John Smith

Ferns

British and foreign: the history, organography, classification, and enumeration of the species of garden ferns with a treatise on their cultivation

John Smith

Ferns

British and foreign: the history, organography, classification, and enumeration of the species of garden ferns with a treatise on their cultivation

ISBN/EAN: 9783337070045

Printed in Europe, USA, Canada, Australia, Japan

Cover: Foto ©berggeist007 / pixelio.de

More available books at **www.hansebooks.com**

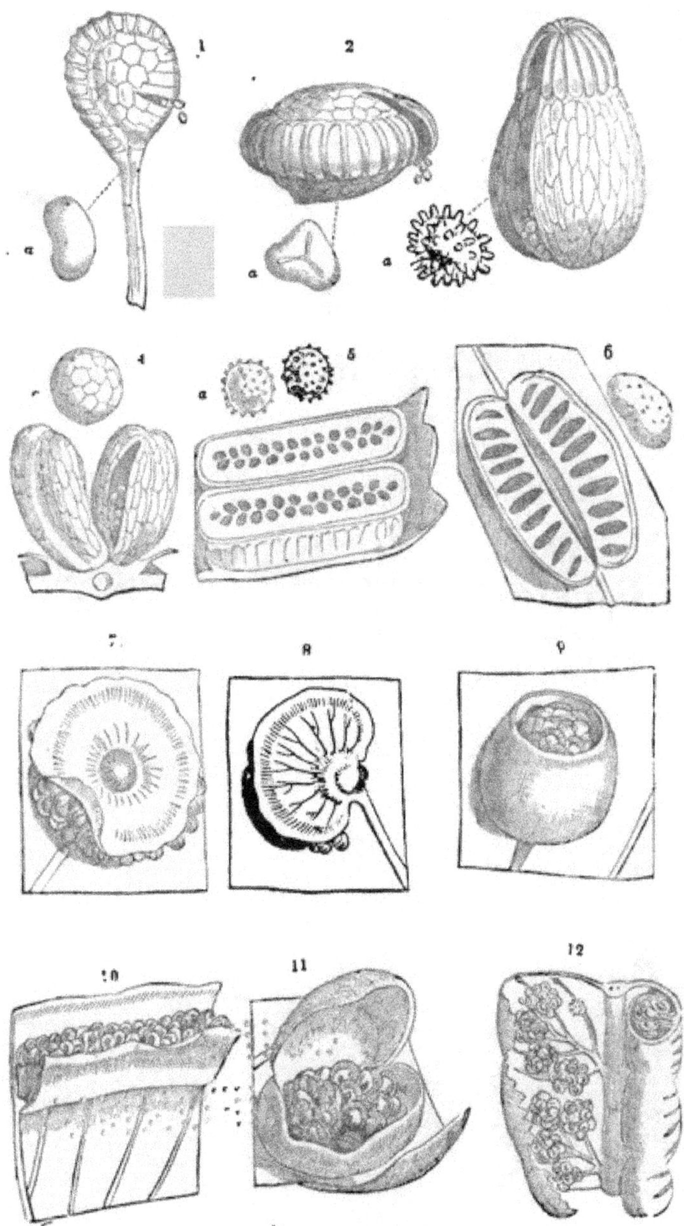

ILLUSTRATION OF ORGANOGRAPHY. (*See p. 55.*)

FERNS: BRITISH & FOREIGN.

THE HISTORY, ORGANOGRAPHY, CLASSIFICATION, AND ENUMERATION OF THE SPECIES OF GARDEN FERNS.

WITH

A TREATISE ON THEIR CULTIVATION,

ETC. ETC.

By JOHN SMITH, A.L.S.

EX-CURATOR OF THE ROYAL BOTANIC GARDENS, KEW;
AUTHOR OF
"DOMESTIC BOTANY," "HISTORIA FILICUM," ETC.

NEW AND ENLARGED EDITION.

London:

DAVID BOGUE,

3, ST. MARTIN'S PLACE, TRAFALGAR SQUARE, W.C.

1879.

CONTENTS.

PREFACE TO THE SECOND EDITION.

TEN years have now elapsed since the publication of "Ferns: British and Foreign," of which the following pages are a corrected reprint. Since then, a number of exotic species have been introduced, making considerable additions to the cultivated collections in this country, of which the names of many have from time to time been noticed in the Horticultural Journals and in Nurserymen's Catalogues. These I have collected and arranged under their respective genera and tribes, so as to form an Appendix to the present edition. I, however, deem it necessary to state, that shortly after the publication of the first edition in 1866, my sight entirely failed, and consequently I have not been able to follow up my rule, which was, not to enter a species on the list of living plants without first having seen it growing, or had specimens sent me taken from plants cultivated in this country. It being impossible for me now to do so, I have consequently availed myself of the great knowledge of Ferns possessed by Mr. William Gower, formerly foreman of the Fern collection at Kew, whose name is already noticed in the preface of the first edition, and who being well acquainted with the principal Fern collections in this country, makes it a special point to obtain a knowledge of all new introductions.

For my knowledge of the additions to the Kew collection since 1864, I am indebted to the kindness of Dr. Hooker for having mentioned to Mr. Baker, the Assistant Curator in the Kew Herbarium, my desire to obtain a list of the new introductions, which he readily furnished me; and whom I have also further to thank for special information regarding certain species. The number of new species introduced since the last enumeration amount to 234, of which the names of about one-half are derived from Mr. Baker's list, some of which are specially interesting as constituting several genera new to this work, which will be found noticed under their respective tribes.

I continue to follow Sir William Hooker's "Species Filicum" * in the adoption of specific names and synonyms as far as possible; at the same time availing myself of certain corrections, made by Mr. J. G. Baker in a work entitled "Synopsis Filicum," being an abridgment of the "Species Filicum,"—this work was commenced by Sir William Hooker, the first part of which was published in 1865, only a few months before the death of that celebrated Pteridologist; it was, however, taken up and completed by Mr. Baker, forming a volume of 482 pages, containing brief

* A work in five volumes, being a description of all known ferns.

descriptions of 2,228 species. A second edition of this work was published in 1874, with an Appendix, which contains the descriptions of 438 new species; thus making the total number of known ferns, as identified by the Kew Herbarium, to amount to 2,646. In this work Mr. Baker has, however, made considerable alterations in the nomenclature and synonyme, as given in the "Species Filicum," which adds additional testimony to what I have stated at page 56. The propriety of making such, whether right or wrong, is not necessary for me to comment on here,* for to do so would lead to this Appendix being much enlarged, and only burden genera and species with additional synonyms, which, for the sake of amateur cultivators of Ferns, I deem it prudent to avoid as much as possible; it is only in a few special cases that I have thought it necessary to change or add synonyms to names in the original enumeration.

I felt desirous to state the name by whom each individual species was introduced, but I found this to be a difficult task, and therefore deem it sufficient to state, that the principal introducers were the following nurserymen:—Messrs. Backhouse, York; Bull, King's Road, Chelsea; Henderson, Pine-apple Place, Edgware Road; Jackson, Kingston; Standish, Ascot; Stansfield, Todmorden; Veitch, King's Road, Chelsea; and Williams, Holloway.

* For full particulars on this point see my "Historia Filicum," a work of 429 pages, with 29 lithographic plates, published by Macmillan & Co., 1875.

In consequence of trade collectors and importers of new plants being desirous of bringing them into early public notice, names are given them without having first taken the precaution to ascertain whether they are not already named and described in Botanical works; thus names frequently appear in Nurserymen's Catalogues, as new, without descriptions, or even their native country given.* Many of such introductions are, however, from time to time described in the *Gardener's Chronicle* by Mr. T. Moore, whose knowledge and writings on ferns are sufficient to warrant their adoption as new species. I have, therefore, in the present addenda, omitted many of these provisional names.

In the plant catalogues of Nurserymen who make ferns a special object of trade, besides the enumeration of specific names, a great number of what are called varieties are also recorded, and their prices affixed, of which Mr. Stansfield's Catalogue contains the names of nearly 500. These consist of abnormal forms of a few British species, principally of *Asplenium Filix-fœmina, Lastrea Filix-mas, Polystichum aculeatum, Scolopendrium vulgare, Lomaria Spicant,* and *Polypodium vulgare,* to which numbers of new forms are yearly

* It should be made a special rule that all importers or possessors of supposed new plants, before offering them for sale, should have them examined by some competent authority, for which there is now ample means in the National Botanical Establishment of Kew, either by examining the living plants in the garden, or in the Herbarium, or by books in the library, or the Herbarium in the British Museum, which now contains my Fern collection.

being added by cultivation, or found wild. The exhibition from time to time of these so-called varieties at the meetings of the Royal Horticultural Society, and the awarding of first-class certificates, are indicators of the great interest in which the curious forms are held by amateurs. Varieties are also found amongst exotic species, especially in the genera *Gymnogramma, Adiantum*, and *Pteris;* but these are comparatively few to those native of this country.

That ferns still continue to be in favour with the plant-loving public, is manifest by the frequent public sales of large importations from abroad, especially of tree ferns, some of which are of large size, and valued as ornamental plants for conservatories, and at public or private banquets or balls, the prices varying according to height, from £5 to £50, or even more; high prices are also given for species of certain genera, such as *Adiantum, Gleichenia, Todea*, &c.

With regard to cultivation, I have nothing to add to what I have already given in the first edition.

J. SMITH,

Ex-Curator, Royal Botanic Gardens, Kew.

July, 1876.

PREFACE TO THE FIRST EDITION.

NEARLY ten years have now elapsed since my "Catalogue of Cultivated Ferns" was published. During that period many new and fine species have been introduced to the gardens of this country. I have therefore been induced to draw up the following enumeration, including in it all the species that have come under my notice in a living state, either by the evidence of plants in the Kew collection or by specimens from living plants sent me from various sources.

Thinking it might be interesting to Fern growers, I have drawn up a brief history of the introduction of Exotic Ferns from the earliest records to the present time.

In order to assist students in the study of Ferns, an explanation is given of the principal organs and terms used in describing them, as well as remarks on their classification. I have also given an alphabetical list of the generic names, accompanied with the derivation of each name.

Great complaints are frequently made, and not without good reason, of the many names given to the same plant; to assist in correcting this, I have drawn up a complete Index of the species and synonymes noticed in this work; therefore, by turning to the page referred to in the Index, the name will be found either with a number before it showing it to be the adopted name, or following the same as a synonyme, or what it has been and is still called by some writers and cultivators. Although this Index contains nearly three thousand names, yet, it must be understood, I have only taken up the synonymes that I consider most essential; those who desire to enter more fully into this subject, may consult the "Index Filicum" of Mr. T. Moore—a work which, when completed, will be a treasure to pteridologists.

With regard to the wood-cuts, I have to explain that about half of the number were not specially prepared for this work; these, in general, represent a portion of the fronds of their natural size, in some cases without fructification, the venation only being given. The drawings for the other half have been made principally from specimens in my herbarium or from living plants. In many of them a magnified portion of the frond is given, showing the character of the genus more distinctly.

As a companion to the scientific enumeration, I have given a treatise on their cultivation, which occupies a considerable space in this book, and it is hoped will be of service to the Fern-growing public.

In drawing up this treatise I have been greatly assisted by Mr. Henry Prestoe and Mr. William Gower (both recently foremen in the Fern department at Kew), two highly scientific and nature-observing practical cultivators.

In conclusion, I have to thank Dr. Berthold Seemann and R. Heward, Esq., for looking over the proofs as they passed through the press. Several causes have occurred to delay the publication of this work; the principal one being a partial failure of my sight, incapacitating me from much use of the pen and examination of new garden Ferns, which I trust will be accepted as my excuse for any errors or omissions that may be found in the following pages.

J. SMITH,

Ex-Curator, Royal Botanic Gardens,

Kew.

KEW, *May*, 1866.

FERNS: BRITISH AND FOREIGN.

HISTORY OF INTRODUCTION OF EXOTIC FERNS.

THOUGH Ferns now occupy a conspicuous place in our gardens, and are in high favour with cultivators, it is only in comparatively recent times that they have been brought into notice. During the last century certain classes of plants came into fashion, and after a season of popularity again fell into disrepute. Thus: Tulips were once the rage. At the time of the establishing of the several provincial Botanic Gardens, all of which were founded upon a strictly botanical footing, though many of them have now, to a greater or less extent, degenerated into places of amusement, the plants in greatest demand were those of our New Holland and Cape colonies, principally the Heaths, *Proteas*, *Aloes*, and their kindred. In after-years dealers obtained large prices for Cactuses; but, with the exception of a few of the easily-grown and most showy kinds, these are now scarcely saleable. Still more recently the magnificent-flowering Orchids were promoted to the first place in our gardens; and though these may still be said to maintain their position, the expense attending their cultivation is so great that they are for the most part confined to the gardens of the wealthy. Ferns, on the contrary, may, as a general rule, be grown in a comparatively inexpensive manner. The discovery made by Mr. N. B. Ward, that these plants can be grown to great perfection in small ornamental closed cases (now well known as "Wardian Cases"), suitable not only for the drawing-rooms of the wealthy, but for humbler dwellings, renders it possible for amateurs to indulge their love of Ferns without going to the expense of erecting hothouses and employing a staff of gardeners; and it is to be hoped that this will be the means of retaining them in favour and spreading them still wider.

The enumeration in the following pages shows that at the present time above nine hundred exotic species of Ferns are cultivated in the various public and private gardens in this country; and of these by far the greater number have been introduced during the last quarter of a century. A very large, indeed almost a complete, collection of them may at present be seen in the Royal Botanic Garden at Kew, where, during forty years, I devoted attention to their cultivation, and to the study of their generic and specific distinctions, using every endeavour in my power, assisted by the extensive influence of the eminent Director, Sir W. J. Hooker, to introduce new species, both by raising them from spores taken from my herbarium, and through correspondence with persons residing in our Colonies and the Superintendents of Botanic and other gardens on the Continent. Being well acquainted with the latter branch of the subject,—the introduction of new species,—I have thought that a few pages devoted to the history of these plants in its connection with our gardens might prove acceptable. My own personal knowledge dates from about the year 1822; and for information respecting those in cultivation previous to that time, I have taken for my guide the most important and most reliable of the garden catalogues.

The first work upon which dependence can be placed for the early-introduced species is the so-called second edition of the "Hortus Kewensis," published by the younger Aiton, in 1813; I say the so-called, for there were in reality two previous editions—one by Dr. John Hill, in 1768, and a second by William Aiton, in 1789. In the first of these only ten exotic species are recorded as being in the garden at Kew, which then belonged to the Princess of Wales, but the dates of their introduction are not given; and in the second, that of the elder Aiton, thirty-four, all of which have the name of their supposed introducer attached to them, and the date at which they were introduced. I here give preference to the last edition, as I know, from personal intercourse with the compilers, that great care was exercised in its preparation, particularly with regard to the rectification of dates, and I shall take it as the basis of my remarks. But, in the first place, it is necessary to say a few words in defence of that work. It has been asserted that a considerable number of the plants enumerated in the "Hortus Kewensis" never existed in a living state, either in the garden at Kew or elsewhere in this country. This statement, however, is merely supposition, and rests upon questionable authority. In the early days of Kew Gardens, large sums of money were devoted to the payment of botanical collectors, and great exertions were made, under the patronage of Sir Joseph Banks, to stimulate the introduction of new and rare plants, by inducing the commanders of ships of war and East-Indiamen to take an interest in the subject. By these means a very large number of plants were actually introduced into the country in a living state; but the Department under whose charge the garden then was, took no steps to provide proper accommodation; and this, together with the very frequent change of foremen, led, as a natural consequence, to the death of the plants.

Taking, then, the third "Hortus Kewensis," and excluding our indigenous species, I find that the total number of "Garden Ferns" introduced previously to the year 1813 amounts to as many as eighty-three. The merit of being the first introducer of these plants belongs to Mr. John Tradescant* the younger, who in early life made a voyage to Virginia; and I find it recorded in Parkinson's "Theatrum Botanicum," published in 1640, that upon his return from that country in 1628 he brought with him, amongst other rare plants, the *Cystopteris bulbifera* and *Adiantum pedatum*. These, therefore, must be regarded as the nucleus of our present large collections. At first the progress seems to have been exceedingly slow, for between the time of Tradescant and the close of the seventeenth century, only five additional species were introduced; viz., *Asplenium rhizophyllum* and *Onoclea*

* John Tradescant had a Botanic Garden and Museum at Lambeth.

sensibilis from North America, the former in 1680 and the latter in 1699, in which year also *Adiantum reniforme* and *Davallia Canariensis* were brought from Madeira, while the fifth, *Blechnum australe*, was, according to Plukenet, who figured it in the second part of his "Phytographia," cultivated in the King's garden at Hampton Court as early as the year 1671, when his work was published, the garden there containing a considerable collection of rare plants. During the next forty-two years no additions appear to have been made, excepting the *Phlebodium aureum*, which was introduced by Lord Petre at some time prior to 1742, the date of his death, the precise year being unknown. Progressing onwards, I do not find any more recorded until the year 1769; but between that year and the commencement of the following century sixty-eight species were added to the eight already existing in our gardens. Out of this large number, no less than thirty-seven were brought home by Rear-Admiral Bligh, in H.M.S. *Providence*, on his return, in 1793, from his second voyage, undertaken for the purpose of introducing the Bread-fruit and other useful trees into our West Indian colonies. And, finally, during the first few years of the present century, up to 1813, the date of the publication of the "Hortus Kewensis," seven others were introduced.

A summary of the foregoing shows that upwards of one-half of the Ferns known at the last-mentioned date were West Indian species, forty-four having been received at various times from those islands, the majority through Bligh's expedition. North America and Madeira, with the neighbouring islands, stand next as the largest contributors, fifteen having been introduced from the former and twelve from the latter country; while of the remaining sixteen, four appear to have come from the East Indies, four from the Cape of Good Hope, three from New Holland, and one from St. Helena,—making in all eighty-three species.

The next catalogue of garden plants worthy of notice is the "Hortus Suburbanus Londinensis," published in 1818, only five years after the "Hortus Kewensis," by Mr. Sweet, the Superintendent of the then celebrated nursery of Mr. Colville, at Chelsea. In it I find an enumeration of one hundred and eight exotic ferns; but this work, like the similar more important "Hortus Britannicus," brought out by the indefatigable Loudon in 1830, and which contains no less than three hundred and thirty exotic Ferns, includes not only a considerable proportion of bad species, but also a large number that did not really exist in British gardens, many having been entered without authentic evidence, and others added upon the mere expectation that they might shortly be introduced,—expectations which, in many cases, have not been realized to this day. No reliance can therefore be placed upon either of these works, and I cannot accept them as authorities.

During the latter part of the eighteenth century and the commencement of the nineteenth, the only private individuals who turned their attention, with any amount of energy, to the introduction of new and rare plants, were the long- and far-famed nurserymen at Hackney, the Messrs. Loddiges; and to them we owe the greater part, if not the whole, of the

Ferns existing at that period in British gardens, and not included in the "Hortus Kewensis." Speaking from my personal recollection of the important collection of plants in the Hackney Nursery, as it existed in the year 1825, I think it contained about a hundred good species of exotic Ferns; but I can obtain no earlier catalogue than one published in 1849, in which two hundred and fifty-one are enumerated.

In the year 1822 I found the collection of Ferns at Kew extremely poor, especially as regards Tropical species, very many of those introduced in previous years having been lost, and very few new ones added. Any person who remembers the hothouses in existence forty years ago, will have but little difficulty in accounting for the falling off of the Fern collection. In those days hot-water pipes were unknown, and the houses were exclusively heated by means of brick flues, too often imperfectly constructed, and the excessively dry and ungenial atmosphere thus induced was quite unsuited for the good cultivation or even for the mere preservation of these moisture-loving plants. Nearly all the North American species enumerated in the "Hortus Kewensis" were growing very finely in a north border, and most of the Madeira species were also in existence; but, including these and the few added since 1813, I cannot estimate the entire Kew collection of exotic Ferns at that period at more than forty species.

Between 1813 and 1846, when my first Catalogue of the Ferns at Kew appeared, no reliable list was published in this country. Several, however, were brought out by Continental botanists, which are of sufficient importance to be worthy of a passing notice. The first of these in point of date is the "Enumeratio Plantarum Horti Regii Botanici Berolinensis," by Wildenow, published in 1809, with a Supplement by Schlechtendal, bringing it down to 1813. We are thus enabled to compare the numbers in the Kew and Berlin gardens at the same period; which were eighty-three in the former, and thirty in the latter, including eleven not known at Kew. During the succeeding nine years more attention appears to have been given to Ferns at the Berlin garden; for Link, in his first "Enumeratio," in 1822, describes ninety-one exotic species, which is more than double the number then existing at Kew. After this the increase in number was still more rapid; for in the second edition of Link's "Enumeratio," published in 1833, no less than two hundred and thirty-nine are described; and in the third, in 1841, two hundred and fifty-eight, exclusive of varieties.

By this time, however, the collection at Kew had received large additions, both through importations of living plants and by raising from spores. In 1845 it was so extensive that I was induced to draw up a classified enumeration, which was published as an appendix to the *Botanical Magazine* for 1846. The number of exotic species there enumerated is three hundred and forty-eight, and I do not think many were to be found in other gardens in this country which were not at Kew, so that the Kew list may be taken as a guide to the number then in British gardens generally.

Four years later, Kunze, of Leipzig, contributed to

the 23rd volume of the "Linnæa" an alphabetical index of the Ferns cultivated in European gardens, and in this the large number of eight hundred and forty-three exotic species are enumerated; but the authorities upon which a very considerable portion of these were inserted cannot be relied upon, many names having been taken from such catalogues as those of Sweet, Loudon, &c., and I am therefore obliged to conclude that the number given by Kunze as living in Europe in 1850 is greatly exaggerated. This conclusion, too, is confirmed by the fact that in 1857, after I had by correspondence become acquainted with the collections in the principal gardens on the Continent, and after that at Kew had obtained most of their novelties by means of exchange, I could, in my "Catalogue of Cultivated Ferns," enumerate only five hundred and sixty exotic species as known in British gardens. Since the last-mentioned year, the constantly increasing demand for Ferns consequent upon their wider spread cultivation, has greatly stimulated the introduction of new ones, and our collections have increased at the rate of about fifty species a year.

It now remains to say a few words regarding the means by which these plants have been obtained, and the persons who have been most active in introducing them, taking the Kew collection as a foundation. Firstly, with reference to the paid collectors employed in various parts of the world, directly or indirectly in the service of the Royal Botanic Garden, Kew, and to whom that garden is indebted for additions to its Fern collection. It would appear that so long back as the year 1775 Mr. Francis Masson, one of the earliest, if not the earliest, collector sent out from Kew, and who succeeded in introducing large numbers of Cape *Proteaceæ* and *Ericaceæ*, sent home several Ferns from the Cape of Good Hope and Madeira. This collector proceeded to the Cape in 1774, and came home by way of Madeira about the year 1778, returning again in 1786, and remaining at the Cape during the nine following years. Early in the present century Mr. George Caley, who was originally a horse-doctor, residing near Birmingham, but acquired a love for plants through collecting herbs, was sent out by Sir Joseph Banks to New South Wales, and to him we owe *Platycerium alcicorne*, *Doodia aspera*, and *Davallia pyxidata*, the first introductions from Australia, received about the year 1808. The next collectors to whom the garden was indebted for Ferns, are the Messrs. Allan Cunningham and James Bowie. They left Kew in 1814, on a botanical expedition to Brazil, where they remained exploring the country and sending home large collections till 1816, when the former proceeded to New South Wales, and the latter to the Cape of Good Hope. No living Ferns appear to have resulted from the Brazilian expedition; but several Australian species and one or two from Norfolk Island were received from Mr. Cunningham, and two or three from Mr. Bowie from the Cape.

Several other collectors were employed in the service of these gardens, when under the Directorship of Mr. Aiton, such as Messrs. Barclay and Armstrong; but I can trace no Ferns to them, nor, with certainty, to David Lockhart, a gardener from Kew, who accompanied the ill-fated expedition of Captain Tuckey up the Congo, in 1816, though I believe it possible that some of the first-known Western-African species are attributable to him. Lockhart was one of the few who did not fall a victim to the deadly climate of that country. After lying a long time in the hospital at Bahia, he returned to England, and shortly afterwards received the appointment of Superintendent of the Botanic Garden in Trinidad, where he died in 1845, after a service of a quarter of a century in that island, during which period he sent many fine plants to Kew.

The principal collectors employed directly or indirectly in the service of Kew during the Directorship of Sir William Hooker, and to whom the garden is indebted for any additions to its Fern collection, are Messrs. Purdie, Seemann, Milne, Barter, and Mann. The first of these, Mr. William Purdie, was engaged as collector in 1843, in which year he was despatched to Jamaica, and after spending several months in the exploration of that island, and forwarding many valuable plants to Kew, he proceeded to Santa Martha, and was employed for more than two years in various parts of New Granada. To him we are indebted not only for numerous showy flowering plants, but for a considerable number of our garden Ferns.

Upon the death of Mr. Thomas Edmonston,* Mr. (now Dr.) Berthold Seemann was appointed his successor, as botanist to H.M.S. *Herald*, and joined that vessel at Panama, in January, 1847, remaining with her until her return to England in June, 1851. Previously

* Mr. Edmonston was shot by the accidental discharge of a rifle, at Sua Bay, Ecuador, January 24, 1846.

to joining the *Herald*, he explored several parts of Panama and Veraguas, and while on board the *Herald* visited various parts of the western coast of America, between Lima on the south and California on the north, occasionally making long journeys inland, particularly in Peru, Ecuador, and Mexico. He likewise accompanied the *Herald* in her three voyages to the Arctic regions in search of the ill-fated Sir John Franklin; visited twice the Sandwich Islands; and returned home by way of Hong-Kong, Singapore, and the Cape of Good Hope. Notwithstanding the few facilities afforded by surveying expeditions for collecting living plants, Dr. Seemann succeeded in introducing some interesting ones to our gardens; and among Ferns we may mention two very remarkable ones, *Deparia prolifera* from the Sandwich Islands, and *Dictyoxiphium Panamense* from Panama.

On the *Herald* being recommissioned in 1852, Mr. William Milne was appointed assistant-botanist, for the special purpose of collecting plants for the Royal Gardens. During the six years he was attached to the *Herald* he visited New Caledonia, the Fiji, and other Polynesian islands, as well as many other places in the southern hemisphere.

The Admiralty having, early in 1857, determined upon sending out a second exploring expedition up the Niger, under the command of Dr. Baikie, R.N., Mr. Charles Barter, a zealous and intelligent young gardener, was appointed to accompany him in the capacity of botanist, and to collect plants for Kew. The officers of the expedition proceeded by way of Sierra Leone, and joined the *Dayspring*, a small vessel fitted for river navigation, at Fernando Po; but, unfortunately, after they had penetrated up the river as far as Nupe, their ship was wrecked, and they were obliged to form a camp on the banks. Poor Barter, however, fell a victim to the deadly climate in July, 1859. As a collector he was indefatigable in the discharge of his duties, and discovered a large number of undescribed plants, including several new genera, one of which has been dedicated to his memory under the name of *Barteria*. But, owing to the mishaps attending this expedition, and the difficulties of transit, no living plants were received from the interior of the country, but several Ferns were transmitted from Sierra Leone and Fernando Po.*

No sooner had the news of Barter's death arrived, than Mr. Gustav Mann, undeterred by his fate, volunteered to fill the vacancy, and was appointed accordingly. Being unable, from want of means of communication, to join Dr. Baikie, he was employed for three years in exploring the island of Fernando Po and the African coast, in the neighbourhood of the Camaroons and Gaboon rivers. In spite of the difficulties and dangers attending the undertaking, he made several ascents of the lofty Clarence Peak of Fernando Po and of the Camaroons mountain on the African mainland. On the former, at an elevation of 5,000 feet, he found a fine new *Cyathea*, forming groups, with stems rising to a height of 30 feet. We are likewise indebted to him

* Since the above was written, news has reached this country of the death of Dr. Baikie at Sierra Leone, in January, 1865, whilst on his homeward voyage.

for many other rare and interesting living Ferns, and fine sets of dried specimens.

Besides contributions from special collectors, the Fern collection at Kew has been largely increased through the instrumentality of the officers of our numerous Colonial Botanic Gardens, and likewise by many private individuals residing in foreign countries; and as the exertions of these gentlemen are worthy of being recorded, I give a few particulars respecting them, adopting a geographical arrangement for the sake of brevity.

The first Colonial garden from which, so far as I am aware, Ferns were received at Kew, was that of Ceylon, Mr. Alexander Moon, the Director, having, in 1824, sent home a collection of plants, amongst which was *Niphobolus costatus*. But the first person who forwarded any considerable number from that island was the lamented Mr. George Gardner, well known to botanists as a botanical traveller in Brazil, who was Director of the Ceylon garden between 1844 and his death in 1848. This gentleman was succeeded by Mr. (now Dr.) G. H. K. Thwaites, the present able Director, and to him also the Kew collection is greatly indebted for a large number of rare and beautiful species; such as *Schizocæna sinuata*, *Asplenium radiatum*, *Actinostachys radiata*, *Helminthostachys Zeylanica*, and many others.

Though Continental India is extremely rich in Ferns, it has, singularly enough, contributed very few to our gardens, no persons in that country having devoted themselves specially to the subject; indeed most of those received thence have been accidentally imported along with Orchids, including the half-dozen species recorded in my first Enumeration as coming from Dr. Wallich, the only Superintendent of the Calcutta garden who has the credit of having introduced any. Indeed, with the exception of those from Ceylon, Kew has received very few Ferns from Asia and the adjacent islands, most of those now in cultivation having been introduced by nurserymen or through Continental gardens. Two or three were brought from Hong-Kong, in 1850, by Mr. J. C. Braine, including one which proved to be a new genus, and to this I gave the name of *Brainea* in honour of its introducer.

Another tropical island in the Eastern hemisphere, whence large additions have been made to the Fern collection at Kew, is the Mauritius. The Botanic Garden in that island has long enjoyed the reputation of possessing a fine set of plants; but until the year 1852, when it came under the able management of the present Director, Mr. James Duncan, very little correspondence was kept up with the gardens of Europe. Mr. Duncan has, at considerable risk, ransacked the forests of the island in quest principally of Ferns, and has been very successful in transmitting living plants to this country, enriching our gardens with many fine species.

The "Synopsis Filicum Capensis" of Pappe and Rawson shows that the Fern Flora of Southern Africa is extremely rich; but up to the present time we possess scarcely a dozen Cape species in our gardens, and most of them have been raised from spores. *Alsophila Capensis* and *Lomaria Capensis* were introduced in 1845 by Mr. Charles Zeyher, as also were *Lastrea athmantica* and *Cyathea Dregei* from Natal by Mr. J. Plant.

Progressing westwards we come to the two strangely isolated islands in the Southern Atlantic, St. Helena and Ascension. From the former of these we have *Asplenium compressum*, introduced by Mr. Thomas Fraser in 1825, and *Asplenium reclinatum*, brought home by Dr. J. D. Hooker on his return from Sir John Ross's Antarctic expedition in 1844; together with *Lomaria alpina* and *L. Magellanica* from the Falkland Islands; while from Ascension Mr. Wren sent numerous fine plants of *Marattia purpurescens* in 1848.

From Australia several individuals have been contributors. *Grammitis Australis* was received from the Sydney garden in 1833, when under the direction of Mr. Richard Cunningham; and Mr. Charles Moore, the present Director of that garden, has also introduced several, including *Trichiocarpa Moorei*, from New Caledonia, while to Mr. Bidwill we owe the curious *Platycerium grande*. But some of the most beautiful of the Australian Ferns, such as the *Gleichenias*, were transmitted to this country by Mr. Walter Hill, the able Director of the Botanic Garden of Brisbane, in the rapidly rising colony of Queensland, who obtained them during his stay in Sydney in 1850. Two species of *Gleicheniaceæ* were, however, previously known in our gardens,—the *Gleichenia microphylla* and *G. flabellata*, both of which, together with several other Ferns, were sent from Tasmania, in 1845, by Mr. Ronald Gunn.

About the year 1841 or 1842, some very fine Ferns, including two Tree-Ferns, the *Dicksonia squarrosa* and *Cyathea medullaris*, were brought from New Zealand, where they had been collected by Mr. J. Edgerly, a gardener, who had proceeded to that country on speculation, and who was the first to introduce the beautiful *Veronica speciosa.* Others have also been sent from New Zealand by the Rev. William Colenso, and by the late Dr. Sinclair, R.N., the beautiful little *Trichomanes reniforme* being one of those due to the latter gentleman.

Turning next to the West Indies, we commence with the island of Jamaica, whence more Ferns have been received at Kew than from any other part of the Western hemisphere. The person to whose energy and perseverance this is mainly due is Mr. Nathaniel Wilson, the Island botanist and Director of the Botanic Garden. He has been a resident in the island for upwards of twenty years, and during that time has thoroughly explored the Blue Mountains and other districts rich in Ferns, liberally forwarding to Kew the results of his numerous journeys. Among his earliest contributions was the beautiful Tree-fern, *Cyathea arborea,* which, though recorded in the "Hortus Kewensis" as having been brought home by Admiral Bligh in 1793, had long been lost to our gardens. Within the last few years he has succeeded, after many failures, in transmitting numerous species of *Trichomanes* and *Hymenophyllum,* which now form so conspicuous a feature in the present rich collection. The other contributors from the same island are, in 1851, Mr. George Manson, and in 1854 and following years, W. T. March, Esq., the latter gentleman sending several arborescent species as well as *Hymenophylleæ,* and others.

Numerous fine species, including several *Cyatheas* and *Alsophilas,* were received in 1855 and 1856 from the French island of Martinique, where they had been collected by the Director of the Garden, M. Belanger. A considerable number of rare species were likewise sent from Dominica in 1853, and several following years, by Dr. Imray; such as *Hemitelia Imrayana, Neurocallis præstantissima, Elaphoglossum undulatum,* and the true *E. longifolium,* &c. While from Trinidad we are indebted to the Island botanist, the late Dr. Crüger, for *Schizæa elegans, Saccoloma elegans, Amphidesmium rostratum, Hymenostachys diversifrons, Trichomanes pennatum,* and other equally rare species. His successor, Mr. Henry Prestoe, who previous to his appointment last year was foreman of the collection of Ferns at Kew, has already transmitted a large collection of rare Ferns, and in the finest condition, proving that they had been collected and put up by one who perfectly understood their nature. The situation he now holds will enable him to be of great service in transmitting new plants to this country.

Comparatively few, considering the richness of its Fern Flora, have been received from tropical America; Dr. Gardner and J. Wetherall, Esq., in Brazil; H. Cadogan Rothery, Esq., in Guiana; Mrs. Colonel McDonald, in Honduras; and Mr. Wagener, in Venezuela, being the principal contributors.

From the above it will be seen that a large number of Ferns have been introduced through the agency of the garden at Kew; but, besides these, a very considerable number are due to the exertions of some of our leading nurserymen, who, in consequence of the great demand for, and the large prices realized by, the finer and rarer species, have imported them, either direct from their native countries, or from the Continent, where, as will be presently noticed, a good many species not previously known in the gardens of this country have been introduced.

I have already alluded to the Messrs. Loddiges, of Hackney, as having at an early period turned their attention to Ferns, and as being the earliest to form a collection of them. But the only nurserymen whose names are recorded in the second edition of the "Hortus Kewensis" are the old-established firm of Messrs. Lee & Kennedy, of Hammersmith, who are stated to have introduced *Polypodium asplenifolium* and *Asplenium monanthemum* in 1790: in later times the Messrs. Lee have imported several from New Zealand. Other New Zealand species have been brought into notice by Mr. Standish, of Bagshot, they having been collected in New Zealand by Mr. J. Watson, now a nurseryman at St. Alban's, and who still continues to import. Several sent from Japan by Mr. Fortune have likewise been sent out from Mr. Standish's nursery. To the Messrs. Low & Sons, of the Clapton nursery, we are indebted for some rare Bornean and Malayan species, collected by Mr. Hugh Low, jun., and amongst others for the remarkable *Arthropteris obliterata*, called *Lindsæa Lowii* in the gardens, and the little curious *Leucostegia parvula;* but more especially many rare species of *Hymenophyllum* and *Trichomanes*, as well as the rare *Thyrsopteris elegans*, collected by Mr. Thomas Bridges in Juan Fernandez. But to the Messrs. Veitch & Sons, of Exeter and Chelsea, among nurserymen, must be assigned the credit of having introduced the greatest number of these plants, the collectors employed by them in Chili and other parts of the American continent, in India, the Malayan continent and islands, and in Japan, having sent home numerous fine species, while through other sources they have obtained many additions from Australia, New Zealand, and other countries. Messrs. Rollisson, of Tooting, have likewise succeeded in enriching our collections with a considerable number, received principally from Mr. John Henshall, their collector in Java and the neighbouring islands. Mr. Robert Sim, of Foot's Cray, has an extensive collection, which he increases by importations, and is very successful in raising plants from spores; and the Messrs. Backhouse & Son, of York, the principal nurserymen cultivators of exotic Ferns in the provinces, have introduced a good many fine species of *Trichomanes*, and others from Chili and the West Indies. A few have likewise been introduced by other nurserymen, but the above are the principal of those who have obtained them from their native countries. I may, however, mention Messrs. Osborn & Sons, of Fulham, as having introduced one or two from Tasmania; Messrs. Jackson & Son, of Kingston, the *Angiopteris Assamica*, from Assam; and Mr. B. Williams several from various parts; while all these and some others have also imported considerable numbers from the Continental gardens and nurseries.

In addition to all these sources, occasional introductions have taken place through several other Botanic Gardens in this country, as well as through some of the numerous amateur cultivators; but Fern amateurs have increased so largely during the last few years, that it is obviously impossible to mention them in detail. With respect to the former, the first provincial Botanic Garden in which Ferns were brought into notice was that of Liverpool, under the Curatorship of the late Mr. John Shepherd, more than thirty years ago; and the collection there has lately been considerably augmented by Mr. Tyerman. At Birmingham, too, Mr. David Cameron in early times formed a good collection, which, however, has now given place to gaudy florist flowers. A good deal of attention is at the present time paid to Ferns at the Glasgow Botanic Garden by Mr. Peter Clarke; and also by Dr. David Moore, Director of the Botanic Garden of the Royal Society of Dublin, at Glasnevin, who has likewise introduced several new species from Trinidad and New South Wales. The Royal Horticultural Society of London must be mentioned as having introduced *Cibotium Schiedei*, and one or two other Mexican species, received from their collector, Mr. Theodore Hartweg.

Want of space precludes my particularizing the numerous private growers of the present day; but I cannot altogether pass over one or two of the earlier ones who formed large collections of species, and did much to stimulate the taste for these plants. I more particularly allude to Mr. James Henderson and John Riley, Esq. Under the patronage of the Earls of Fitzwilliam, to whom he had long been gardener, Mr. Henderson has for upwards of thirty years been a zealous cultivator of Ferns, and has been very successful in raising them from spores, adding by that means a good number of species to our collections. One of the earliest amateurs distinguished for his love of Ferns, was John Riley, Esq., of Papplewick, near Nottingham, who was also a successful raiser, and brought together a collection containing nearly 300 species, which, upon his death in 1846, was purchased by Dr. Forbes Young, of Lambeth, who considerably augmented it; but unfortunately the death of its second owner, in 1859, caused its dispersion. Few amateurs at the present day study Ferns scientifically, or form collections numerically large in species, the principal of those in the neighbourhood of London devoting their attention to a select number of the most beautiful ones, such as are well known to the frequenters of our metropolitan flower-shows. I must, however, except E. J. Lowe, Esq., of Beeston, near Nottingham, who formed a considerable collection, and published an illustrated work upon them in nine octavo volumes.

In the public and private gardens on the Continent Ferns claimed a large share of attention, and many of these possess fine collections of them, containing numerous species not yet known in British gardens, though our nurserymen are constantly on the look-out for novelties, and import a great number from these sources. I have mentioned above that the directors of the Botanic Garden of Berlin, at an early period, possessed an extensive collection, and many species are reputed to have been raised in this establishment. The publication of the "Ferns of the Leipzig Garden," an illustrated work, in folio, by Dr. Mettenius, shows that the garden under his direction is exceedingly rich in Ferns, and the University fortunate in having a Professor so well able to do justice to the collection. At Vienna, also, a collection of Ferns has long existed under the direction of the late Dr. Schott. Several other German gardens, as those at Gœttingen and Herrenhausen, also possess a considerable number, some very interesting novelties in the latter having been obtained by M. Hermann Wendland, during a journey through Central America; and in many of these gardens species have been obtained by means of spores taken from dried specimens, while through the Dutch Botanic gardens, numerous rare Ferns have been introduced from Java, Surinam, and other Dutch colonies.

Among private individuals on the Continent who have made large additions to our collections, by the introduction of species from their native countries, I cannot omit to notice M. J. Linden, of Brussels, who himself travelled in the West Indies, Venezuela, and New Granada, and who employed several enthusiastic collectors in various parts of the same and neighbouring countries, by whom a great number of the new plants were brought into cultivation. But besides these M. Linden has also received several species new to our gardens, from New Caledonia and the Philippine Islands. About six years ago another private traveller in Venezuela and New Granada, Dr. Karsten, likewise enriched continental gardens by the introduction of numerous fine species of tree and other Ferns, some of which have not yet been imported to this country.

The total number of Ferns cultivated in our gardens at the present day may be regarded as forming about one-third of all the species known to botanists by means of dried specimens, and described in the numerous works of pteridology. Among the remaining two-thirds are very many fine species, equal or superior in merit, as garden plants, to any of those we already possess. It may be worth while to mention a few of the more striking of these, together with the countries in which they are found, in order to draw the attention of some of our enterprising nurserymen to them and induce them to take steps for their introduction. Assuredly in this fern-loving age many would prove of great commercial value. First, there is the magnificent *Matonia pectinata*, found only on Mount Ophir, in Malacca; a Fern resembling the *Gleichenias* in habit, but rising to a height of five or six feet, with beautiful fronds, divided, like those of fan-palms, into numerous pectinate segments. Two other Ferns of much the same habit, the *Dipteris Wallichii* and *D. Horsfieldii*, are likewise worthy a place in our gardens. The former of these is found in the mountains of Silhet, and the latter in Java, Borneo, the Philippines, Fiji, and neighbouring islands. In the Philippine Islands, where the Fern Flora has about 250 representatives, there are numerous fine species, such as *Dryostachium splendens* and *Aglaomorpha Meyeniana*, both somewhat resembling *Drynaria quercifolia* in the general aspect and mode of growth of their barren fronds, both having rhizomes equally tenacious of life; *Lomagramme pteroides*, with large pinnate fronds three feet high, having long linear, lanceolate articulate pinnæ, bearing amorphous sori; *Photinopteris Horsfieldii*, the glistening sterile fronds of which are pinnate and between two and three feet high, and have very broad elliptic-lanceolate pinnæ, similar to the common laurel, while the fertile ones are very much contracted; *Gleichenia excelsa*, a very strong-growing species with fronds five or six feet high, having spreading pinnæ two to three feet in length. The beautiful *Schizocæna Brunonis* of Penang and

Malacca would also form a striking addition to our tropical ferneries, its pinnate fronds being from three to four feet long on stipes about half as long again, rising from an arborescent caudex. In Blume's "Enumeration of the Ferns of Java" alone, no less than 460 species are described, of which about 300 are regarded as new. Many of them are, however, not distinct as species, but are fine and showy and well worth the cultivator's notice.

Remarkable for their structural characters, there are *Sphæropteris barbata* of Nepal, and *Diacalpe aspidioides* of Eastern Bengal and Java; the fronds of the former resembling those of a *Lastrea dilatata*, but having globose sori with cup-shaped indusia elevated on distinct pedicels, while the latter has very similar sori not elevated. Another Fern of Eastern Bengal worthy of notice is the *Acrophorus nodosus*, a species with large decompound fronds remarkable on account of their pinnæ standing out almost horizontally, or at right angles with the main rachis. The same district, including the Khasaya and Silhet hills, Assam, Bootan, Sikkim, &c., is extremely prolific in fine Ferns, which, though familiar enough in a botanical point of view, are still unknown in our gardens: they would yield a rich harvest to a collector of living plants; and it is not a little remarkable that so few of them have as yet been introduced through the Botanic Garden of Calcutta. The total number of known species of Indian Ferns may be stated in round numbers to be 400; and what we have of these have been received from their other habitats. I cannot, of course, attempt to give a list of Indian desiderata; but, in addition to the two or three above alluded to, I may mention *Kaulfussia Assamica*, a remarkable Marattiaceous Fern found in Assam, having ample trifoliate somewhat fleshy fronds, from eighteen inches to two feet in height, with the fructification, which consists of hollow circular sporangia, scattered irregularly on the under surface. It is allied to the *K. æsculifolia* from the Malayan islands, included in the following enumeration, and which is extremely rare, if indeed it be not altogether lost from our gardens: *Lomaria glauca*, a Fern of Khasaya, with pinnate fronds, fine glacous underneath, two feet high; and, finally, the *Alsophila gigantea*, a native not only of Silhet, Nepal, and other parts of India, but of Ceylon, Penang, and other Indian islands, a magnificent Tree-fern with a trunk fifty feet high, bearing a crown of large bi- or tri-pinnate fronds, the segments of which are very variable. Dr. Thwaites, in his "Flora of Ceylon," enumerates 214 Ferns, of which about one-half are embodied in the following pages, but many are yet rare, and several have failed to become established. This is especially the case with *Asplenium* (*Actiniopteris*) *radiatum*, *Actinostachys radiata*, and the singular *Polystichum anomalum*, a most remarkable Fern, which bears perfect sori on both sides; it is found at an elevation of from 5,000 to 6,000 feet, and no doubt our want of success in not keeping it, as well as the unhealthy look of other Ceylon Polystichums, is owing to their being placed in the tropical house,—their elevation and their resemblance to the European *Polystichum aculeatum* indicates that they would be more at home in the temperate house. Several interesting species have yet to be introduced, such as the small pinnatifid *Poly-*

podiæ, including *P. contiguum* and *P. Emersoni*, which, on account of their sporangia being seated in a deep cyst, and protruding outwards, are referred by some authors to *Davallia*. Two fine Tree-ferns are also worthy of notice, *Alsophila crinita* and *Cyathea Walkeri*. Many of the islands of the Eastern Archipelago likewise offer a fertile field for the fern collector, being rich in species of *Hymenophyllum*, and *Trichomanes*, *Asplenium*, *Lomaria*, as well as in Tree-ferns of the genera *Cyathea* and *Alsophila*. The one or more Sumatran Ferns yielding the singular styptic drugs brought to this country of late years under the uncouth names of Penghawa, Djambi, and Pakoe Kidang, would be of interest to cultivators from their beauty, and to pharmaceutists from their properties. *Cibotium djambianum*, *Dicksonia chrysotricha*, and two other species of *Alsophila* (*A. lurida* and *tomentosa*) have been mentioned as the sources of these drugs; but very little is known respecting any of them, and more information is desired. Two singular and peculiar Ferns widely spread throughout this region are *Tænitis blechnoides* and *Osmunda Javanica*, which, although long known in herbaria, have not yet found a place in our living collections. The Moluccas and Celebes, especially, possess large numbers; and among those in the former is the rare *Cystodium sorbifolium*, known only from a few imperfect specimens from these islands, and from the island of Honimœ. Mauritius contains several which would be acceptable in our gardens; but, with such an enthusiastic explorer as Mr. Duncan in that island, it is to be hoped that they will not long be classed among our *desiderata*. I would particularly call attention to *Ochropteris pallens*, *Antrophyum Boryanum*, *Ophioglossom palmatum*. *Cyathea canaliculata*, and *C. excelsa*, as well as *Adiantum asarifolium* and *A. Mauritianum*,—all found in that island,—are still scarce in, if not altogether lost to our gardens. I have already alluded to the paucity of species from Southern Africa at present in our gardens, although in the "Synopsis" of Pappe and Rawson, published in 1858, no less than 165 are described, and the localities where they are to be found given in detail. Notwithstanding that few of these are remarkable or striking in appearance, many would be prized on account of their small size and neatness, and they would be very suitable for Ward's cases. Even in European countries there are several Ferns which we do not yet possess in British gardens; for instance, the *Lastrea fragrans* of the Arctic and sub-Arctic regions, said by Sir W. J. Hooker to be "one of the most beautiful of all ferns," is, as far as I am aware, known only from dried specimens; while *Asplenium fissum*, found in several parts of Southern Germany and Italy, is rare even in herbaria, and altogether unknown in the gardens of this country. On the south-eastern confines of Europe, in the Caucasus, there is also the *Woodsia Caucasica*, an interesting species, closely allied to *W. elongata*, of Northern India.

From the Western hemisphere, also, there are numerous fine species yet to be introduced. Fee's "Catalogue of Mexican Ferns" shows that that country contains upwards of 300 not known in our gardens, though many of them would be very acceptable. Among these are several Tree-ferns, such as the remarkable *Cyathea Mexicana*, found in the neighbourhood of

Jalapa, Cordova, and Oxaca, while the little tufted *Schaffneria nigripes,* found between Vera Cruz and Orizaba, would be an interesting acquisition to growers whose space is limited, being only three or four inches high, and having intensely black glossy stipes, and broadly obovate-cuneate fronds. There are also many very pretty species of *Cheilanthes,*—the *Cheilanthes speciosissima,* with broad lanceolate multifid coriaceous fronds, measuring as much as two feet in length; and several very desirable species of *Gleichenia, Trichomanes,* and *Hymenophyllum;* of the latter genera, in particular, there are still many beautiful western species yet to be introduced. From the more northern countries of the American continent our hardy fern growers might obtain many additions. For example, it would be interesting to have the American *Cryptogramme acrostichoides* side by side with our British *C. crispa;* and if to these the Himalayan *C. Brunoniana* were added, we might then be able to ascertain whether they be really distinct species or merely forms of one and the same plant. The closely allied *Pellœa gracilis,* found in many parts of North America and also in Northern India, would be worth having on account of its remarkable resemblance to our *C. crispa.* *Polystichum munitum,* a Fern distributed over Western America, from California to as far north as Nutka, and, therefore, probably hardy, would be a fit companion for our own *Polystichum Lonchites,* though considerably larger. Lovers of golden Ferns would be glad of the *Gymnogramme triangularis,* a species resembling *Pellœa argentea* in appearance, but much larger and covered with golden farina on the under-side; and as this is found as far north on the western coast as the Columbia river, it will, it may be presumed, prove hardy with us. Did space permit, this list of *desiderata* from North America might be greatly extended; but I must pass on to the countries of the South, which offer a rich field to the fern-collector.

In my enumeration of the Ferns of Panama, in Seemann's "Botany of the Voyage of H. M. S. Herald," I have described a very pretty Fern from Southern Darien under the name of *Glyphotænium crispum*, which would be an acquisition to those who grow Ferns in a natural manner, though not suited for pot culture. It is found on trees, from the branches of which its tufts of long and narrow wavy fronds hang down in a very graceful manner. In Darien, Panama, and the adjacent Pacific islands, also, there are several Tree-ferns which we have not yet got; such as *Hemitelia petiolata*, a distinct species, with large pinnate fronds, having widely-separated petiolated pinnules; and *Alsophila elongata*, a very robust species. New Granada, Venezuela, and other countries north of the equator, though explored by several collectors, would still yield a good many desirable novelties to our gardens. In the former country I may indicate the several species of the extremely curious genus *Jamesonia*, with their very narrow, erect, rigid fronds, continuously developing little orbicular, concave, imbricated pinnæ, and densely clothed, while young, with ferruginous hairs; and also *Dryomenes Purdiei*, a magnificent Fern with very deeply pinnatifid fronds from four to five feet long, having extremely broad segments, covered with numerous small sori, which may probably possess indusia; but younger specimens than those in my herbarium are required to settle this point. In Venezuela there is the remarbable *Amphiblestra latifolia*, a Fern resembling some of the larger species of *Aspidium* in habit and appearance, but having a line of confluent sori on the margin like the *Pteridiæ*, to which tribe it is generally referred, some authors retaining it under Humboldt's name, *Pteris latifolia*, and also a species of *Trichomanes* of extraordinary size, *T. Kunzeanum*, nearly allied to our own Irish species *T. radicans*, but with rather rigid fronds, from two to three feet in length. One or two species of *Lindsæa* are also found in Venezuela, particularly the neat *L. stricta*, with fronds varying from pinnate to tri-pinnate, though most commonly bi-pinnate. But the head-quarters of the genus *Lindsæa* in the Western hemisphere are Guiana, where is found the rare *L. reniformis*,* resembling in the general appearance of its fronds the well-known *Adiantum reniforme*, and the yet to be introduced *Gymnogramme reniformis* of Brazil, a rare plant even in herbaria. The exceedingly beautiful *L. trapeziformis*, which has bi-pinnate fronds two feet or more in height, is also found here, as well as in other parts of tropical America and the West Indies. I may remark that, notwithstanding that some of the loosely compiled garden catalogues in common use among gardeners mention as many as twenty-six species of this genus being in cultivation, I know of only two, and I have made many inquiries upon the subject. The genus contains upwards of sixty described species, dispersed over the tropical and sub-tropical countries of

* Lately introduced by Messrs. Backhouse of York, but yet rare.

both hemispheres, and many of them would be highly prized by fern-growers. Amongst other Guiana Ferns worthy of notice there is one to which I would wish particularly to draw attention, not only on account of its singularity, but of the little that is known of it by pteridologists. I allude to the *Danæa simplicifolia* of Rudge, of which I have only seen two specimens, one in Rudge's herbarium, and the other in Schomburgk's Guiana collection. In general appearance the sterile fronds of this Fern resemble those of *Elaphoglossum latifolium*, being about eight inches in length (including the stipes) and of an ovate-lanceolate form, attenuated to the base, while the fertile ones are narrower, and still more attenuated downwards. Nor must I omit to notice the very remarkable *Hewardia adiantoides* of French Guiana, still very rare in herbaria. It would be a noble addition to our large species of *Adiantum*, its fronds being two or three feet high, very broad, and irregularly bi-pinnate, with remote, alternate, petiolate pinnules from three to five inches long, and about two inches wide, and borne upon glossy black stipes. Closely allied to this is the *Hewardia dolosa* of Eastern Brazil, Surinam, and Ecuador, with much longer but comparatively narrower pinnules and rough hairy stipes. There is also in Dutch and British Guiana, as well as in Brazil (in the neighbourhood of Rio Janeiro), a species of the curious *Schizæaceous* genus, *Actinostachys* (*A. pennula*, Hook.), resembling the Ceylon *A. digitata*, already in our gardens, though extremely rare. While the beautiful *Schizæa flabellum*, with its fern-shaped fronds, cleft into two to form broad wedge-shaped segments, and upon stipes a foot or so high, is found in British Guiana, and also on the banks of the Orinoco, Rio Negro, and Yapura rivers. Several other species of Schizæa are likewise worthy of a place in our gardens, such as the pretty *Schizæa pectinata* of the Cape of Good Hope, and *Schizæa dichotoma*, which is found not only in Guiana and Venezuela, but widely dispersed through the Pacific islands as far south as New Zealand, occurring also in Java, Mysore, the Mauritius, and other parts of the Eastern hemisphere. Allied to these, also, are the two Brazilian species of *Coptophyllum* described by Dr. Gardner, and likewise the *Trochopteris elegans* of the same author, all of which some pteridologists include under the genus *Anemia*, and perhaps rightly so with respect to the former, for they have the same relationship with true *Anemia* that *Osmunda cinnamomea* has with *O. regalis*, their barren and fertile fronds being distinct. Both species are found in the province of Goyaz; one being named *C. millefolium* and the other *C. buniifolium*, from the general resemblance in the divisions of their barren fronds to the leaves of *Achillea millefolium* and Bunium. The *Trochopteris elegans* is an exceedingly curious little Fern, with flat, radiating fronds of a somewhat spathulate form but more or less five-lobed, the two lower lobes being deeper and bearing the sporangia, the entire plant resembling a rosette, and growing on rocks like a lichen. Dr. Gardner found it on the Serra de Natividad, in the province of Goyaz. Amongst other Brazilian Ferns worth being looked after, I may mention two species of *Antigramme*—*A. Brasiliense* and *A. Douglassii*, the former having oblong-lanceolate fronds about a span long, tapering downward to a short stipe; and the latter ovate fronds of the same length, but usually cordate at the base and upon long stipes. *Lomaria zamioides* of Gardner, a plant with a trunk four feet high, resembling a *Zamia*, found by Gardner in boggy places near the summit of the Organ Mountains, would also be a valuable addition to our small-growing Tree-ferns.* Brazil is rich in Tree-ferns, but only a few of them have as yet been introduced. I will mention only one or two. *Dicksonia Sellowiana*, found on the Organ Mountains, is, like the *Lomaria* above mentioned, remarkable for its resemblance to an extreme southern species, dried specimens being scarcely distinguishable from the *Dicksonia antarctica*, though most probably if the two were cultivated side by side they would prove very distinct. *Cyathea vestita* and *C. Schamschin* appear to be very plentiful throughout Brazil, and both are very fine species, the former having a trunk from twenty to thirty feet high. The two species of *Trichopteris*—*T. excelsa* and *T. elegans*—are also very graceful trees, found in Southern Brazil, and although the latter is included in the following enumeration, it is still very rare in our collections. Several special localities in Brazil may be mentioned as abounding in Ferns, such as the Organ Mountains and St. Catherine's, in the east; on the eastern slopes of the Andes, where at elevations of from fifteen hundred to four and five thousand feet, in some localities, they flourish in great luxuriance. At Tarrapota, in Peru, Dr. Spruce, in a diameter of fifty miles, collected no less than two hundred and

* Fine plants of this Fern have been recently imported to this country by Mr. Low of the Clapton Nurseries.

fifty species, twenty of which were Tree-ferns, and many new and interesting species.

Before leaving tropical America I must say a few words respecting the West Indies, the Fern Flora of which is to a great extent identical with that of the countries on the Atlantic coast of South America. A tolerably accurate idea of the number of species indigenous to the West Indies may be obtained from Grisebach's Flora of the islands belonging to Great Britain, where three hundred and forty are described, and their particular localities noted. Out of these, two hundred and twenty will be found enumerated in the following pages as already in our gardens; and as our intercourse with most of these islands is now so frequent, and the voyage accomplished with such rapidity, we may expect ere long to receive all the most striking types of the remaining ones. Indeed, the West Indian correspondents of the Royal Gardens at Kew, as well as those of several nurserymen, and other private individuals, are continually forwarding Ferns to this country; and under these circumstances I do not think it worth while to mention any particular species; but it is worthy of remark that among our desiderata is the numerous group represented by *Polypodium trichomanoides*.

Passing westwards to Ecuador and Peru, I might give a long list of desiderata, particularly of pretty little Alpine species from the Andes, belonging to *Cheilanthes*, *Notholæna*, *Asplenium*, and *Polypodium*, but want of space compels me to confine my remarks to a few of the most desirable ones. In his second century of Ferns, Sir W. J. Hooker has figured a beautiful *Polybotrya*, named *P. Lech-*

leriana, after its discoverer, Dr. Lechler. It has large, finely divided, somewhat membranaceous fronds, three or more feet in height, resembling a species of *Darea*, and thick scandent rhizomes. *Cyathea microphylla*, found by the same collector, and figured in the same work, appears to be a neat little Tree-fern, with stems four feet high and finely divided fronds, two or three feet long, ferrugineous from hairs on the under side. Some species of *Gymnogramme* are worthy of note, such as *G. elongata*, with narrow pinnate fronds a foot or more long, something like those of the well-known *Nothølæna trichomanoides*, and clothed with copious longish hairs; *G. flabellata*, the fronds of which are about a foot high, bipinnate, with dark shining stipes, and little flabelliform, dichotomously divided, green pinnules, and extremely neat; *G. incisa*, which has bipinnate fronds a span or more high, and scarcely more than an inch wide, with the pinnules deeply incised. These Gymnogramms are also found in Venezuela, New Granada, and countries north of the equator, where there is also a remarkable scandent species, *G. refracta*, the finely cut fronds of which continuously increase to a great length, and ramble over the branches of trees. Our collections of *Gleicheniæ* might be also enriched with several species from Peru and Chili, particularly *G. simplex* from the former, and *G. pedalis* from the latter. *G. simplex* having simple, pectinately pinnatifid fronds a foot and a half in length, with short stipes; and *G. pedalis* fronds of the ordinary form, something like *G. furcata*, but smaller, neater, and more compact in its mode of growth. And, finally, the two singular *Polypodiæ*, with dimorphous fronds, would be very pretty additions to our ferneries. One of these, *Polypodium heteromorphum*, Hook., was found by Dr. Jameson "upon the top of the mountain face of dripping rocks;" and has simple fronds like those of *Asplenium Trichomanes*, mixed in the same tuft with others which are repeatedly branched in a regular dichotomous manner like the *Gleicheniæ;* while the other, *Polypodium bifrons*, Hook., found by the same botanist in Ecuador, growing on branches of trees partially immersed in water, has sterile fronds resembling oak leaves in their general outline, and narrow wavy fertile ones. To the creeping rhizomes of the specimens collected by Dr. Jameson there were attached curious bodies, resembling small potatoes; but these were most probably adventitious, and caused by some insect. Dr. J. W. Sturm, in his little work on the Fern Flora of Chili, enumerates one hundred and sixty-one species as found in that country and the adjacent island of Juan Fernandez; but very few of these have as yet been introduced, though many of them would prove acceptable additions to our half-hardy collections.

The numerous islands of the Pacific Ocean are, as a general rule, rich in Ferns, and worthy of being visited by a collector of living plants. The Hawaiian or Sandwich Islands, for example, would afford three fine species of *Cibotium*. One of them, which has the stipes densely clothed with beautiful golden silky moniliform hairs, is so abundant that these hairs are collected as an article of commerce and are largely exported to California and Australia for the purpose of stuffing cushions, &c.; *Polypodium pellucidum*, a creeping species, allied to our *P. vulgare*, but differing in having pellucid striæ between the fascicles of veins, and varying so much in the more or less compound division of its fronds, that one state of it was described as a distinct species by Sir W. J. Hooker, under the name *P. myriocarpon*; *Asplenium Sandwichianum*, with large tripinnate fronds three feet high, with numerous small segments, bearing some resemblance to a *Mimosa* leaf, and others too numerous to mention.

The Galapagos, although not rich in Ferns, are worthy of notice, on account of a very rare and remarkable species, first described and figured by Sir W. J. Hooker, in the "Icones Plantarum," under the name of *Acrostichum* (*Neurocallis*) *aureo-nitens*, and more recently in the fifth vol. of the "Species Filicum" as *Acrostichum* (*Chrysodium*) *aureo-nitens*. Judging by either of the sectional names, it might be supposed to have some resemblance to the well-known *Acrostichum aureum*, but such is not the case, reticulated venation and apparent amorphous sori being the only characters that place it in that alliance; in habit it is totally distinct, just as distinct from *Acrosticum* (*Chrysodium*) *aureum* as *Ceterach officinarum* is from *Asplenium* (*Hemidictyon*) *marginatum*. The plant has simple barren and pinnate fertile fronds 6 to 10 inches in length, the whole plant being densely clothed with shining scales. To me its relationship seems to be with *Hemionites vestita*, a beautiful Fern of India, and also with another little-known species, the *Gymnogramme* (*Eugymnogramme*) *Muellerii*, a native of north-eastern Australia, described and figured by Sir W. J. Hooker, in the fifth vol. of the "Species Filicum," which also seems to me to be closely related to

Hemionites vestita, but described as having free veins. Whatever difference then may actually be in the character of the venation of these three species there can be no doubt but that they are closely allied and constitute a very natural group. I have always considered that *Hemionites vestita* does not well associate with true *Hemionites,* but now, having found two companions for it, I view them as forming a natural genus, to which I apply the name *Chrysodium.* I hope that ere long we may have the opportunity of becoming better acquainted with them; their silky appearance renders them worthy of being added to our living collections.

In the Fijis again, Ferns form a conspicuous feature in the vegetation. During a visit of only six months, recently paid to these islands by Dr. Seemann, for the purpose of exploring them and investigating their Flora, he collected specimens of about 800 species of plants, and of these one-seventh were Ferns, very few of which are yet known in our gardens. A few of the desiderata are worth mentioning, particularly the graceful *Todea Wilkesiana,* found by the collectors attached to the United States' Exploring Expedition, and named by Mr. Brackenridge in compliment to Commodore Wilkes, who was in command of the expedition. It is spoken of by Brackenridge as the "Little Tree-fern," and as being not more than three or four feet high; but Seemann found it in the mountains of Somosomo, where it grows as underwood, attaining seven feet in height, and often with several crowns. The stem is as slender as a walking-stick, and the fronds bipinnate, and about two feet in length, with the ultimate pinnules thin, but not so delicate or so finely cut as *T. hymenophyloides*. *Davallia Fejeensis* is a species with highly decompound fronds, a foot or so high, having the segments so narrow that they bear only a single sorus upon each. A species of *Hemonites*, *H. lanceolata*, and *Syngramme pinnata*, are found in these islands; the latter having, on old plants, large pinnate fronds about one to two feet high, including the rather long stipes, the first simple lanceolate fronds from a foot to eighteen inches high, but it is questionable whether these simple fronds be not merely a state of the latter plant, for other species of Syngramme are known to have simple fronds as well as pinnate. Allied to *Syngramme* is the long and well-known *Tænites blechnoides*, which has a wide geographical range, but is not yet introduced alive; the form usually seen in herbaria from the Malayan islands has large simply pinnate fronds, with long tapering pinnæ, like *Blechnum orientale*.

There is also another Fijian Fern, desirable as much on account of its botanical character as from the singularity of its appearance, viz., *Diclidopteris angustissima*, which grows epiphytically on trees, chiefly the Tahitian chestnut (*Inocarpus edulis*), in the manner of *Vittaria*, and has narrow, thin, grass-like fronds, varying from six inches to a foot in length. In all the Fijian specimens I have seen, the fructification is seated in a groove upon a vein running along the side of the midrib, and parallel with it, though in the generic character drawn up by Brackenridge, it is said to be normally in two rows, one on either side of the midrib; but, as Brackenridge alludes to its being occasionally on one side only, I am not disposed to consider it as a distinct species without further evidence, though it is worthy of remark that the specimens seen by that author were partly from the Samoan, and partly from the Fijian group; and it is possible that the two forms are separated geographically, as well as by their technical characters. Did space permit many others might be indicated—not only from these islands, but from other Polynesian groups; though, as far as it is at present known, the Fern Flora is very uniform in species throughout. New Caledonia, the Soloman Isles, and others, have not, however, yet been well explored, either by botanical or horticultural collectors.

As might be expected from the great intercourse that has been carried on of late years between this country and New Zealand, the greater part of the Ferns indigenous to that colony are now to be found in our half-hardy ferneries, only about 20—a small number—out of the 120 species described by Dr. Hooker in his "Handbook of the New Zealand Flora," remaining to be introduced. One especially I should be glad to see in a living state: viz., the remarkable as well as handsome and very rare *Loxsoma Cunninghamii*, found by Cunningham on the Keri Keri River, Bay of Islands, and by Sinclair on the Wangarei River, in the Northern Island. This Fern possesses the habit of a *Microlepia*, and has broadly triangular decompound fronds, two to three feet high, glaucous below, with sori intermediate in character between *Trichomanes* and *Davallia*.* The Tasmanian Ferns are likewise nearly all intro-

* This Fern was introduced, but has not become established.

duced; only one half-dozen out of the 52 species described by Dr. Hooker in his "Flora Tasmania" being unknown in our gardens, while of the Australian ones about a third are still wanting to complete our collection, and one of these is the extremely rare *Platyzoma microphylla*, found by R. Brown on the borders of the Gulf of Carpentaria during Flinder's voyage—an extremely neat little Fern, with rigid pinnate fronds a foot long, and hardly one-eighth of an inch broad, having minute oval pinnules, with revolute edges and powdery beneath, growing in tufts from short creeping rhizomes.*

I have now traced the progress of the introduction of exotic Ferns to the gardens of this country, and shown that many novelties have yet to come. No doubt, more or less of them will from time to time be introduced, as they are eagerly sought after by numerous amateurs. Select private collections are thus formed, in many cases consisting of rare and unique plants; but, in the course of time, changes in private establishments take place, and thus collections of Ferns get dispersed, and species are often lost to the country. It is, therefore, only to such public establishments as that of Kew that we have to look to for the preservation of special collections. As there is no law or rule defining what kinds of plants should or should not be grown in public Botanic Gardens, the matter resting entirely with the Director or Curator, some families of plants are often more favoured than others, although all are of equal merit

* Since the above was written about a dozen of the species named have been introduced, and will be found in the Appendix to the Second Edition.

in a botanical point of view. To a certain extent, the Fern collection at Kew is a proof of this; it so happens that both Sir W. J. Hooker and myself had an early predilection for Ferns, which has led to the gradual increase of the fine collection at Kew; and, although I am now* incapacitated, by failing sight, from doing more in support of this collection, still, happily, it remains under the direction of Sir W. J. Hooker, who, doubtless, will not allow it to deteriorate, either in number of species or otherwise. One great means towards assisting in their preservation is continuing to view them as a scientific collection. Scientifically-arranged collections are presumed to be the leading features of all Botanic Gardens. Unfortunately it is not the most showy or attractive. My long experience has shown me that as soon as a scientific arrangement in any family of plants is lost sight of, and showy cultivation made the first consideration, a rapid loss of species is the sure consequence. For their proper maintenance it is most essential that the cultivator should view even the most humble species with a scientific and conservative eye. It is also much to be desired that an official rule should be made, requiring an inventory of the collections to be taken every few years, and the publication of a general catalogue; or, in order to meet the various tastes of the public, separate catalogues of special families, like the one I now publish of the Ferns, might be issued.

* May, 1864.

II.—ORGANOGRAPHY.

FOR the purpose of rendering the technical descriptions occurring in the following pages intelligible to those not well acquainted with botany, I have thought it necessary to devote a preliminary chapter to organography, being the explanation of the various terms in common use among pteridologists. I adopt this course in preference to giving an ordinary glossary, because I think a better idea of the structure of the plants, and the relation of one organ to another, and of the relation of the terms to the organs themselves, may be conveyed by it; but for convenience of reference I append an alphabetical list of the terms, paged so that they can be easily found in the explanatory chapter. In the generic characters I have endeavoured to avoid needless technicalities, though I have not attempted to frame them in what is commonly called a "popular" style, and I hope that with the aid of the following explanations, persons of ordinary abilities who have not made botany their study, will be able to understand them. In many cases, especially in describing the form and shape of the fronds, the same terms are employed as in flowering plants; but as now and then they have special significations, I have briefly explained all that occur in this work.

Ferns (*Filices*) are flowerless plants, and form the highest order of the division of the vegetable kingdom termed *Cryptogamia*. Their most evident organs consist of the stem and the leaves, the latter of which are always called *fronds*, and are variously traversed by *veins*, ramifying in a determinate manner in the different genera. Upon certain definite parts of these veins, generally on the under side of the frond, termed the *receptacles*, clusters or lines of free one-celled spore-cases (*sporangia*) are produced, or occasionally many-celled ones (*synangia*), and in these cases the reproductive *spores* are contained. The clusters are called *sori*.

VERNATION.

The word *vernation*, as employed by me, designates the mode of growth of Ferns, or, in other words, the manner in which their fronds are developed and connected with the stem.

VERNATION is either—

Articulated when the fronds are attached to the stem by a joint, and leave a clean scar when they fall away ; or,

Adherent when no such joint exists, and the bases are continuous with the stem.

And it is either—

Uniserial when the fronds are produced one after the other, in a single lineal series, sometimes close together (*contiguous*), and at other times far apart (*distant*) ; or,

Fasciculate when they surround a central axis, upon the top of which they form a crown.

STEM.

In a large number of Ferns the stem is not at first sight very evident ; and even when plainly visible, it is frequently confounded with the root by the unlearned (as, for example, the underground stems of *Pteris aquilina*) ; but in others, as in Tree-ferns, it is very marked. It is an organ of considerable importance for classifying purposes, and often affords valuable distinctive characters.

The principal modifications of the stem are the—

Rhizome, a brittle, fleshy, prostrate stem, producing roots along its under side, mostly growing above ground (*epigæous*), and then furnished with scales (*squamose*), but occasionally under ground (*hypogæous*), and then destitute of scales. It varies greatly in length, and is either simple or branched; when very short and branched it forms tufts (*cæspitose*), and when very long (*surculose*) it usually climbs on trees (*scandent*). Very rarely it is erect (*subfrutescent*). Its point of growth is always evidently (often considerably) in advance of the undeveloped fronds; and the fronds themselves are produced singly from special, more or less distant, points on its sides, termed nodes, at which they are articulated.

Sarmentum, a tough slender running stem, rooting like a rhizome, and either epigæous or hypogæous, but differing in having the bases of the fronds adherent and continuous with it, and in its point of growth being coincident with, or scarcely ever in advance of, the undeveloped frond.

Caudex, an *erect* or reclining (*decumbent*) stem, either simple or tufted (*cæspitose*), through the growth of offsets, or rarely sending out long running shoots, which root at their extremity (*stoloniferous*). It is often very small, scarcely rising above the earth, but generally more or less elevated, and sometimes forms a cylindrical trunk (*arborescent*), occasionally 50 or more feet high, which, in many species, is thickened by the growth of numerous aërial, outgrowing, wiry roots. And it bears a crown of usually adherent fronds, developed in a spiral series, upon its apex.

FRONDS.

The fronds of Ferns are either *barren* or *fertile*. In the great majority the latter do not differ very much from the former, though they are generally rather narrower in all their parts. But sometimes they are very evidently different on the same plant, the barren presenting the ordinary leafy appearance, and the fertile being decidedly contracted, occasionally so much so that the leafy part is entirely absent, or in some the two kinds are combined in the same frond, the fertile portion being contracted, and the barren leafy.

When young the fronds are involutely coiled, in the manner of a watch-spring, and gradually uncurl during the period of growth (circinate); rarely straight, as in *Ophioglosseæ*.

Fully developed fronds vary in size from less than an inch to 15 or 20 feet in length, and from a line, or even less, to 10 or 15 feet in breadth. They also vary in form, in circumscription, and in texture; and they are either furnished with a leaf-stalk (*stipes*) or are leafy to the base (*sessile*).

In describing the form, circumscription, texture, and surface of the *fronds* of Ferns, the same terms are employed as in the case of the leaves of flowering plants. They vary from simple entire to decompound-multifid. In compound fronds the primary divisions are termed *pinnæ*, and when more than once divided, the ultimate ones *pinnules;* and the terms applied to simple fronds are equally applicable to these divisions. The divisions or branches of their stipes also are termed the *rachis.*

Their texture is very different in different species. Some being thin, membranous, and even pellucid, while others are thick and coriaceous, or fleshy, rigid or flaccid.

The surfaces of the fronds are either quite smooth, or furnished with different kinds of hairs, glands, or scales (the latter have received the name of *ramenta*, and are generally membranous and deciduous), or they are covered, particularly the under surface, with white or yellow farina.

The plants called Fern Allies differ entirely in habit and mode of growth from true Ferns; that the word fronds is not applicable; but as the genus *Selaginella* is called "fern-like plants," I therefore apply the term "*frondules*" to the species with distinct stems, and to the main branches of the surculose species.

VEINS.

In Ferns the mode in which the veins are disposed in the substance of the fronds, or the *venation*, as it is termed, is of more importance than in flowering plants, the characters relied upon for distinguishing the genera depending more or less upon it, and there are numerous terms applied to it.

The midrib of simple fronds, or of the pinnæ or pinnules of compound fronds, is called the *costa*, and is in the former a continuation of the stipes, gradually decreasing in thickness towards the apex, or altogether disappearing (*evanescent*), and in the latter a continuation or branch of the ultimate rachis with which it is either *adherent* or *articulated.* It is generally central; but is sometimes excentric, or even quite on one side (*unilateral*), or sometimes there is no costa at all. From the sides of the costa veins are produced at more or less distance from each other, generally equal on each side, except when the costa is excentric or the frond or segment has a radiating axis. The direction of the first or primary veins is, as in the leaves of other plants, towards the margin and apex of the frond or segment, forming a more or less acute or obtuse angle, or sometimes nearly a right angle with the costa.

In describing venation the words *veins*, *venules*, and *veinlets* are employed, each successive one of which is intended as a diminutive of the preceding; "*veins*" being applied to the first ramification of the midrib, "*venules*" to the branches, and "veinlets" to the branches of the venules. Some fronds have veins only, others veins and venules, and others again all three.

Terms are occasionally employed to express the relative distinctness of the venation, particularly when any marked peculiarity exists: thus it is said to be—

Elevated, or external, when they are so thick that they are readily seen and felt on the under surface of the frond; and—

Internal when very much sunk in the substance of the frond.

The primary veins are—

Costæform when very strong and well defined, more or less resembling the costa in general appearance;

Undefined when of the same size as and not distinguishable from the venules and veinlets; and—

Evanescent when they gradually disappear towards the margin.

Veins are spoken of as—

Free when each vein proceeding from the midrib, however much it may be divided, is entirely unconnected with the neighbouring ones; and—

Anastomosing when the venules of one vein are in some way connected with those of the next.

A fascicle comprehends a single vein with all its venules and veinlets.

Free veins are—

Simple when each vein proceeds from the costa to the margin without branching (83).

Forked when they divide at an acute angle into two or more branches after leaving the costa (51).

Simply forked, or *dichotomous*, when the division is into two branches (96).

Pinnately forked when the primary veins are scarcely defined, and branch several times one after the other on both sides (75).

Pinnate when the primary veins that run from the costa to the margin are distinctly defined, and produce venules in regular order on both sides, so that the fascicles have a feather-like appearance (121).

Radiate when the veins spread out from a definite point at the base of the frond or segment (93).

The simplest form of anastomosing venation is when the apices of the veins are combined or connected by means of a marginal vein (113). In the more complicated forms it is spoken of as—

Angularly anastomosing when the venules of one vein join those of the next, and form an angle at their point of junction (65); when the angle is very acute the term *acutely anastomosing* is employed, or sometimes called cathedrate.

Arcuately anastomosing when the venules of one vein join those of the next, and together form an arch or curve (63).

Transversely anastomosing when the venules of one vein join those of the next, and together form a nearly straight line (104).

Distantly anastomosing when the venules are parallel with the costa, close together, and joined at long intervals by short cross veinlets.

Compoundly anastomosing when the venules are irregularly connected in a more or less net-like manner, and have variously directed free or conniving veinlets in the areoles (21–28, 43).

Reticulated when the veins, venules, and veinlets are all connected together in a more or less net-like manner: *uniform* is used in reference to reticulated venation when there is no apparent difference between the veins, venules, and veinlets (31, 55).

Areoles are the spaces formed by the anastomosing of veins, and are of various shapes and sizes: those next the costa are called *costal areoles.*

In speaking of the venules of forked and pinnate veins it is sometimes necessary to indicate a particular one in the fascicle: thus, the—

Anterior venules are those on that side of the vein next the apex of the frond or segment; and the

Posterior venules those on the opposite side farther from the apex.

Venules and veinlets are likewise said to be—

Excurrent when directed towards the margin of the frond or segment; and

Recurrent when directed from the margin;

And their apices are said to be

Clavate when thickened like a club.

FRUCTIFICATION.

As a general rule, what is called the fructification of Ferns is seated on more or less regularly arranged points or lines on the under surface or margin of the fronds, and is usually of well-defined form. There are, however, some variations from this. For example, in *Acrosticheæ* it either covers the whole under surface of the fronds, or is in irregular undefined patches, and in some other cases, as *Botrychium, Osmunda,* &c., where the fertile fronds are much contracted, it assumes a spike-like or racemose form.

The terms used in describing the fructification may be classed under four heads:—1st. Those relating to the *receptacle;* 2nd. those relating to the *sporangium* and *synangium;* 3rd. those relating to the *sorus;* and 4th. those relating to the *indusium.*

1. *Receptacle.*

The receptacles are the sites upon which the sporangia are seated, and are generally either thickened points on, or long thickened portions of, some part of the venation.

In position they are—

Terminal when on the points of the veins or their branches (5, 7).

Basal when close to the costa (1).

Axillary when on the point where the veins fork (131).

Compital when on the angular crossings or points of confluence of two or more venules or veinlets.

Medial when in none of the above positions, but some intermediate part of the veins or the branches (21, 28).

They are *superficial*, or *immersed* in the substance of the frond, or *elevated* above its surface, and then columnar (plate I. fig. 9) or globose.

In form they are—

Punctiform when small and dot-like.

Elongated when long and line-like.

Amorphous when of no defined form (46).

2. *Sporangium.*

The spore-cases, or *sporangia*, are the organs which contain the reproductive spores, and are borne in masses upon the receptacles. They are thin and transparent, or horny and opaque, unilocular and globose, oval or pyriform, usually pedicellate, which is articulate, but sometimes sessile, and either furnished with a more or less complete articulated elastic ring (*annulate*) (plate I. fig. 1), or destitute of a ring (*exannulate*) (plate I. fig. 4). In annulate sporangia the ring is said to be

Vertical when it rises immediately from the apex of the pedicel (of which is a continuation), and passes vertically over the apex of the sporangium (plate I. fig. 1).

Horizontal when it passes horizontally round the sporangium either at or about its middle (plate I. fig. 2), or at the apex (*apical*) (plate I. fig. 3.)

Oblique when it has neither of the above directions, but passes round the sporangium in some direction intermediate between them.

When the sporangia arrive at maturity and are under certain favourable conditions as to dryness, the elasticity of the ring causes them to burst open with force and sound sufficient to be heard, and this takes place in a direction at or very near to a right angle with the direction of the ring. In exannulate sporangia the opening takes place by a simple slit or pore (plate I. figs. 4 and 5).

Synangium.

The *synangia* are formed by the union of a greater or lesser number of exannulate sporangia, arranged side by side, forming a series of cells, disposed in a circle, or in two rows side by side, united in one mass, which either remain united (plate I. fig. 5), or separate longitudinally in two valve-like lobes (plate I. fig. 6). The cells open for the escape of the spores by a slit on their inner side or by a pore at their apex.

In Lycopodiaceæ and Marsileaceæ there are two kinds of sporangia, the one containing numerous small spores, the other only a few—considerably larger. Some authors consider them to represent different sexes, and therefore named the first Antheridangia, the other Oophoridangia. The large spores are known to vegetate, and some say the small ones also; the large ones are called *Corpuscules.* In the genus *Marsilea* the sporangia are called conceptacles, because they contain free vesicles of two kinds, one containing small spores, Antheridangia, the other large ones, Oophoridangia.

3. *Sorus.*

The *sori* are the masses of sporangia borne upon the receptacles, and are either *naked* or furnished with variously shaped hairs and scales, or with membranous or rarely coriaceous covers of various forms (*indusia*); their form and position correspond with and are dependent upon those of the receptacles, which are their foundations. Thus, when the receptacles are punctiform, the sori are always round (5) or globose, while elongated receptacles bear sori of many forms, *oblong ovate, oval, elliptical, arcuate, linear* (50), *reticulated* (54), &c. When situated on the margin of the frond or segment (*marginal*), a little within the margin (*antemarginal*), somewhere between the margin and the midrib (*intramarginal*), close to the midrib (*costal or basal*), or sometimes on a pedicel, and projecting slightly beyond the margin (*exserted or extrorse*) (73). In some cases they are irregularly scattered, but in others they are arranged either in rows (*serial*) (7) or in continuous lines, and when these diverge at an angle from the midrib they are said to be *oblique* (110); and when parallel with either the margin or the costa, *transverse* (96 and 100). As a general rule, each sorus is distinct and well-defined, but in many cases the receptacles are so very close together that one sorus runs into another (*confluent*), or sometimes the receptacles themselves are joined and form a more or less perfectly united simple sorus, or when not perfectly joined (as in *Cryptogramme* and *Platyloma*) a compound linear sorus.

4. *Indusium.*

As stated above, the sori of some ferns are naked while those of others are furnished with a kind of cover, to which the name *indusium* is given by some authors, and *involucre* by others.

The indusia present many well-marked forms, and often afford valuable characters for distinguishing genera, though they are by no means constant. Three kinds are distinguishable: *special*, *accessory*, and *universal*.

True or *special* indusia are of a cellular membranous nature, and are produced from the receptacles to which they are attached in different ways. In some cases they are in the form of an orbicular disk, and then rise from the centres of the receptacles to which they are attached by their own centres, their edges being free all round; this form is called peltate or central (plate I. fig. 7). More frequently, however, the indusia are more or less elongated, and are then attached to the sides of the receptacles (*lateral*) (plate I. fig. 8). In this case their attachment is either on the side next the costa (*interior*), or on that next to or at the margin (*exterior*), and is either by a point or sinus on their side, in which case their form varies from *reniform* to *oval* and *oblong*, or it is by the entire length of one side, when they are *linear* (110). Their surface is flat (*plane*), arched (*vaulted*), or hood-like (*cucullate*), and their edges are either entire or variously laciniated or fringed.

Besides these two modes of attachment, there is a third kind where the indusia are attached all round the base of the receptacle, and they are at first globose and entire, but ultimately their apex opens, and then they assume a cup-like (*calyciform*) form with the margin more or less entire (plate I. fig. 9); sometimes the attachment is only half round the receptacle (*semi-calyciform*).

Accessory indusia, sometimes in addition to the true indusia, portions of the margin of the frond are changed in texture and form, what are here termed *accessory indusia*, and which resemble the true indusia in appearance. These connive more or less with the true indusia, which in these cases are always attached on the interior side of the receptacles, and the two combined indusia form continuous or interrupted grooves, or urceolate, bilabiate, or tubulose cysts, open exteriorly and containing the sporangia (plate I. figs. 10 and 11).

Universal indusia occur in cases when the segments of the fertile fronds are contracted. They consist simply of the margins of the segments being more or less changed in texture, and rolled inwards so as to include all the sori upon the segment (plate I. fig. 12).

There is also another kind of indusium, called "indusoid scales;" they only occur in a few species of the division Eremobrya. In *Pleopeltis* this consists of orbicular, peltate, glistening imbricate disks, covering the sporangia; in *Hymenolepis* they are very thin and membranous; in *Schellolepis* they are very irregular in form, and seem to be imperfect sporangia; their deformity being caused by the excessively crowded immersed sporangia; they are also found in *Tænitis* and *Vittaria*, and have received the name of paraphyses. I however do not use this term in describing those genera. The orbicular disks of *Pleopeltis*, however, seem to be more special organs, particularly in the smooth-fronded species.

I have now explained the terms of the chief organs and structure of Ferns made use of for their classification. I fear a beginner will say it is quite enough to deter any one from entering upon the study of Ferns; but he should bear in mind that it is quite as impossible to read a language without first learning the alphabet as to understand botanical descriptions without first mastering the technical terms employed in them. He will be further impressed with the difficulty of study when he finds that the very first point of investigation is to determine whether the fern before him has or has not a ring to its spore-cases. He presumes that a microscope is required to determine this first starting-point; but such is not actually the case, for with the aid of a pocket lens he will be able to detect the presence or absence of a ring, and as annulate and exannulate Ferns in cultivation in this country are in proportion to one another as one to forty-five, he may soon become aware that the great majority of Ferns belong to the annulate section. But the best way for a beginner is to procure a few correctly-named species of each tribe, and carefully compare them with the characters given in the following pages. He will soon overcome the dread of technical phrases, and before long will be able to refer his unnamed species to their respective tribes and genera.

EXPLANATION OF THE PLATE.

Annulate Sporangia—

FIG. 1. Sporangium with a vertical ring, mag. 100 diameters (sub-order Polypodiaceæ).
2. Sporangium with a horizontal ring, mag. 100 diameters (sub-order Gleicheniaceæ).
3. Sporangium with an apical ring, mag. 100 diameters (sub-order Osmundaceæ).
(*a.*) Spores of each highly magnified, 200 and 300 diameters.

Exannulate (Order Marattiaceæ)—

FIG. 4. Sporangia (two) free, opening by a vertical slit, mag. 25 diameters (Angiopteris).
5. Sporangia united (synangium), opening by pores, mag. 7 diameters (Danæa).
6. Sporangia united (synangium), opening by slits, mag. 9 diameters (Marattia).
(*a.*) Spores of each highly magnified, 300 diameters.

Indusia—

FIG. 7. Indusium peltate orbicular, slightly magnified (Aspidium).
8. Indusium lateral reniform, slightly magnified (Nephrolepis).
9. Indusium calyciform, slightly magnified (Cyathea).
10. Indusium linear, interiorly attached, slightly magnified (Asplenium).
11. Indusium valvate, slightly magnified (tribe Dicksoniieæ).
12. Indusium universal, slightly magnified (Struthiopteris).

ON THE GENERA OF FERNS AND THEIR CLASSIFICATION.

THE systems for the classification of Ferns are almost as numerous as pteridologists themselves; indeed, nearly every author, from Linnæus downwards, who has written upon the subject, has propounded his own views, and these have generally differed both from his predecessors and from his contemporaries. But the point upon which pteridologists appear to differ most, and on which their only agreement seems to be an agreement to differ, is the definition of genera and their limits. I say emphatically appear to differ, for in the works of those most at issue, the differences are not so much in the limits of the groups themselves as in the relative importance assigned to them. For example, while some, as Presl, Moore, and myself, break up the old Linnæan genera, *Polypodium*, *Aspidium*, &c., into a greater or lesser number of smaller genera, others, as Hooker and Mettenius, prefer adhering to the Linnæan genera, without greatly altering their characters, and adopting the modern generic names as sectional ones for such divisions as they find themselves compelled to make. It would occupy too much space to enter fully upon this subject, and I must leave it for a more extensive work upon the genera of Ferns, long contemplated by me,* contenting myself here with a brief mention of the organs more or less employed by pteridologists in establishing and classifying genera.

* See "Historia Filicum."—Macmillan & Co. 1875.

An examination of the works of Linnæus shows that he was acquainted with about one hundred and eighty species, and these he classed under eleven genera (viz., *Osmunda, Onoclea, Acrostichum, Hemionitis, Polypodium, Asplenium, Pteris, Blechnum, Lonchitis, Adiantum,* and *Trichomanes*), which were founded upon purely artificial characters, derived solely from the shape and position of the fructification. This system was amply sufficient for the limited number of species then known; indeed, the proportion of genera to species was much larger in Linnæus's days than in our own; but when the number of species had been greatly augmented, it became obvious that, in order to avoid genera of unwieldy dimensions, if not for other reasons, additional characters must be sought for; and these have gradually been introduced. It is a remarkable fact, however, that although the number of species now known exceeds by about twenty-fold that known to Linnæus, it is quite possible to arrange them all under the eleven genera established by that author.

After the time of Linnæus, the first additional organ relied upon for generic characters was the *indusium*, which was employed by Sir J. E. Smith and Professor Roth, and afterwards more fully by Swartz, who divided the twenty-five genera known to him into "naked" and "indusiate." Linnæus noticed the fact of the sori following the course of the veins in his character of *Hemionitis*, but, in 1810, Robert Brown first specially employed characters taken from the *position of the sori upon the veins*. This was the next important step in advance. The same learned botanist was also before anybody else to point out the importance of venation as an aid to classification, but the credit of being the first to employ characters from venation upon a large scale is due to Professor Presl, who, in 1836, published his celebrated "Tentamen Pteridographiæ," where he described one hundred and fifteen genera of Polypodiaceæ alone, in the characters of all of which the venation holds the most prominent place. Several years before seeing Presl's "Tentamen," I had been engaged in working out, and had completed, a treatise upon the same subject, which, with a few necessary alterations in nomenclature, I afterwards published.* My views for the most coincided with those of Presl, but I had paid more attention to forming natural groups and bringing together species agreeing in their mode of growth, and vegetative organs; for it appeared to me that pteridologists did not give sufficient importance to that point, and even now it is not taken into consideration as much as it deserves to be. With the exception of my own more recent efforts to obtain characters from the mode of growth presently to be explained, the only further suggestion of any importance remaining to be noticed is that of M. Fée, who, in his work on the *Polypodiaceæ*, introduced characters taken from the form and structure of the sporangia, the number of articulations in their rings, and the form of their spores. The form of the sporangia, and direction of their rings, had previously been adopted by Presl and myself for distinguishing the main orders or sub-orders of Ferns, and I, in common with all modern pteridologists, still rely upon those organs for that purpose; but I cannot consent to their introduction into generic and specific characters, as proposed by Fée. Even were the dif-

* Hook. Journ. Bot., 1841.

ferences pointed out by him constant, which they are not, the organs themselves are so minute that the study of Ferns would be impeded rather than facilitated by the laborious microscopic examination demanded. The spores also vary at different ages, and are thus apt to mislead. No practical advantage is gained by the introduction of such characters; and natural groups and alliances can be established without them, by employing such tangible characters as do not require much aid from the microscope for their observation.

I now come to consider the characters taken from mode of growth. My long connection with the Royal Botanic Garden at Kew, where an unrivalled collection of Ferns exists, has given me abundant facilities for the observation of growing plants, and after an attentive study and close examination of many years I am induced to attach a higher value for systematic purposes to the different modes of growth than my contemporaries may be disposed to do. My views upon this subject were first published in Seemann's "Botany of the Voyage of H.M.S. *Herald*" (p. 226), and subsequent observations have but confirmed them.

Ferns present two very distinct modes of growth, the one of which I term *Eremobrya,* and the other *Desmobrya,* and these are comparatively as distinct as the primary divisions of flowering plants; but I do not, as has been suggested, consider that there is any analogy between the structure of the stems of *Eremobrya* and *Endogens,* and *Desmobrya* and *Exogens,* that their respective modes of development are identical, or that *Eremobrya* and *Desmobrya* are of equal value in a general systematic point of view with *Exogen* and

Endogen. The terms equivalent to the two latter are *Pleurogen* and *Acrogen*.

In *Eremobrya* the fronds are produced singly from the sides of a rhizome, which has its growing-point always evidently in advance of the young developing frond. Each frond springs from a separate node, more or less distant from its neighbour, and is there articulated with the rhizome, so that when it has passed its maturity it separates at the node, and leaves behind a clean concave scar. The rhizome is solid, fleshy, and brittle, and when young always densely covered with scales (excepting in hypogeous rhizomes), which seldom, except in the very few scaly-fronded species, extend higher than the node; but it varies in some respects, being in some cases long and slender, and either simple or branched, and in others short and thick. The essential distinction between *Eremobrya* and *Desmobrya* rests in the fronds of the former being *articulated with the axis*, while those of the latter are *adherent and continuous with the axis.*

In *Desmobrya* the fronds are developed in two modes. In a large number of Ferns belonging to this division they come out from the apparent apex of the axis in a spiral series, and form a fascicle or corona. In this case the axis or stem is an erect or decumbent *caudex*, very variable in size, being sometimes scarcely elevated above the ground, and sometimes, in extreme cases, rising to the height of fifty or more feet. Almost an equally large number, however, have their fronds developed in a single alternate series, and their stem forms a *sarmentum*, in which the point of growth is in most cases scarcely at all in advance of the developing frond, and would appear to be coincident with it, though sometimes the prelongation is evidently in advance, and then the mode of growth appears to agree with *Eremobrya; but the non-articulation of the stipes at once distinguishes it.* Whatever the character of the stem of *Desmobryous* Ferns, it is always formed of the united and adherent bases of the fronds, and increases by the successive evolution of fresh fronds, each succeeding one of which is produced on the interior side of the bases of the preceding ones.

All Ferns are referable to one or other of these two divisions, and in general the difference between them is readily seen, particularly when living plants are examined; but, as in all attempts to generalize from special organs or structures, there are exceptions. For example, in *Elaphoglossum* the fronds are neither strictly adherent nor strictly articulate, but have a swelling some distance up the stipes, at which point, though there is no change in structure, the vascular bundles are so weak that the fronds ultimately separate there; and hence I regard the genus as an aberrant form of Desmobrya (?). In *Woodsia*, again, the stipes has an elevated articulation; but the axis is a *caudex* formed of the adherent bases of the stipes, and this, together with its fasciculate frond, indicates its true affinity to be in *Desmobrya*. A few also occur, as in the section *Ctenopterideæ* of the tribe *Polypodiæ*, in which the articulation is obscure, and a careful examination is required to detect it.

Notwithstanding these few exceptions, there can be no doubt that the two modes of growth above described are widely distinct, and the two groups into which Ferns are thereby divided are quite distinct in habit and appearance. The plants too seem to be endowed with very different natures, for the vitality and tenacity of life is much greater in the *Eremobryous* than in the *Desmobryous* division; and it is not a little remarkable that so far as observations upon cultivated plants enable me to ascertain, the latter are freely reproduced from spores, while the former are in proportion rarely reproduced by that means. In confirmation of this tenacity of life in *Eremobrya* I may mention that in importations of Ferns from distant countries those belonging to that division generally arrive in a living state, while *Desmobryous* ones, particularly those with *sarmentum*, are often killed in the transport.

I have now briefly reviewed in chronological order all the organs or structures upon which pteridologists rely for the formation of genera. Unfortunately, scarcely two can be found who agree as to the principles upon which genera of Ferns should be founded, or as to the value of the several organs for generic purposes. Some apply to Ferns the principles which characterize the genera among flowering plants, depending for the most part upon characters taken from the organs of reproduction. Others place great reliance upon the different modifications of venation; whilst I believe I stand alone in endeavouring to obtain natural genera, that is, genera having species associated by general habit and appearance, and by employing auxiliary characters taken from the modes in which the plants grow. Habit is not excluded from generic characters of flowering plants; indeed numerous instances might be quoted in which it is allowed by eminent botanists to constitute the chief distinction between allied genera, and by introducing it into the characters of Fern genera, more natural groups and sequences are obtained than by a strict adherence to the artificial characters afforded by the fructification and venation. Among Ferns no single organ alone affords characters sufficient for general systematic purposes. Were the principle upon which Linnæus acted—that is, a strict adherence to the fructification alone—applied in its integrity to the enormous mass of Ferns now known,—and it would be quite possible to do so, the most incongruous plants would be associated under one genus, and the magnitude of the genera would be quite overwhelming. The same would be the case were venation alone or habit alone to be taken into consideration. In some instances, however, a marked difference in one set of characters indicates well-defined groups; but as a general rule a combination of differences in two or more sets is requisite. Great difference of opinion exists as to what is and what is not a genus; but so long as plants are distributed into well-circumscribed groups of not too great an extent, it appears to me that it is a matter of little importance whether those groups be termed genera, sub-genera, or sections. For my own part I prefer regarding them as genera.

It has not been without due consideration that I have arrived at this conclusion. It also saves a great deal of unnecessary trouble, both in speaking and writing about Ferns, it being more easy to say and write *Elaphoglossum conforme*, than *Acrostichum* (*Elaphoglossum*) *conforme*, or *Gymnogramme tomentosa*, than *Gymnogramme* (*Eugymnogramme*) *tomentosa*, &c. Also by studying the character of the smaller groups individually, and treating them as genera, their nature is at once brought to the mind, without having to think of their association with a host of species of quite distinct characters. It also leads to investigation, and, accordingly, to a better knowledge of the structure of Ferns.

With regard to the characters that define the limits of species, as much uncertainty prevails amongst authors as with genera. This is owing to several causes; such as many species being normally heteromorphous, presenting at the same time different forms, which again vary at another period of growth; and in many instances authors have described the different states as distinct species; and in some cases different fronds of the same plant, and even portions of the same frond, have been placed under separate genera. It also frequently happens that two or more presumed species present so many intermediate gradations of form, that only the most extreme states appear sufficiently distinct to warrant their adoption as species, the numerous intermediate forms seeming to set specific distinctions at defiance. It therefore becomes a question what is the limit of form or of structure that constitutes a species. Generally understood, a species is an organized structure endowed with an essence or quality peculiar to itself, and possessing the power of multiplying and transmitting its type to new generations without change, *ad infinitum*. Admitting this definition as correct, it seems to be beyond human power to ascertain whether the serial gradations of form are genuine descendants of original creations, or only deviations from one original, brought into existence during the lapse of ages by the different climatic and local influences they have been subjected to. It is well known that phænogamous plants assume different forms and aspects, effected by the agency of man and by various natural causes; the difference from the original types being often so great, that if evidence of the change were not on record, the botanist of the present day would be justified in describing them as *distinct originally created species.* With Ferns we possess but little evidence of new forms having come into existence, the chief examples being found in several intermediate states in the genus *Gymnogramme,* which of late years have made their appearance in gardens, and seem to have as good right to be regarded as species as the original typical forms first known. If such changes do actually take place, and we are to deduce from them that races of intermediate forms originate in the progress of time and through the causes above alluded to, then great difficulty must attend any attempt to define species of Ferns. This is especially applicable in determining species from extensive suites of herbarium specimens. The number of species will be diminished or increased in accordance with the botanist's idea of specific differences: he will either amalgamate a number of allied forms under one specific name, or separate more or less of them as distinct species. On inspecting living examples of allied forms, the latter view seems to claim adoption; for although words often fail to convey the differences between individuals, still the eye readily detects them, and knowing that each maintains its own peculiar phase or habit from year to year, the scientific observer considers himself justified in naming them distinct species. It is a botanical rule to retain the names under which species are first described, whether continued in their original genus, or in whatever genus they may afterwards be placed. But as many species of Linnæus, Swartz, and other old, as well as modern authors, are but indifferently described, many being derived from imperfect specimens, and with nothing but the meagre description left us for their identification, it frequently happens that some modern author detects, or supposes he has found out, that the new species of his contemporary is one of the Linnæan or Swartzian doubtful species, and faith in his decision being admitted, familiar names become changed, thus burdening the science with additional synonyms, and rendering it in many cases impossible to reconcile one author's views with another. As an instance of the different views of authors on the identification of species and their synonyms, the genus *Asplenium* is a good example, it having within these few years, and near about the same time, been *revised* by Dr. Mettenius, Sir W. J. Hooker, and Mr. Moore. The two latter had the advantage of profiting by Dr. Mettenius's views, but in a great many cases I find it quite impossible to reconcile or agree with the views of either. As an example of the different views, I will cite the plant known in gardens for the last forty years by the name of *Asplenium Shepherdii.* The above-mentioned authors place it as a synonym, each under a different species and with different synonyms. To show the impossibility of reconciling one with the other, it will be sufficient to notice that in the *Index Filicum* it is found as one of twenty-three synonyms under *Diplazium radicans.* Believing as I do that these synonyms represent several distinct species, and the plant in question being one of them, I deem it best to retain it under the name it has been so long known by, and which is very well represented in "Lowe's Ferns," vol. v. p. 47.

These observations briefly explain a few of the causes of the plurality of names possessed by most Ferns, also the difficulty of arriving at satisfactory conclusions respecting their generic and specific distinctions, affording little hope of an early unanimity amongst authors, and fully justifying every one who has studied Pteridology in giving his own views.

This being the case, I have to explain that some important changes in the relative position of tribes and genera have been made in the following enumeration, in order to bring natural allied genera together; thus *Oleandra* and the articulated *Davallia* are now placed in *Eremobrya*, which is their proper place. The tribe *Aspidieæ* I now make a section of the tribe *Phegopterideæ*, their former separation being entirely dependent on the presence or absence of indusia, an organ not to be depended on in this tribe, when in many cases I am doubtful even of its value as a generic distinction, such as between *Dictyopteris* and *Aspidium*, *Goniopteris* and *Nephrodium*, *Phegopteris* and *Lastrea*, these genera containing species perfectly analogous to one another in general habit. The fugaceous nature of the indusium also makes it an organ of less importance than it is generally considered. In many species it is very small, and is soon lost or obliterated by the swelling of the sporangia; it is therefore only by watching living plants while the sori are yet young, that many species can be proved to be indusiate or non-indusiate.

I have long been dissatisfied with the position of *Hymenophylleæ* as a section of the tribe *Dicksonieæ*, it having no natural affinity with the typical representative of that tribe. I have, therefore, characterized them as a distinct sub-order. Mettenius, in his work on *Hymenophylleæ*, published in 1864, removes them from the position they have hitherto held between *Cyatheæ* and *Gleicheniceæ*, and places them before *Polypodieæ*, assigning to them the lowest rank amongst the Ferns; in their downward relationship they would border on mosses. It, however, appears that as far back as the year 1828 the elder Reichenbach regarded the *Hymenophylleæ* as the lowest group of Ferns, and indicated their relationship to be with *Hepaticæ*. But to discuss the views of these two authors on this subject would require more space than this work will allow.

These, with a few others, are the principal changes I have introduced; more might be made, but as, without being accompanied with full explanations showing my reasons, they might be considered unnecessary, I defer my views on the subject for another and more general work on the genera of Ferns, already alluded to.

The limited size of this book does not permit me to give descriptions of the species; but in order to assist in referring species to their respective genera, I have given the general characters and a woodcut of each genus, and also the principal synonyms, with references to one or more published figures. The native country of each species I have given only in its widest sense, as many species have a wide geographical distribution, and to state their precise localities would require much space, and is the less necessary, as the special localities of each species are given by Sir W. J. Hooker in his great work, the "Species Filicum," now happily brought to a close after twenty years' arduous and patient study. Another work has also been compiled during the last few years: I allude to "Lowe's Ferns." It consists of nine volumes, with 550 plates, containing figures of about two-thirds of the species in cultivation, with vague descriptions and many erroneous synonyms. This is a remarkable work in its way, but devoid of scientific merit; the figures being the only part worthy of notice; many of them are good representations of species—all such I have quoted; others are not to be relied upon, and tend rather to mislead.

In the following pages I have classified Ferns and certain other Cryptogamic plants, called Fern allies, under five orders, viz.:—

Order I. Filices. Annulate, or true *Ferns*.
„ II. Marattiaceæ. Exannulate. *Ferns*.
„ III. Ophioglossaceæ. Adder's-tongue. *Fern Ally*.
„ IV. Lycopodiaceæ. Lycopods. *Fern Ally*.
„ V. Marsileaceæ. Rhizocarps. *Fern Ally*

The two first of these orders agree in having circinate unfolding fronds, but differing essentially in habit and nature of their spore-cases; in the first, the spore-cases being membranous, and girded by an articulate ring, and the other firm and coriaceous, and destitute of a ring; they also differ in the nature of their roots, true Ferns having slender filiform, often soft, mossy roots, or they are hard and wiry, whereas in *Marattiaceæ* they are thick and fleshy, indicating quite a distinct habit of growth from that of true Ferns. The third order, *Ophioglossaceæ*, seems to possess some affinity to *Marattiaceæ* in the nature of its roots and spore-cases, but its straight vernation marks it as quite distinct. With *Lycopodiaceæ* it is connected through *Phylloglossum Drummondii*, a singular little plant, having the appearance of a small plant of *Ophioglossum Lusitanicum*, but with a spike formed of small bracts containing sporangia in their axis, analogous to *Lycopodiaceæ*; otherwise the family of *Lycopods* stands quite isolated, appearing to have no very evident transition forms connecting it with any other except the extinct order *Lepidodendreæ*: the same may be said of the last order, *Marsileaceæ*.

The most important of the above orders is *Filices*. Sir W. J. Hooker, in the "Species Filicum," describes two thousand five hundred species of annulate Ferns, which, with those described since the first publication of that work, twenty years ago, may now be considered to amount to no fewer than three thousand. To arrange and classify this mass of species is no easy task. The chief writers on Ferns adopt the difference in the position and direction of the ring, as the first important character for subdividing the order. This, however, divides it very unequally, the greater mass having the ring of the spore-case vertical, which characterizes the sub-order *Polypodiaceæ*; this I have in the following arrangement subdivided into eleven tribes, as follows:—

Conspectus of Arrangement of Orders, Sub-Orders, and Tribes.

1. Annulatæ.—*Sporangia* furnished with an articulate elastic ring.

Order I.—*Filices*.

Frond circinately unfolding. *Sporangia* furnished with vertical, horizontal, or sub-oblique ring.

Sub-Order I.—*Polypodiaceæ*.

Ring vertical.

Division I.—*Eremobrya.*

Fronds articulated with the rhizome.

Tribe I. Oleandreæ.—*Sori* round, medial, intra-marginal. *sium* lateral, interiorly attached, or sometimes central plane.

II. Davalleæ.—*Sori* round, terminal, marginal. *Indusium* lateral, interiorly attached, vertically urceolate.

III. Polypodeæ.—*Sori* round or linear, naked.

Division II.—*Desmobrya.*

Fronds adherent to the stem.

Tribe IV. Acrosticheæ.—*Sori* amorphous, naked.

V. Grammiteæ.—*Sori* oblong or linear, simple, forked, or reticulated, naked.

VI. Phegopterideæ.—*Sori* round, rarely linear, naked or indusiate. *Indusium* lateral, interiorly attached or central, or rarely calyciform.

VII. Pterideæ.—*Sori* marginal, round, or linear and transverse. *Indusium* lateral, exteriorly attached on the margin.

VIII. Blechneæ.—*Sori* intra-marginal, linear, transverse. *Indusium* lateral, exteriorly attached.

IX. Apleneæ.—*Sori* linear, oblique. *Indusium* lateral.

X. Dicksoneæ.—*Sori* marginal, round, or linear and transverse. *Indusium* lateral, interiorly attached, conniving with the changed margin, forming a groove or urceolate sub-bivalved cyst.

XI. Cyatheæ.—*Sori* round, intra-marginal. *Receptacles* elevated.—*Indusium* calyciform, or lateral and interiorly attached or absent.

Sub-Order II.—*Gleicheniaceæ.*

Ring horizontal. (*Sori* intra-marginal.)

Sub-Order III.—*Hymenophyllaceæ.*

Ring horizontal or oblique. (*Sori* marginal.)

Sub-Order IV.—*Osmundaceæ.*

Ring apical, often rudimentary only.

Tribe I. Schizææ.—*Sporangia* produced on contracted racemes, or on terminal or marginal spike-like appendices, ring complete.

II. Osmundeæ. — *Sporangia* globose. *Ring* rudimentary only.

2. Exannulatæ.—*Sporangia* coriaceous, destitute of a ring.

Order II.—*Marattiaceæ.*

Fronds circinate. *Sporangia* dorsal, free, or connate, opaque, coriaceous.

Order III.—*Ophioglossaceæ.*

Vernation straight, the fronds rising from a root-stock, consisting of a fascicle (more or less according to age) of fleshy roots. *Sporangia* homogeneous, connate on spikes, or free and paniculate.

Order IV.—*Lycopodiaceæ.*

Plants with indefinite prolonging, erect or pendulous, stems furnished with acerose rusciform, or jungermania-like leaves (sometimes very small), bearing 1–3-celled sporangia in their axes, or on catkin-like spikes.

Order V.—*Marsileaceæ.*

Plants floating or growing in water, consisting of grass or trefoil-like leaves, or branched with imbricate leaves, bearing 1–3, or many-celled sporangia at their base or otherwise (see the characters of the respective genera).

AN ENUMERATION

OF

CULTIVATED FERNS.

Order I.—FILICES.

Fronds circinately unfolding, uniform and leafy, bearing sporangia on their under side or margin (rarely on both sides); or of two forms, one leafy and sterile, the other wholly, or some portion of its segments more or less contracted and fertile. *Sporangia* membraneous, one-celled, free, furnished with a vertical, horizontal, or oblique articulated elastic ring.

Sub-Order I.—POLYPODIACEÆ.

Sporangia globose or oval, unilocular, pedicellate or sessile, membraneous, furnished with a vertical ring, and opening at a right angle to the direction of the ring.

Division I.—Eremobrya.

Fronds in vernation lateral, solitary, attached to the axis (rhizome) by a special articulation.

* *Sori indusiate.*

Tribe I.—OLEANDREÆ.

Sori round, medial. *Indusium* lateral, interiorly attached or sometimes central, plane.

1. OLEANDRA, *Cav.*

Rhizome surculose or erect, subfrutescent and ramose; node of articulation sessile, or more or less elevated on the stipes. *Fronds* simple, entire, linear-lanceolate, 1—1½ foot long, smooth or pilose. *Veins* simple, or once or twice forked; venules free, parallel, their apices curved outwards, forming a narrow cartilaginous margin. *Receptacles* punctiform, medial, or basal on the anterior venules. *Sori* round, transversely uniserial, or irregular. *Indusium* reniform, or rarely orbicular.

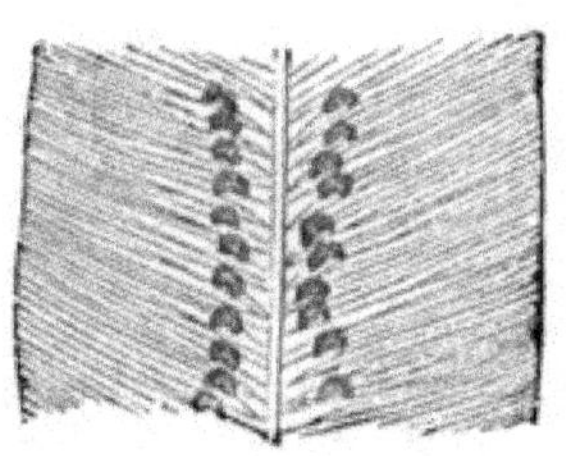

Genus 1.—Portion of mature frond—under side. No. 4.

1. **O. nodosa,** *Presl.; Hook. Sp. Fil.* 4, *p.* 157. *Lowe's Ferns*, 7, *t.* 17. Aspidium nodosum, *Willd.* (*Plum. Fil. t.* 136); *Hook. Exot. Fil. t.* 117. Aspidium articulatum, *Schk. Fil. t.* 27.—West Indies and Guiana.

2. **O. articulata,** *Presl.* Aspidium articulatum, *Sw.* (*excl. Syn. Plum. et Schk.*).—East Indies, Mauritius, and Natal.

3. **O. Wallichii,** *Presl.; Hook. Sp. Fil.* 4, *p.* 158. Aspidium Wallichii, *Hook. Exot. Fil. t.* 5. *Kunze, Fil. t.* 19. Neuronia Asplenioides, *D. Don.*—East Indies.

4. **O. neriiformis,** *Cav.; Hook. Fil. Exot. t.* 58; *Lowe's Ferns*, 7, *t.* 16. Aspidium neriiforme, *Sw.; Kunze, Fil. t.* 18. Ophiopteris verticillata, *Reinw.*—Var. hirtella, *Moore.* Oleandra hirtella, *Miq.; Kunze, Fil. t.* 129. Oleandra pilosa, *Hook. et Bauer, Gen. Fil. t.* 45 *B.*—East Indies, Malayan Archipelago, and Tropical America.

TRIBE II.—DAVALLIEÆ.

Sori round or oblong, terminal, marginal. *Indusium* lateral, interior, plane, or its sides more or less adnate, forming a vertical cyst, open exteriorly.

2. HUMATA, *Cav.*

Rhizome surculose, slender, squamiferous. *Fronds* linear-lanceolate, entire, sinuose, pinnatifid or deltoid bipinnatifid, smooth, coriaceous. *Veins* simple or forked; venules free, often thickened and clavate. *Receptacles* terminal, punctiform, on all or only on the anterior venules of each fascicle. *Sori* marginal or anti-marginal. *Indusium* sub-rotund or reniform, coriaceous, interiorly attached by its base only, shorter or equal with the margin, and forming with it a bilabiate vertical or sometimes oblique cyst.

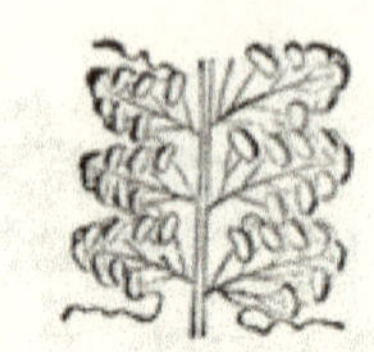

Genus 2.—Portion of fertile frond—under side. No. 1.

1. **H. heterophylla,** *J. Sm.; Hook. et Bauer, Gen. Fil. t.* 114. Humata ophioglossa, *Cav.* Humata pinnatifida, *Cav.* Davallia heterophylla, *Hook. et Grev. Ic. Fil. t.* 230; *Hook. Fil. Exot. t.* 27; *Lowe's Ferns*, 8, *t.* 19; *Hook. Sp. Fil.* 1, *f.* 152.—Malayan Archipelago.
2. **H. pedata,** *J. Sm.* Davallia pedata, *Sm.; Hook. Sp. Fil.* 1, *t.* 45 *A; Hook. Gard. Ferns, t.* 7. Pachypleura pedata, *Presl.*—Malayan Archipelago.
3. **H. Cumingii,** *J. Sm.* Davallia Cumingii, *Hook. Sp. Fil.* 1, *t.* 45 *B.*—Philippine Islands, Ceylon.

3. DAVALLIA, *Sm.*

Rhizome surculose creeping, or sub-erect and sub-frutescent. *Fronds* generally deltoid, pinnate, bi-tripinnate, or multifid, smooth, often coriaceous. *Veins* forked; venules free, the fertile ones often very short. *Receptacles* punctiform, terminal. *Sori* sub-rotund or vertically oblong, marginal. *Indusium* scariose, its sides adnate, forming an urceolate or tubular vertical cyst, open exteriorly.

Genus 3.—Pinnule of fertile frond—under side. No. 7.

* *Fronds pinnate, pinnæ entire or lobed.*

1. D. **pentaphylla,** *Blume; Hook. Fil. Exot. t.* 37; *Kunze, Fil. t.* 108. Scyphularia pentaphylla, *Fée.* Stenolobus pentaphyllus, *Presl.* Davallia tryphylla, *Hook. Sp. Fil.* 1, *t.* 46 *A; Lowe's Ferns,* 8, *t.* 18.—Malayan Archipelago.

** *Fronds bi-tripinnately compound.*

2. D. **bullata,** *Wall.; Hook. Sp. Fil. t.* 50 *B.*—East Indies.

3. D. **dissecta,** *J. Sm.; Moore in Gard. Chron.* 1855, *p.* 469; *Lowe's Ferns,* 8, *t.* 20.—Malayan Archipelago. β. decora, Davallia decora, *Moore in Sim's Cat.*—Java.

4. D. **Canariensis,** *Sm.; Hook. Sp. Fil. t.* 56 *A; Lodd. Bot. Cab. t.* 142. Trichomanes Canariense, *Linn.* Polypodium Lusitanicum, *Linn.*—South of Europe, Madeira, and Canary Islands.

5. D. **ornata,** *Wall.* Stenolobus ornatus, *Presl.* Davallia solida, β. latifolia, *Hook. Sp. Fil. t.* 42 *B; Hook. Fil. Exot. t.* 57.—Singapore.

6. D. **solida,** *Sw.; Schk. Fil. t.* 126.—Malayan and Polynesian Islands.

7. D. **pyxidata,** *Cav.; Hook. Gen. Fil. t.* 27; *Hook. Sp. Fil. t.* 55 *C.*—Australia.

8. D. **Lindleyi,** *Hook. Sp. Fil. t.* 58 *B.*—New Zealand?

9. D. **elegans,** *Sw.; Hook. Sp. Fil. t.* 43 *A B; Lowe's Ferns,* 8, *t.* 22. Davallia bidentata, *Schk. Fil. t.* 127.—Malayan Archipelago.

10. D. **divaricata,** *Blume.* Davallia polyantha, *Hook. Sp. Fil. t.* 59 *A; Lowe's Ferns,* 8, *t.* 23.—Malayan Archipelago.

11. D. **elata,** *Sw.; Schk. Fil. t.* 127 *B; Hook. Sp. Fil.* 1, 166, *t.* 55 *A.*—Society Islands, Malayan Archipelago, &c.

12. D. **nitidula,** *Kunze; Schk. Supp. Fil. t.* 37, *f.* 2; *Hook. Sp. Fil. t.* 44 *A.* D. Kunzii, *Hort.*—South and West Africa.

13. D. **Vogelii,** *Hook. Sp. Fil. t.* 59 *B.*—Fernando Po.

4. LEUCOSTEGIA, *Presl.*

Rhizome thick, short, surculose, sometimes hypogæous. *Fronds* deltoid, tripinnatifid, or multifid, sometimes lanceolate and bipinnatifid. *Veins* forked; venules free, the anterior ones often very short. *Receptacles* terminal, superficial, or immersed on the exterior venules. *Sori* round. *Indusium* sub-reniform, oblong, or nearly orbicular, plane, interiorly attached by its broad base, equal with or shorter than the margin, thin, scariose.

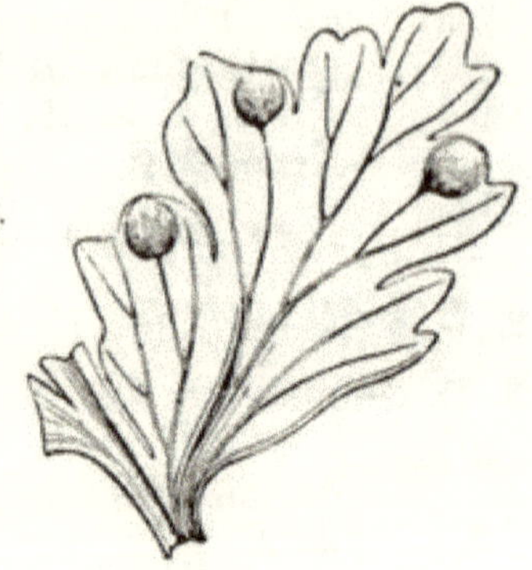

Genus 4.—Fertile pinna.

* *Rhizome epigæous squamose.*

1. L. **hirsuta,** *J. Sm. En. Fil. Philipp.* Microlepia hirsuta, *Moore.* Davallia ciliata, *Hook. Sp. Fil.* 1, 184, *t.* 60 *A.*—Luzon.

2 L. **Borneensis,** *J. Sm.;* Nephrodium (Lastrea) Borneense, *Hook. Sp. Fil.* 4, p. 111; *Hook. Ic. Pl. t.* 993.—Borneo.

3. L. **parvula,** *J. Sm.* Davallia parvula, *Wall.; Hook et Grev. Ic. Fil. f.* 138.—Malayan Islands, Singapore.

4. L. **pulchra,** *J. Sm.* Davallia pulchra, *D. Don.* Acrophorus pulchra, *Moore Ind. Fil.* (*excl. syn.* Davallia chærophylla).—Nepal.

5. L. **chærophylla,** *J. Sm.* Davallia chærophylla, *Wall.; Hook. Sp. Fil.* 1, 157, *t.* 51 *A.* Acrophorus chærophyllus, *Moore.* Humata chærophylla, *Mettin. Fil. Hort. Lips. t.* 27, *f.* 9, 10.—East Indies. T.

6. L. **affinis,** *J. Sm.* Davallia affinis, *Hook. Sp. Fil.* 1, 158, *t.* 52 *B.* Acrophorus affinis, *Moore.* Humata affinis, *Mett. Fil. Hort. Lips. t.* 27, *f.* 5, 6.—Ceylon, Singapore, Philippine Islands.

** *Rhizome hypogæous. Fronds deciduous.*

7. L. **immersa,** *Presl.; J. Sm.; Hook. Gen. Fil. t.* 52 *A.* Davallia immersa, *Wall.; Hook. Fil. Exot. t.* 79. Acrophorus immersus, *Moore.* Humata immersa *Mettin.*—East Indies.

** *Sori naked.*

TRIBE III.—POLYPODIEÆ, *J. Sm.*

Sori round, oblong, or linear, destitute of a special indusium.

5. POLYPODIUM, *Linn. in part.*

Rhizome generally short and thick, sometimes sub-hypogæous. *Fronds* pinnatifid, pinnate, or bi-tripinnatifid, rarely simple, smooth, villose, or squamiferous, from 6 inches to 2-3 feet high. *Veins* forked, very rarely simple; venules free. *Receptacles* punctiform, superficial, terminal on the lower anterior venules. *Sori* round or rarely oval, transversely uniserial or solitary on laciniæ.

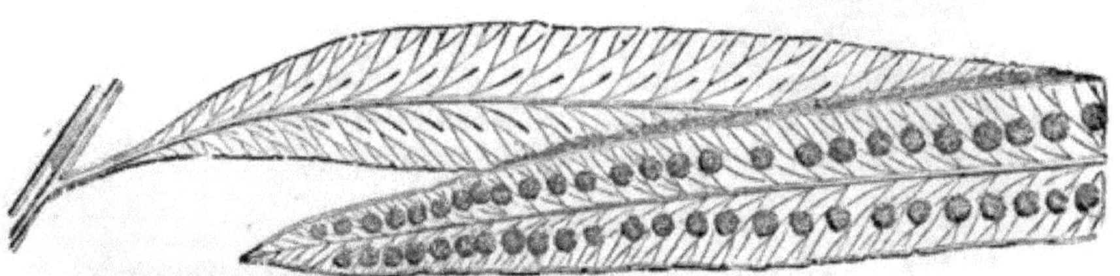

Genus 5.—Portion of mature frond. No. 7.

* *Fronds pinnatifid.*

1. P. pectinatum, *Linn.; Plum. Fil. t.* 83; *Hook. Gard. Ferns, t.* 10; *Lowe's Ferns,* 2, *t.* 21. — Tropical America.
2. P. Paradiseæ, *Lang. et Fisch. Ic. Fil. t.* 11; *Lowe's Ferns,* 2, *t.* 1. P. Otites, Hort. (*non Linn.*).—Brazil.
3. P. Schkuhrii, *Radd. Fil. Bras. t.* 27. P. pectinatum, *Schk. Fil. t.* 17 *C* (*excl. syn.*). P. plumula, *Moore and Houlst.* (*non Humb.*). P. plumosum, Hort.—Brazil.
4. P. Martensii, *Mett.* P. affine, *Mart. et Gal. Fil. Mex. t.* 8, *f.* 1 (*not Blume*).—Mexico.
5. P. vulgare, *Linn.; Hook. Brit. Ferns, t.* 2; *Eng. Bot.* 1149; *Lindl. and Moore, Nat. Print. Ferns, t.* 1, *f. A, B, C, D; Bolt. Fil. Brit. t.* 18; *Sowerby, Ferns of Gr. Brit. t.* 1.

Var. Cambricum, *Willd.; Bolt. Fil. Brit. t.* 2, *f.* 5 *A; Lindl. and Moore, Nat. Print. Ferns, t.* 3, *f. A.* P. Cambricum, *Linn.*

Var. semilacerum, *Link.; Lindl. and Moore, Nat. Print. Ferns, t.* 2 *A* (*bis*). P. vulgare, *var.* Hibernicum, *Sowerby, Ferns of Gr. Brit. t.* 10.

Var. acutum, *Lindl. and Moore, Nat. Print. Ferns, t.* 1 *E.*

Var. serratum, *Willd.; Lindl. and Moore, Nat. Print. Ferns, t.* 2 *B* (*bis*).

Var. crenatum, *Lindl. and Moore, Nat. Print. Ferns, t.* 3 *B.*

Var. bifidum, *Lindl. and Moore, Nat. Print. Ferns, t.* 1 *F.*

Var. cristatum, *Linn.; Lowe's New Ferns, t.* 26 *B.*

6. **P. plebejum,** *Schlecht.; Hook. Gard. Ferns, t.* 48; *Lowe's New Ferns, t.* 33. P. Karwinskianum, *A. Braun; J. Sm. Cat. Cult. Ferns,* 1857.—Tropical America. **T.**

** *Fronds pinnate.*

7. **P. Henchmanii,** *J. Sm.; Moore and Houlst. in Mag. of Bot.; Lowe's Ferns,* 1, *t.* 30. P. fraternum, *J. Sm. Cat. Cult. Ferns,* 1857 (? *Schlecht.*).—Mexico.

8. **P. subpetiolatum,** *Hook. Ic. Pl. t.* 391, 392. P. biserratum, *Mart. et Gal. Fil. Mex. t.* 9, *f.* 1.—Mexico.

9. **P. sororium,** *H. B. K.*—West Indies and Tropical America.

---

6. LEPICYSTIS, *J. Sm.*

Rhizome short and rigid, or slender and surculose. *Fronds* pinnatifid, 6—18 inches high, densely covered with round or elongated ciliated scales. *Veins* pinnately forked, anastomosing,

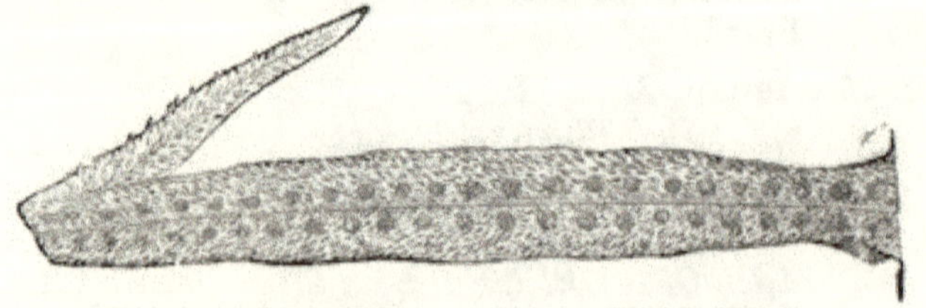

Genus 6.—Portion of fertile frond, under side. No. 3.

lower exterior venules free. *Receptacles* punctiform, terminal on the free venules in the costal areoles. *Sori* round, transverse, uniserial, protruding through the dense scales.

1. **L. incana,** *J. Sm.* Polypodium incanum, *Sw.* P. velatum, *Schk. Fil. t.* 11 *B.*—Tropical America and Southern United States.

2. **L. sepulta,** *J. Sm.* Polypodium sepultum, *Kaulf.; Lowe's Ferns,* 1, *t.* 34 *A.* P. rufulum, *Presl.* P. hirsutissimum, *Rad. Fil. Bras. t.* 26. Acrostichum lepidopteris, *Lang. et Fisch. Ic. Fil. t.* 2.—Tropical America.

3. **L. squamata,** *J. Sm.* Polypodium squamatum, *Linn.* (*Plum. Fil. t.* 79); *Lowe's New Ferns, t.* 34.—West Indies.

4. **L. rhagadiolepis,** *J. Sm.* Goniophlebium rhagadiolepis, *Fée, Mem. Polypod. t.* 19, *f.* 3. Polypodium thysanolepis, *A. Braun.*—Tropical America. **T.**

---

7. GONIOPHLEBIUM, *Presl.; J. Sm.*

Rhizome thick and fleshy, or slender and sub-hypogæous. *Fronds* pinnatifid or pinnate, rarely simple, uniform, 1—3 feet high, smooth or slightly pubescent, segments and pinnæ adherent with the rachis. *Veins* once or more times forked, or equally pinnate, the lower anterior venule always free, the rest angularly anastomosing, and generally producing an excurrent free veinlet from the junctions. *Receptacles* punctiform, superficial, terminal on the anterior free venules and also often on the

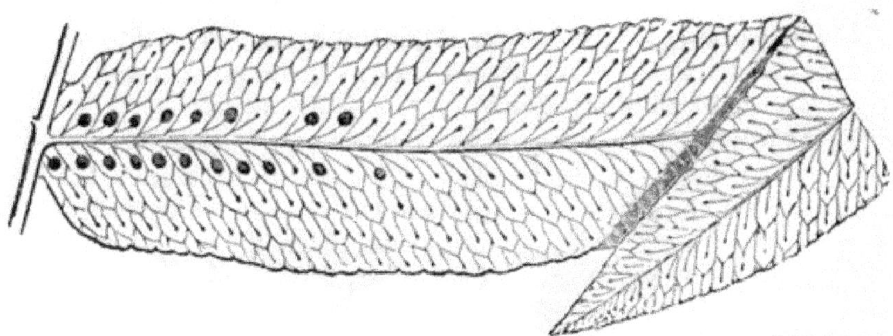

Genus 7.—Portion of mature frond. No. 12.

excurrent veinlets. *Sori* round, or rarely oblong, solitary in the areoles, or transverse, 1–6-serial, naked.

* *Fronds pinnatifid.*

1. G. **appendiculatum,** *Moore in Gard. Chron.* (1856). Polypodium appendiculatum, *Linden; J. Sm. Cat. Cult. Ferns*, 1857, *p.* 2; *Hook. Fil. Exot. t.* 87. P. scriptum, *Hort.* P. sculptum, *Hort.*—Venezuela and Mexico.

2. G. **plectolepis,** *Moore.* Polypodium (Goniophlebium) plectolepis, *Hook. Sp. Fil.* 5, *p.* 30.—Dominica, Mexico.

3. G. **loriceum,** *J. Sm.* Polypodium loriceum, *Linn.; Plum. Fil. t.* 78. Polypodium gonatodes, *Kunze.* Goniophlebium latipes, *Moore and Houlst.* P. latipes, *Lang. et Fisch. Ic. Fil. t.* 10.—Tropical America.

4. G. **Catharinæ,** *J. Sm.* Polypodium Catharinæ, *Lang. et Fisch. Ic. Fil. t.* 9.—Brazil.

5. G. **glaucum,** *J. Sm.* Polypodium glaucum, *Radd. Fil. Bras. t.* 29, *f.* 1.—Brazil.

6. G. **harpeodes,** *J. Sm.* Polypodium harpeodes, *Link.*—Brazil.

7. G. **colpodes,** *J. Sm.* Polypodium colpodes, *Kunze; Lowe's Ferns*, 2, *t.* 60.—Venezuela.

8. G. **lætum,** *J. Sm.* Polypodium lætum, *Radd. Fil. Bras. t.* 28.—Brazil.

9. G. **vacillans,** *J. Sm.* Polypodium vacillans, *Link.*—Brazil.

** *Fronds pinnate.*

10. G. **fraxinifolium,** *J. Sm.* Polypodium fraxinifolium, *Jacq. Ic. Rar. t.* 639. P. longifolium, *Presl.*—Tropical America.

11. G. **distans,** *J. Sm.* Polypodium distans, *Radd. Fil. Bras. t.* 31. P. polystichum, *Link.* P. deflexum, *Lodd.*—Tropical America.

12. G. **menisciifolium,** *J. Sm.* Polypodium menisciifolium, *Lang. et Fisch. Ic. Fil. t.* 12. P. albopunctatum, *Radd. Fil. Bras. t.* 30; *Lowe's Ferns*, 1, *t.* 36. Goniophlebium albopunctatum, *J. Sm.*—Brazil.

13. G. **dissimile,** *J. Sm.* Polypodium dissimile, *Linn., non Schk.; Lowe's Ferns*, 2, *t.* 35. Goniophlebium chnoodes, *Fée.*—Jamaica.

14. **G. inæquale**, *J. Sm.* Phlebodium inæquale, *Moore.* Polypodium inæquale, *Lowe's Ferns*, 2, *t.* 28. Polypodium (Goniophlebium) Guatemalense, *Hook.*—Guatemala.

15. **G. neriifolium**, *J. Sm.* *Hook. Gen. Fil. t.* 70 *B.* Polypodium neriifolium, *Schk. Fil. t.* 15; *Radd. Fil. Bras. t.* 31 *bis.*—West Indies and Tropical America.

---

8. SCHELLOLEPIS, *J. Sm.*

Vernation contiguous or distant. *Rhizome* slender, sub-hypogeous. *Fronds* pinnate or pinnatifid, generally slender and pendulous, 1½–12 feet long, smooth or nearly so; pinnæ and segments articulated with the rachis. *Veins* once or more times

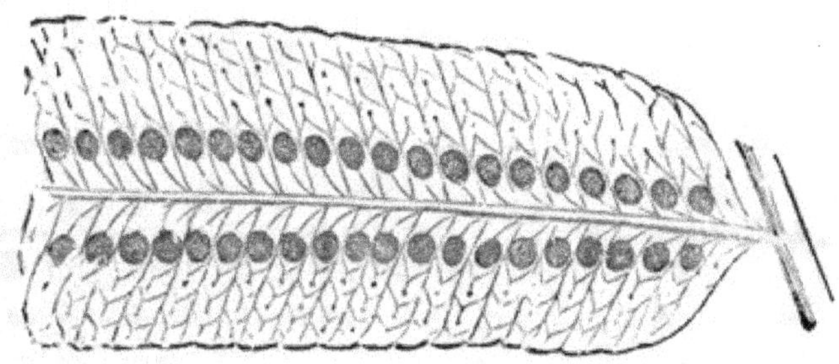

Genus 8.—Portion of pinna of mature frond, under side. No. 3.

forked or pinnate; the lower exterior venules always free, the rest angularly anastomosing. *Receptacles* punctiform, generally immersed, on the apices of the lower free venules. *Sori* round, solitary in the areoles, transverse uniserial, furnished with indusiform laciniate scales.

1. **S. cuspidata**, *J. Sm.* Polypodium cuspidatum, *Bl., not Don.* Goniophlebium cuspidatum, *Presl.* P. grandidens, *Kunze; Metten. Fil. Hort. Leipsic. t.* 23. P. colpothrix, *Kunze.* Goniophlebium argutum, *Cat. Hort. Kew., not* Polypodium argutum, *Wall.*—Java.

2. **S. subauriculata**, *J. Sm.* Polypodium subauriculatum, *Bl. Fl. Jav.* 6, *t.* 83. Goniophlebium subauriculatum, *Presl.* P. Reinwardtii, *Kunze.* P. metamorphum, *Kunze.* Goniophlebium Pleopeltis, *Fée.*—Malayan Archipelago.

3. **S. verrucosa**, *J. Sm.* Polypodium verrucosum, *Wall.; Hook. Gard. Ferns, t.* 41. Marginaria verrucosa, *Hook. Gen. Fil. t.* 14, 10 *B.* Goniophlebium verrucosum, *J. Sm. Cat.* (1857).—Malacca.

---

9. PHLEBODIUM, *R. Br.; J. Sm.*

Rhizome thick and fleshy. *Fronds* large, 2–6 feet high, pinnatifid or subpinnate, membranous, smooth or glaucous. *Veins* pinnate; venules arcuately or angularly anastomosing,

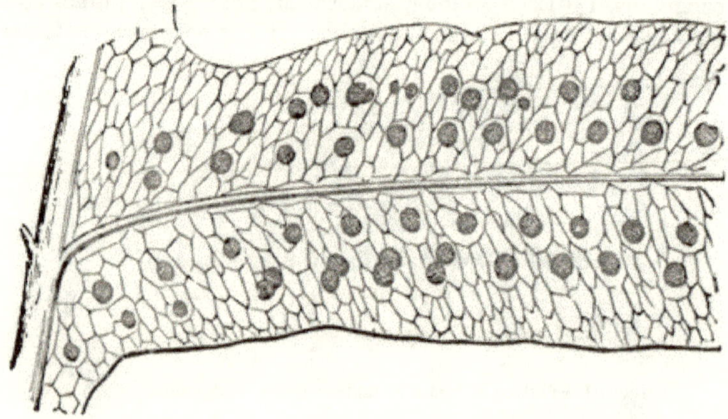

Genus 9.—Portion of pinna of mature frond, under side. No. 1.

producing two or three excurrent veinlets terminating in the areoles; the costal areoles always vacant. *Receptacles* punctiform, on the combined apices of the excurrent veinlets. *Sori* round, transversely 1–6-serial, destitute of scales.

1. **P. aureum**, *R. Br.* Polypodium aureum, *Linn.; Plum. Fil. t.* 76; *Schk. Fil. t.* 12.—Tropical America.

2. **P. sporodocarpum**, *J. Sm.* Polypodium sporodocarpum, *Willd. Lowe's Ferns,* 2, *t.* 6. P. glaucum, *Hort.*—Mexico.

3. **P. areolatum**, *J. Sm.* Polypodium areolatum, *Willd.*—Venezuela.

4. **P. pulvinatum,** *J. Sm.* Polypodium pulvinatum, *Link; Lowe's Ferns*, 2, *t.* 56.—Brazil.

5. **P. dictyocallis,** *J. Sm.* Chrysopteris dictyocallis, *Fée.* Polypodium dictyocallis, *Lowe's Ferns*, 2, *t.* 36. Phlebodium multiseriale, *Moore, Gard. Chron.* (1855).—Tropical America.

---

10. LOPHOLEPIS, *J. Sm.*

Rhizome slender, much elongated. *Fronds* simple, entire, 1–6 inches high, squamose or smooth; the fertile contracted, linear. *Veins* pinnately forked; the lower anterior venules free,

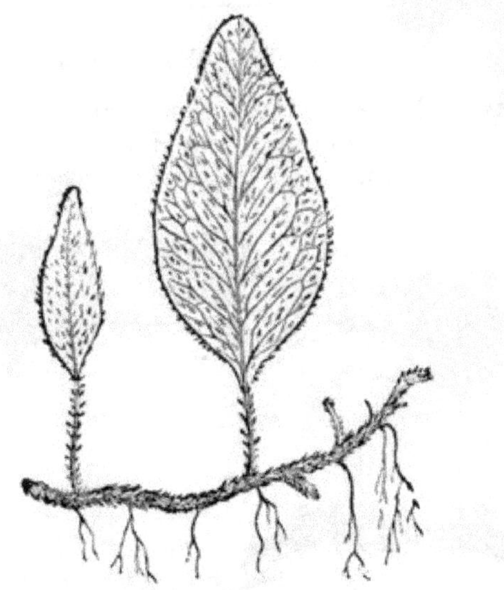

Genus 10.—Portion of rhizome and barren fronds. No. 1.

the rest angularly anastomosing. *Receptacles* punctiform, terminal on the free venules in the costal areoles. *Sori* round, generally confluent, transversely uniserial, furnished with elongated scales, or destitute of scales.

1. **L. piloselloides,** *J. Sm.* Polypodium piloselloides, *Linn.* (*Plum. Fil. t.* 118); *Hook. Gard. Ferns, t.* 18; *Lowe's Ferns*, 1, *t.* 32. Goniophlebium piloselloides, *J. Sm.*

(*olim*). Marginaria piloselloides, *Presl.; Hook. Gen. Fil. t.* 51.—West Indies and Tropical America.

2. L. **ciliata**, *J. Sm.* Polypodium ciliatum, *Willd.;* Goniophlebium ciliatum, *J. Sm.* (*olim*).—West Indies and Tropical America.

3. L. **vaccinifolia**, *J. Sm.* Polypodium vaccinifolium, *Lang. et Fisch. Ic. Fil. t.* 7; *Lowe's Ferns*, 1, *t.* 41. Anapeltis vaccinifolia, *J. Sm. Cat. Cult. Ferns* (1857). Goniophlebium vaccinifolium, *J. Sm. Cat. Kew Ferns*, (1846).—Brazil.

β **albida**, *J. Sm.* Fronds smaller, whitish on the upper surface.—Bahia.

---

11. ANAPELTIS, *J. Sm.*

Rhizome surculose, elongating. *Fronds* simple, 1–6 inches long, the fertile usually contracted and linear, smooth, generally opaque. *Veins* arcuately or angularly anastomosing. *Receptacles* punctiform, produced on the confluent apices of two or more excurrent veinlets terminating in the medial areoles, or sometimes compital. *Sori* round or ovate, transversely uniserial, naked.

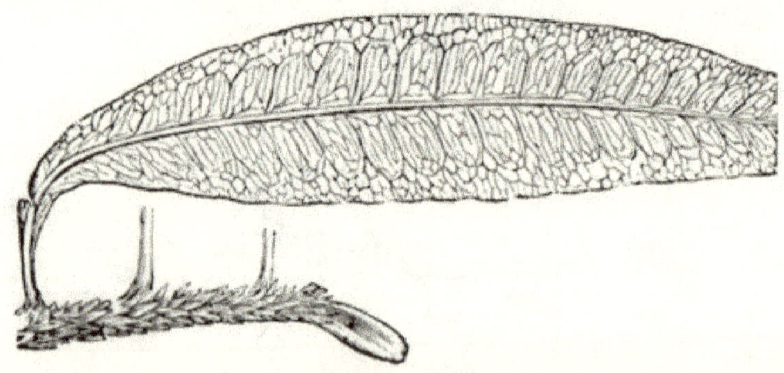

Genus 11.—Portion of barren frond. No. 5.

1. A. **serpens**, *J. Sm.* Polypodium serpens, *Sw.; Plum. Fil. t.* 121. Pleopeltis serpens, *Presl.* Goniophlebium serpens, *Moore.*—West Indies.

2. **A. Owariensis,** *J. Sm.* Polypodium Owariense, *Desv.; Lowe's Ferns,* 2, *t.* 62. Goniophlebium Owariense, *Lodd.*—Sierra Leone.

3. **A. lycopodioides,** *J. Sm.* Polypodium lycopodioides, *Linn.; Plum. Fil. t.* 119. Pleopeltis lycopodioides, *Presl.*—West Indies.

4. **A. nitida,** *J. Sm. En. Fil. Hort. Kew.* (1846). Pleopeltis nitida, *Moore.*—Honduras.

5. **A. stigmatica,** *J. Sm.* Polypodium stigmaticum, *Presl. Rel. Hænk. t.* 3, *f.* 2. Pleopeltis stigmatica, *Presl.* Phlebodium venosum, *Moore et Houlst.* Anapeltis venosa, *J. Sm. Cat. Cult. Ferns* (1857). Polypodium venosum, *Lowe's Ferns,* 1, *t.* 35.—Tropical America.

6. **A. squamulosa,** *J. Sm.* Polypodium squamulosum, *Kaulf.; Lowe's Ferns,* 1, *t.* 50; 2, *t.* 29 *B.* Pleopeltis squamulosa, *Presl.* Polypodium myrtifolium, *Lodd.*—Brazil.

7. **A. geminata,** *J. Sm.* Polypodium geminatum, *Schrad.; Metten.* Polypodium iteophyllum, *Link.*—Brazil.

---

12. PLEOPELTIS, *Humb.; J. Sm.*

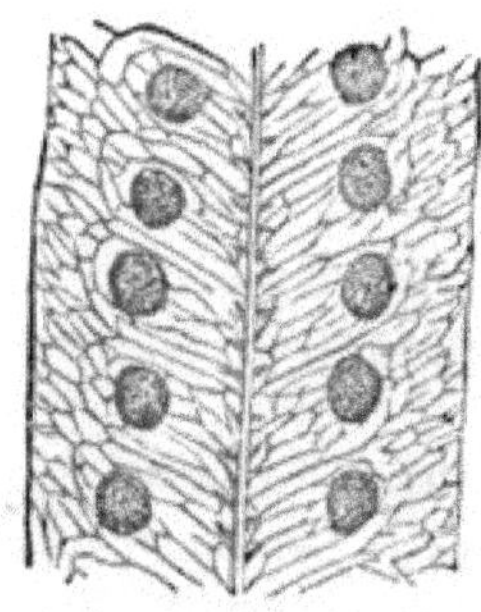

Genus 12.—Portion of mature frond, under side. No. 1.

Rhizome surculose, elongating. *Fronds* simple, sinuose, or pinnatifid, 4–12 inches high, opaque, squamiferous. *Veins* arcuately anastomosing. *Sporangia* produced on the confluent apices of two or more excurrent veinlets, terminating in the medial areoles. *Sori* punctiform, oblong, or (by confluence) linear, transversely uniserial, furnished with indusiform peltate scales.

1. **P. percussa,** *Hook. et Grev. Ic. Fil. t.* 67. Polypodium percussum, *Cav.; Lang. et Fisch. Ic. Fil. t.* 6. Poly-

podium cuspidatum, *Presl. Reliq. Hænk. t.* 1, *f.* 3. Polypodium avenium, *Desv.*—Tropical America.

2. **P. lanceolata,** *Presl.* Polypodium lanceolatum, *Linn.; Plum. Fil. t.* 137. Polypodium macrocarpum, *Willd.* Pleopeltis macrocarpa, *Kaulf.* Pleopeltis lepidota, *Presl.* Pleopeltis Helenæ, *Presl.*—Tropical America, St. Helena, South Africa, and Bourbon.

3. **P. elongata,** *J. Sm.* Grammitis elongata, *Sw.* Synammia elongata, *Presl.* Grammitis lanceolata, *Schk. Fil. t.* 7. —Tropical America.

4. **P. nuda,** *Hook. Exot. Fl. t.* 63 (*non Hook. Gen. Fil.*). Phymatodes (Lepisorus) nuda, *J. Sm. Cat. Cult. Ferns* (1857). Polypodium loriforme, *Wall. Hook. Gard. Ferns, t.* 18. Pleopeltis loriformis, *Presl.;* Drynaria Fortunei, *T. Moore* (*non Link*). Polypodium leiopteris, *Kunze; Metten. Fil. Hort. Leip. t.* 25, *f.* 37.—East Indies.

5. **P. excavata,** *J. Sm.* Polypodium excavatum, *Bory in Willd.* Phymatodes (Lepisorus) excavata, *J. Sm. Cat. Cult. Ferns* (1857). Polypodium scolopendrinum, *D. Don.* Polypodium sesquipedalis, *Wall.* Polypodium phlebodes, *Kunze;* Pleopeltis nuda, *Hook. Gen. Fil. t.* 18 (*non Hook. Exot. Fl.*).—East Indies, Mauritius, and China.

---

13. PARAGRAMMA, *Blume.*

Rhizome short, cæspitose or slender elongated. *Fronds* simple, linear-lanceolate, obtuse, ½ to 1½ foot in length, smooth, coriaceous. *Veins* compound anastomosing, internal, obscure, nearly uniform. *Receptacles* compital, deeply immersed, forming oblong or short linear cysts near to, and parallel with, the margin. *Sori* oblong-linear, marginal, furnished with indusioid stipitate squamae.

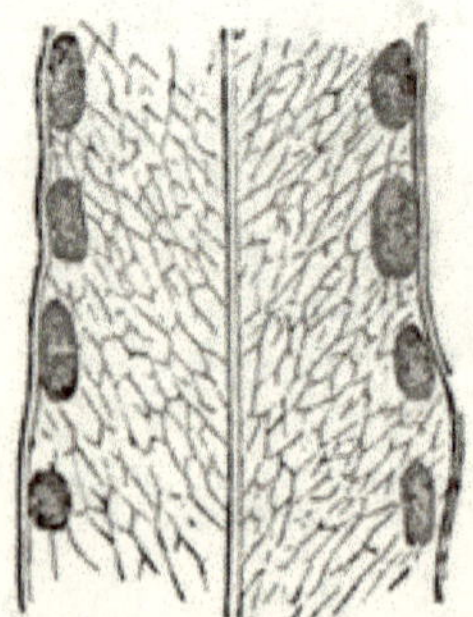

Genus 13.—Portion of mature frond, under side. No. 1.

1. **P. longifolia**, *Moore, Ind. Fil.* Grammatis (Paragramma) longifolia *et* decurrens, *Blume.* Drynaria revoluta, *J. Sm. En. Fil. Phil.* Phymatodes longifolia, *J. Sm. Cat. Cult. Ferns* (1857). Polypodium contiguum, *Wall.*; *Hook. Ic. Pl. t.* 987; *Hook. Fil. Exot. t.* 20. –Malacca, Moulmein, Java, and Luzon.

---

14. NIPHOPSIS, *J. Sm.*

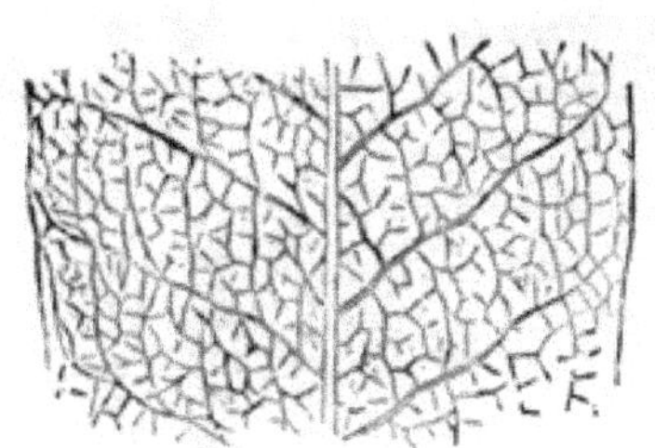

Genus 14.—Portion of barren frond. No. 1.

Rhizome slender, surculose. *Fronds* simple, linear-lanceolate, coriaceous, opaque, densely covered with stellate pubescence, 6 inches to 1 foot in length. *Veins* internal, obscure, compound anastomosing; primary veins indistinct. *Receptacles* compital. *Sori* oval, large, transverse uniserial.

1. **N. angustatus**, *J. Sm. Lowe's New Ferns, t.* 38 *A.* Polypodium angustatum, *Sw.*; *Schk. Fil. t.* 8 *c.* Pleopeltis angustata, *Presl.* Niphobolus angustatus, *Spreng. Hook. Gard. Ferns, t.* 20. Niphobolus sphærocephalus, *Hook. et Grev. Ic. Fil. t.* 94. Polypodium sphærocephalum, *Wall.* Phymatodes sphærocephalus, *Presl.* Niphobolus macrocarpus, *Hook. et Arn.*—Malayan Archipelago.

---

15. DICTYMIA, *J. Sm.*

Rhizomes short. *Fronds* simple, linear or lanceolate, coriaceous, smooth, 6–12 inches long. *Veins* reticulated, uniform, obscure. *Receptacles* punctiform, compital. *Sori* oval, transverse uniserial, destitute of scales.

Genus 15.—Portion of fertile frond. No. 1.

1. **D. attenuata,** *J. Sm. En. Fil. Hort. Kew.* (1846). Polypodium attenuatum, *R. Br.*; *Hook. Gard. Ferns, t.* 30 (*not Hook. Ic. Pl. t.* 409). Dictyopteris attenuata, *Presl.* (*not Hook. Gen. Fil. t.* 71). —New South Wales and Victoria.

---

16. DRYMOGLOSSUM, *Presl.*; *J. Sm.*

Rhizome slender, surculose. *Fronds* simple, entire, 1–4 inches long, of two forms, the sterile subrotund-elliptical, the

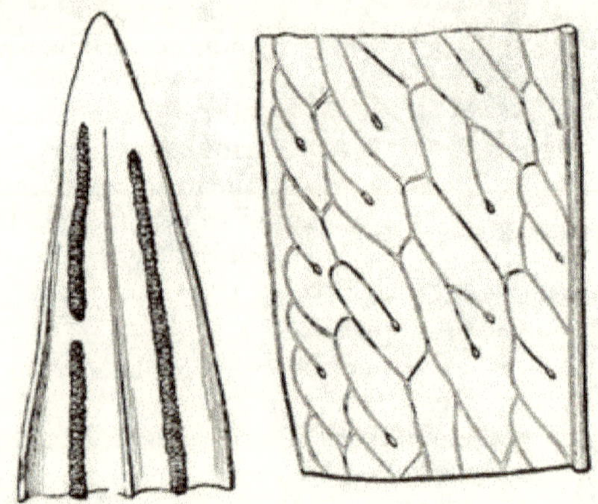

Genus 16.—Barren and fertile frond, slightly enlarged. No. 1.

fertile contracted, linear. *Veins* obscure; venules compoundly anastomosing. *Receptacles* elongated, compital. *Sori* linear, continuous, transverse, intra-marginal, furnished with stellate indusioid scales.

1. **D. piloselloides,** *Presl. Hook. Gard. Ferns, t.* 46. Pteris piloselloides, *Linn. Sw. Syn. Fil. t.* 2, *f.* 3; *Schk. Fil. t.* 87.—East Indies.

---

17. NEVRODIUM, *Fée.*

Rhizome short, cæspitose. *Fronds* simple, entire, 6–12 inches long, lanceolate, thick and fleshy, the fertile portion somewhat contracted. *Veins* obscure; venules compoundly anastomosing. *Receptacles* elongated, compital. *Sori* linear,

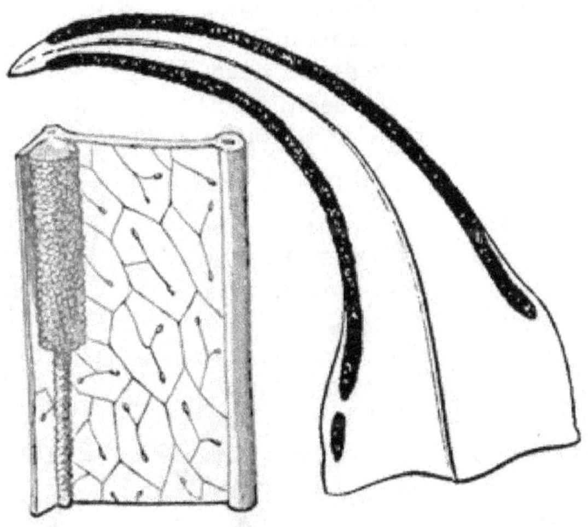

Genus 17.—Portions of fertile frond, natural size. No. 1.

continuous, transverse marginal, on the upper portion of the fronds destitute of scales.

1. **N. lanceolatum,** *Fée, Gen. Fil. t.* 8 *c. Lowe's Ferns,* 2, *t.* 64 *A.* Pteris lanceolata, *Linn.* (*Plum. Fil. t.* 132). Tænitis lanceolata, *R. Br.* Drymoglossum lanceolatum, *J. Sm.* (*olim*). Pteropsis lanceolata, *Desv.; Hook. Fil. Exot. t.* 45.—West Indies.

---

18. DICRANOGLOSSUM, *J. Sm.*

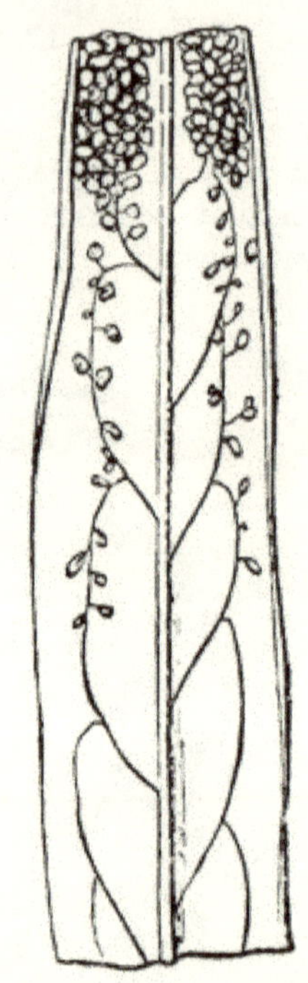

Genus 18.—Portion of fertile frond, under side. No. 1.

Rhizome short, cæspitose. *Fronds* contiguous, furcately-pinnatifid, 6–12 inches high, coriaceous, sparsely squamiferous, segments lanceolate-cuspidate, the fertile slightly contracted. *Veins* obscure, simple, or forked, free, or their apices arcuately anastomosing, forming linear transverse superficial receptacles, which, by contiguity, constitute a continuous or interrupted, linear, intramarginal, naked sorus.

1. **D. furcatum,** *J. Sm.; Bot. Voy. Herald.* Pteris furcata, *Linn.; Plum. Fil. t.* 114. Tænitis furcata, *Willd.; Hook. et Grev. Ic. Fil. t.* 7. Pteropsis furcata, *Presl.; J. Sm. Gen. Fil.* 1841. Cuspidaria furcata, *Fée, Gen. Fil. t.* 8 *A, f.* 2.—West Indies and Tropical America.

---

19. HYMENOLEPIS, *Kaulf.*

Rhizomes short, cæspitose. *Fronds* simple, 6–12 inches long,

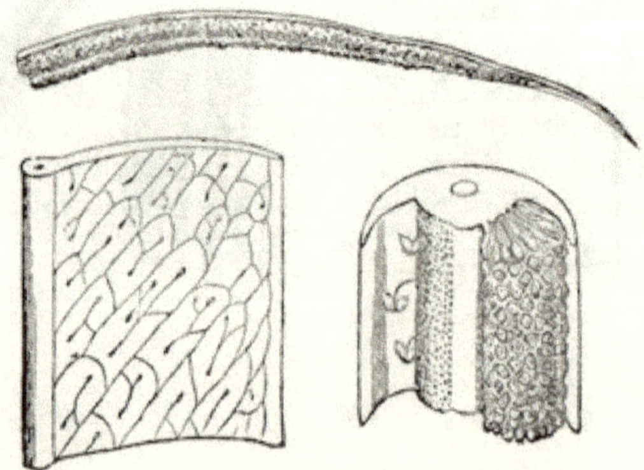

Genus 19.—Portion of fertile frond, natural size; ditto fertile and sterile, enlarged. No. 1.

linear-lanceolate, coriaceous, smooth, the upper portion contracted and fertile, plicate and indusiform, forming a linear spike. *Veins* obscure; venules compoundly anastomosing. *Receptacles* elongated, compital. *Sori* linear, continuous, transverse, on the upper portion of the fronds confluent, furnished with numerous suborbicular hyaline scales.

1. **H. spicata,** *Presl; Hook. Fil. Exot. t.* 78; *Lowe's Ferns,* 2, *t.* 64 *B.* Acrostichum spicatum, *Linn.; Sm. Ic. ined. t.* 49. Lomaria spicata, *Willd.* Gymnopteris spicata, *Presl.; J. Sm. Gen. Fil.* Hymenolepis ophioglossoides, *Kaulf.; Kunze, Fil. t.* 47, *f.* 1. Hymenolepis revoluta, *Bl.; Kunze, Fil. t.* 47, *f.* 2.—Malayan Archipelago.

2. **H. brachystachys,** *J. Sm.* H. spicata, *var.* brachystachys, *Hook. Gard. Ferns, t.* 3. Tænitis ophioglossoides, *Hort. Lips.*—Malayan Archipelago.

---

20. LEPTOCHILUS, *Kaulf.*

Rhizomes short and cæspitose, or long, slender, and surculose. *Fronds* 6–18 inches long, of two forms: the sterile simple, lobed, or pinnatifid, smooth; the fertile contracted, linear-rachiform, its margin revolute and indusiform. *Veins* of sterile frond evi-

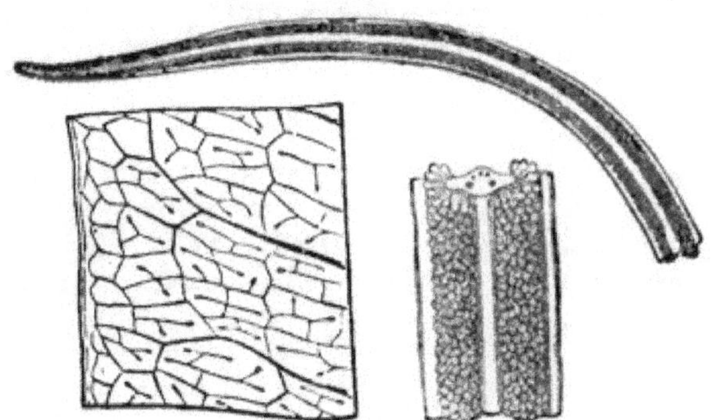

Genus 20.—Portion of fertile and sterile frond, natural size; ditto fertile, enlarged. No. 1.

dent, straight or flexuose, pinnate; venules compoundly anastomosing. *Receptacles* elongated compital. *Sorus* linear, continuous, uniserial, on each side of the costa, ultimately confluent, destitute of scales.

1. L. **decurrens,** *Bl.; Fée, Mem. Acrost. t.* 48, *f.* 1. Anapansia decurrens, *Presl.* Gymnopteris decurrens, *J. Sm.* (*olim*); *Hook. Gard. Ferns, t.* 6.—Ceylon and Malayan Archipelago.

2. L. **axillaris,** *Kaulf. En. Fil. t.* 1, *f.* 10. Acrostichum axillare, *Cav.* Gymnopteris axillaris, *Presl.*—East Indies.

---

21. PHYMATODES, *Presl.; J. Sm.*

Rhizome generally thick, short or much elongated, becoming smooth. *Fronds* simple, pinnatifid or pinnate, smooth, coriaceous or membranous, segments adherent with the rachis.

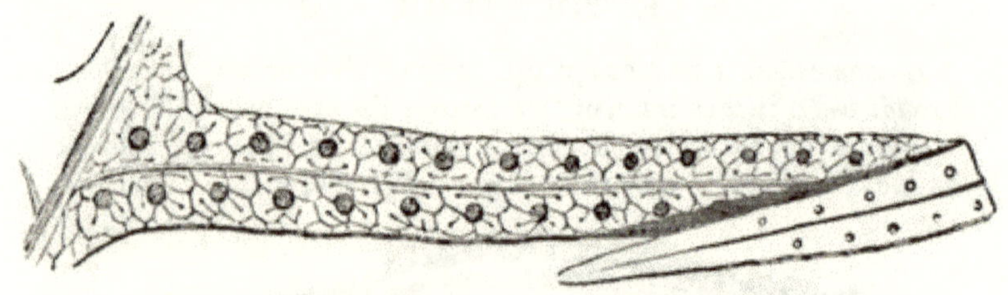

Genus 21.—Portion of mature frond, under side. No. 2.

Veins compound anastomosing, internal, obscure or evident; primary veins generally undefined or evanescent. *Receptacles* compital, generally deeply immersed. *Sori* round or oval, large, transversely uniserial or irregular, naked.

* *Fronds simple or pinnatifid.*

1. P. **pustulata,** *Presl.* Polypodium pustulatum, *Forst. Schk. Fil. t.* 10; *Lowe's Ferns,* 2, *f.* 8. Pleopeltis pustulata, *Moore.*—New Zealand.

2. P. **Billardieri,** *Presl.* Polypodium Billardieri, *R. Br.* Pleopeltis Billardieri, *Moore.* Polypodium scandens, *Labill. Nov. Holl. t.* 240. Polypodium diversifolium,

Willd. Polypodium lepidopodum, *Link.*—Tasmania and New Zealand.

3. **P. terminalis,** *J. Sm.* Chrysopteris terminalis, *Link.*—East Indies.

4. **P. peltidea,** *J. Sm.* Chrysopteris peltidea, *Link.* Polypodium peltideum, *Link; Lowe's Ferns,* 2, *t.* 42. Polypodium phymatodes, *Schk. Fil. t.* 17.—East Indies.

5. **P. nigrescens,** *J. Sm.* Polypodium nigrescens, *Blume, Fil. Jav. t.* 70; *Hook. Fil. Exot. t.* 22. Phymatodes saccata, *J. Sm. Cat. Cult. Ferns* (1857), *p.* 9.—Malayan and Pacific Islands.

6. **P. vulgaris,** *Presl.* Polypodium phymatodes, *Linn.; Jacq. Ic. t.* 637; *Schk. Fil. t.* 9. Pleopeltis phymatodes, *Moore* (*in part*).—Ceylon, South and West Africa, and Mauritius.

7. **P. longipes,** *J. Sm. En. Fil. Hort. Kew.* (1846). Chrysopteris longipes, *Link.* Polypodium phymatodes, *Schk. Fil. t.* 8 *d.*—Malayan Archipelago.

8. **P. glauca,** *J. Sm.* Drynaria (Phymatodes) glauca, *J. Sm. En. Fil. Phil.* Pleopeltis glauca, *Moore.*—Luzon.

9. **P. incurvata,** *J. Sm.* Polypodium incurvatum, *Blume, Fil. Jav. t.* 65. Pleopeltis incurvata, *Moore.*—Java.

10. **P. longissima,** *J. Sm.* Polypodium longissimum, *Bl. Fil. Jav.* 6, *t.* 68. Pleopeltis longissima, *Moore.* Drynaria melanococca, *Moore and Houlst.* Polypodium melanoneuron, *Miq.* Drynaria rubida, *J. Sm. En. Fil. Phil.*—Malayan Archipelago.

** *Fronds pinnate.*

11. **P. leiorhiza,** *Presl.* Polypodium leiorhizon, *Wall.; Hook. Fil. Exot. t.* 25. Pleopeltis leiorhiza, *Moore.* Phymatodes cuspidata, *J. Sm. Cat. Cult. Ferns* (1857), *p.* 10 (*excl. syn. Don.*).—East Indies.

12. **P. albo-squamata,** *J. Sm.* Polypodium albo-squamatum, *Blume, Fil. Jav. t.* 57; *Hook. Gard. Ferns, t.* 47. Pleopeltis albo-squamata, *Presl.*—Java and Borneo.

---

22. PLEURIDIUM, *Fée; J. Sm.*

Rhizome short or elongating. *Fronds* simple, pinnatifid or pinnate, coriaceous, firm, marginate; segments articulated with

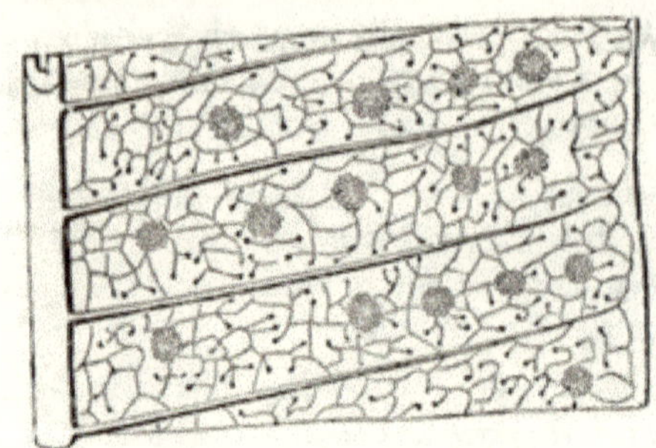

Genus 22.—Portion of fertile frond, natural size. No. 1.

the rachis. *Veins* compound anastomosing; primary veins evident, elevated, costæform, straight. *Receptacles* compital. *Sori* round or oval, or by confluence oblong, transversely uniserial or obliquely 1–2-serial.

* *Fronds simple.*

† *Sori obliquely uniserial.*

1. **P. crassifolium,** *Fée.* Polypodium crassifolium, *Linn.; Plum. Fil. t.* 123. Anaxetum crassifolium, *Schott. Gen. Fil. t.* 1. Polypodium coriaceum, *Radd. Fil. Bras. t.* 25.—Tropical America.

2. **P. albo-punctatissimum,** *J. Sm.* Polypodium albo-punctatissimum, *Linden's Cat.* (1860).—Tropical America.

3. **P. crassinervium,** *J. Sm.* Polypodium crassinervium, *Blume, Fl. Jav. t.* 61.—Java.

†† *Sori obliquely biserial.*

4. **P. rupestre,** *Fée.* Polypodium rupestre, *Blume, Fl. Jav. t.* 55, *f.* 2; *t.* 60, *f.* 1–3.—Java and Luzon.

5. **P. triquetrum,** *J. Sm.* Polypodium triquetrum, *Blume, Fl. Jav. t.* 69.—Java.

** *Fronds pinnatifid or pinnate.*

† *Sori transversely uniserial.*

6. **P. palmatum,** *J. Sm.* Polypodium palmatum, *Bl. Fl. Jav. t.* 64.—Java.

7. **P. oxyloba,** *Presl.* Polypodium oxylobum, *Wall.* Polypodium (Phymatodes) oxylobum, *Hook. Sp. Fil.*—East Indies.

8. **P. angustatum,** *J. Sm.* Polypodium angustatum, *Blume, Fl. Jav. t.* 62. Polypodium Lindleyanum, *Wall.*—Penang, Java.

9. **P. juglandifolium,** *J. Sm.* Polypodium juglandifolium, *D. Don., non Humb.* Polypodium capitellatum, *Wall.* Polypodium Wallichianum, *Spr.*—East Indies. T.

†† *Sori oblique, biserial.*

10. **P. venustum,** *J. Sm.* Polypodium venustum, *Wall.*—East Indies. T.

---

23. SELLIGUEA, *Bory.*

Rhizome slender, elongating epigeous and squamose, or sub-

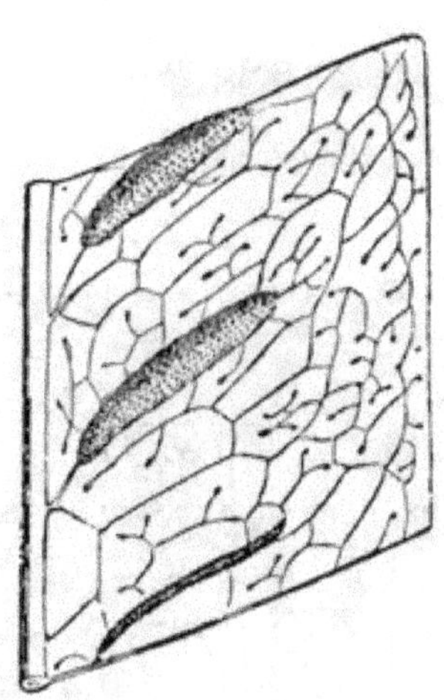

Genus 23.—Portion of fertile frond, natural size No. 2.

hypogeous and naked. *Fronds* stipate, 1-2 feet long, simple, linear lanceolate or broad elliptical, rarely pinnatifid, smooth, opaque, the fertile longer than the sterile, and often sub-contracted. *Primary veins* costæform, straight; venules compound, anastomosing with free veinlets terminating in the areoles. *Receptacles* compital, elongated, oblique, forming a continuous or sub-interrupted linear sorus between the primary veins.

* *Fronds simple.*

1. **S. caudiforme,** *J. Sm.* Polypodium caudiforme, *Blume, Fil. Jav. t.* 54, *f.* 2. Grammitis (Selliguea) caudiformis, *Hook. Bot. Mag. t.* 5328. Gymnogramma (Selliguea) caudiformis, *Hook. Sp. Fil.*—Java.

** *Fronds pinnatifid.*

2. **S. pothifolia,** *J. Sm. in En. Fil. Phil.* Hemionitis pothifolia, *Don.* Grammitis decurrens, *Wall.; Hook. et Grev. Ic. Fil. t.* 6. Gymnogramma (Selliguea) decurrens, *Hook. Sp. Fil.*—India, Japan, Philippine and Fiji Islands.

---

24. COLYSIS, *Presl.; Fée.*

Rhizome short, sub-hypogeous. *Fronds* simple lobed or

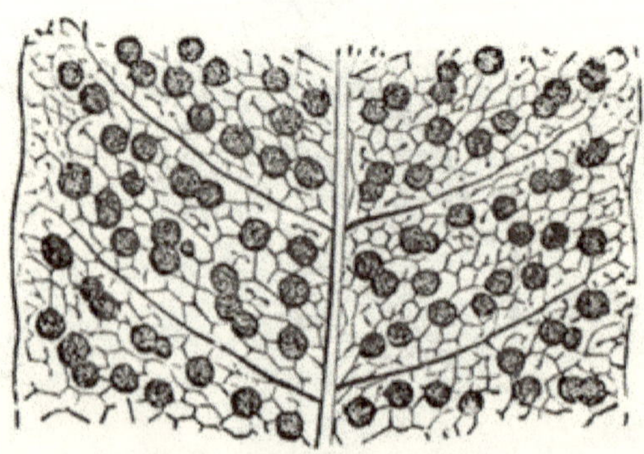

Genus 24.—Portion of mature frond, under side. No 1.

pinnatifid, generally membranous, flaccid, 1–3 feet long. *Veins* compound anastomosing; primary veins costæform, elevated or internal, generally flexuose, sometimes obsolete. *Receptacles* compital, superficial. *Sori* round, or by confluence oblong or linear, irregular or obliquely 1–2-serial.

1. **C. membranacea**, *J. Sm.* Polypodium membranaceum, *Don.* Polypodium hemionitideum, *Wall.; Lowe's Ferns*, 2, *t.* 7. Colysis hemionitidea, *Presl.; Fée.* Hemionitis plantaginea, *Don.* Polypodium grandifolium, *Wall.*—East Indies.

2 **C. Spectra**, *J. Sm.* Polypodium spectrum, *Kaulf.* Polypodium Thouinianum, *Gaud. in Freyc. Voy. Bot. t.* 5, *f.* 1.—Sandwich Islands.

---

25. MICROSORUM, *Link; Fée.*

Rhizome short, subhypogeous. *Fronds* simple, entire or irregularly sinuose, coriaceous, smooth, 1–3 feet long *Veins*

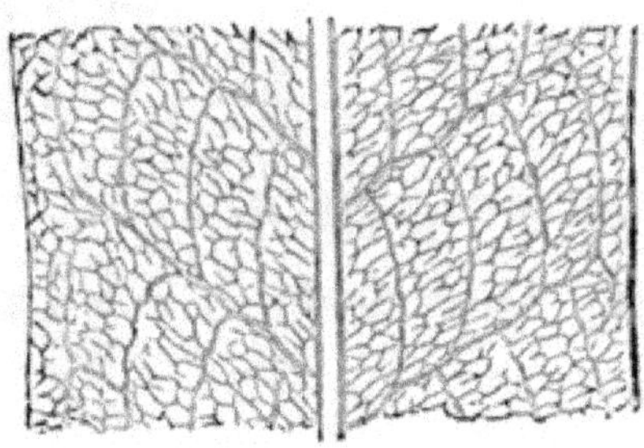

Genus 25.—Portion of mature frond, upper side. No. 1.

compound anastomosing, internal; primary veins obscure. *Receptacles* compital, superficial. *Sori* round, small, numerous, irregular, sometimes subconfluent.

1. **M. irioides**, *Fée.* Polypodium irioides, *Poir.; Hook. et Grev. Ic. Fil. t.* 125. *Hook. Fil. Exot. t.* 4. Polypodium polycephalum, *Wall.* Microsorum irregulare, *Link; Fée.* Microsorum sessile, *Fée.*—β apex of fronds crested.—East Indies, Malayan Archipelago, Australia, and Trinidad.

26. NIPHOBOLUS, *Kaulf.*; *J. Sm.*

Rhizome short or elongated and surculose. *Fronds* simple, linear-lanceolate, oblong-elliptical, or obovate-subrotund, rarely lobed, from less than an inch to three or four feet long, thick and fleshy or coriaceous, covered with sessile or stipulate stellate pubescence; the fertile usually more or less contracted and

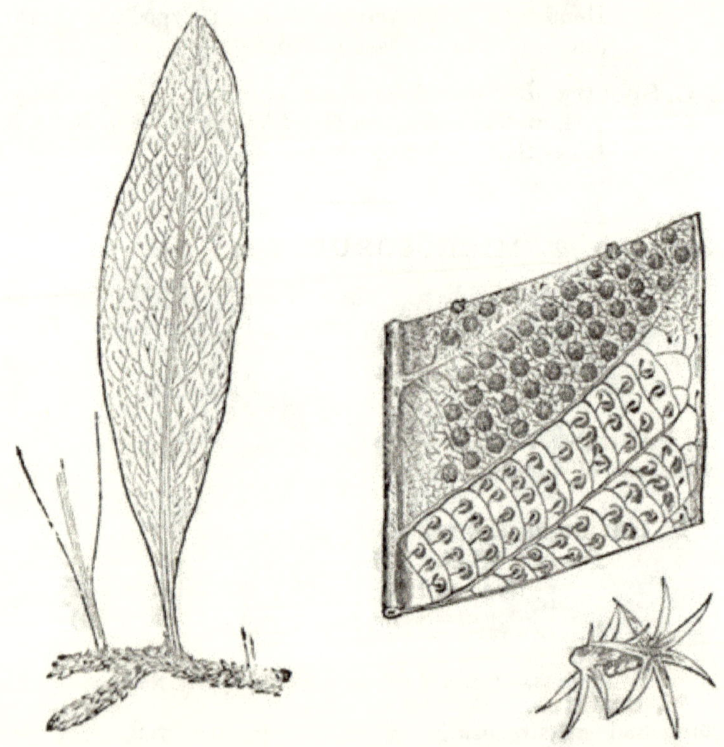

Genus 26.—Portion of rhizome, with a barren frond. No. 3.

longer than the sterile. *Veins* obscure, undefined, or evident and costæform; venules compound anastomosing. *Receptacles* punctiform, immersed, terminal or medial on simple or brachiate free veinlets, or compital. *Sori* round or oval, sub-transverse multiserial between the primary veins, or irregular and confluent, protruding through the dense stellate pubescence.

* *Rhizomes elongated, surculose. Fronds distant. Primary veins undefined.*

1. **N. rupestris,** *Spr.; Hook. et Grev. Ic. Fil. t.* 93; *Lowe's Ferns,* 1, *t.* 20. Polypodium rupestre, *R. Br.* Craspedaria rupestris, *Link.*—Australia. **Tr.**

2. **N. bicolor,** *Kaulf.; Hook. et Grev. Ic. Fil. t.* 44.—New Zealand. **Tr.**

3. **N. adnascens,** *Kaulf.; Hook. Gard. Ferns, t.* 19. Polypodium adnascens, *Sw. Syn. Fil. t.* 2, *f.* 2. Niphobolus pertusus, *Spr.; Lowe's Ferns,* 1, *t.* 21. Polypodium pertusum, *Roxb.; Hook. Exot. Fil. t.* 162.—East Indies.

4. **N. Lingua,** *Spr.; Kunze in Schk. Fil. Supp. t.* 63. *Lowe's Ferns,* 1, *t.* 22. Acrostichum Lingua, *Thunb. Fil. Jap. t.* 33; *Schk. Fil. t.* 1. Polypodium Lingua, *Sw.; Lang. et Fisch. Ic. Fil. t.* 5. Cyclophorus Lingua, *Desv.* Polycampium Lingua, *Presl.* Niphobolus Sinensis, *Hort.*—East Indies and China.

** *Rhizomes short, cæspitose. Fronds contiguous. Primary veins generally evident.*

5. **N. Gardneri,** *Kunze; Hook. Fil. Exot. t.* 68; *Lowe's New Ferns, t.* 38 *B.* Polypodium Gardneri, *Metten. Gen. Polypodium, p.* 129. Niphobolus acrostichoides, *Cat. Fil. Hort. Kew., non* Polypodium (Niphobolus) acrostichoides, *Forst.*—Ceylon.

6. **N. costatus,** *Presl.* Polypodium costatum, *Wall.*—East Indies.

---

27. CAMPYLONEURUM, *Presl.*

Rhizome short and cæspitose or elongated, often subhypogeous. *Fronds* simple or very rarely pinnate, coriaceous, rigid, smooth, 1–2 feet high. *Veins* costæform or undefined, elevated or internal and obscure; venules arcuately or angularly anastomosing, producing two or more excurrent free veinlets. *Re-*

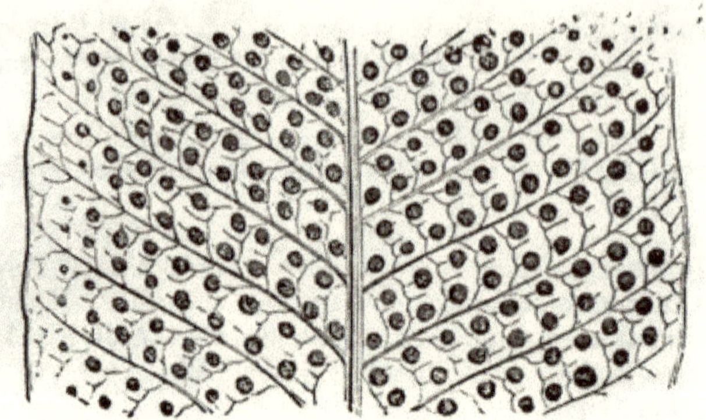

Genus 27.—Portion of mature frond, under side. No. 5.

ceptacles punctiform, terminal or medial on the free veinlets. *Sori* round, obliquely biserial or irregular, destitute of scales.

* *Fronds simple.*

1. **C. ensifolium,** *J. Sm.* Polypodium ensifolium, *Willd.* Marginaria ensifolia, *Presl.* Campyloneurum angustifolium, β tæniosum, *Moore.*—Tropical America.

2. **C. angustifolium,** *Fée.* Polypodium angustifolium, *Sw.; Radd. Fil. Bras. t.* 24, *f.* 2. Marginaria angustifolia, *Presl.* Polypodium dimorphum, *Link.* Polypodium leucorhizon, *Klt.* Polypodium amphostemum, *Kunze.* —Tropical America.

3. **C. fasciale,** *Presl.* Polypodium fasciale, *Humb.* P. lapathifolium, *Radd. Fil. Bras. t.* 24, *f.* 3.—Brazil and Venezuela.

4. **C. rigidum,** *J. Sm. Cat. Cult. Ferns* (1857), *p.* 13. C. lucidum, *Moore.* Polypodium nitidum, *Hook. Fil. Exot. t.* 12 (*excl. syn.*).—Tropical America.

5. **C. repens,** *Presl.; Hook. Gen. Fil. t.* 71 *A.* Polypodium repens, *Linn.; Plum. Fil. t.* 134. C. cæspitosum, *Link; J. Sm. Cat.* (1857). Polypodium cæspitosum, *Link; Metten. Fil. Hort. Lips. t.* 24, *f.* 4, 5.—Tropical America.

6. **C. Phyllitidis**, *Presl.* Polypodium Phyllitidis, *Linn.*; (*Plum. Fil. t.* 130).—Tropical America.

7. **C. nitidum**, *Presl.* Polypodium nitidum, *Kaulf.* Campyloneurum latum, *Moore, Ind. Fil. p.* 225.—Tropical America.

8. **C. brevifolium**, *Link.* Polypodium brevifolium, *Link*; *Mett. Fil. Hort. Lips.*—Tropical America.

** *Fronds pinnate.*

9. **C. decurrens**, *Presl.* Polypodium decurrens, *Radd. Fil. Bras. t.* 33. Polypodium polyanthos, *Hort. Brux.*—Brazil.

---

28. DRYNARIA, *Bory*; *J. Sm.*

Rhizome short, thick, and fleshy. *Fronds* rigid; the sterile (when present) sessile, broad cordate, sinuose or laciniated; the fertile stipitate or sessile, pinnatifid or pinnate, rarely simple, the segments articulated with the rachis; when sessile, the base is similar to the special sterile frond. *Veins* external, elevated,

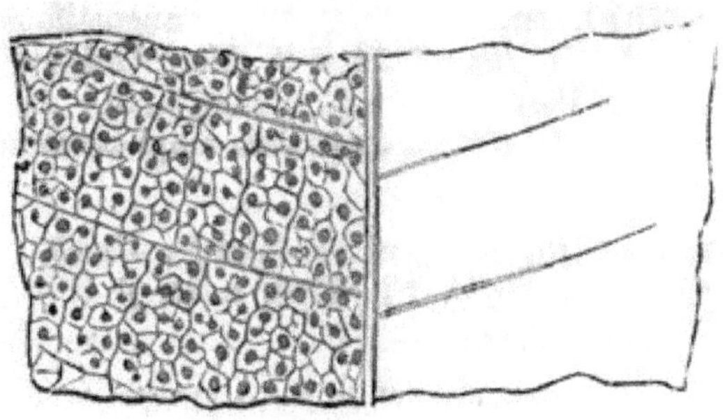

Genus 28.—Portion of mature frond, under side. No. 6.

compound anastomosing, forming quadrate or hexagonal areoles; primary veins costæform or obsolete. *Receptacles* compital. *Sori* round, small, numerous, and irregular, or transversely or obliquely serial, sometimes confluent, forming a linear sorus between the costæform veins.

* *Sori transversely uniserial.*

† *Fronds pinnatifid.*

1. D. **propinqua**, *J. Sm.* Polypodium propinquum, *Wall.* Phymatodes propinqua, *Presl.* Polypodium Willdenowii, *Hook. Gard. Ferns, t.* 35; *non Bory.*—East Indies.

†† *Fronds pinnate.*

2. D. **diversifolia**, *J. Sm.* Polypodium diversifolium, *R. Br.*; *Hook. Gard. Ferns, t.* 5. Polypodium Gaudichaudi, *Bory*; *Bl. Fil. Jav. t.* 57. Drynaria pinnata, *Fée.* Polypodium glaucistipes, *Wall.* Drynaria Hilli, *Hort.*—East Indies, Malayan Archipelago, and Australia.

** *Sori oblique, uniserial.*

† *Fronds pinnatifid.*

3. D. **coronans**, *J. Sm.*; *Fée.* Polypodium coronans, *Wall.*; *Hook. Fil. Exot. t.* 91. Phymatodes coronans, *Presl.*—East Indies and Malacca.

*** *Sori oblique, biserial.*

4. D. **quercifolia**, *Bory*; *Fée.* Polypodium quercifolium, *Linn.*; *Schk. Fil. t.* 13. Phymatodes quercifolia, *Presl.*—East Indies, Mauritius, Malayan Archipelago, and Australia.

**** *Sori numerous, irregular.*

† *Fronds simple.*

5. D. **musæfolia**, *J. Sm.* Polypodium musæfolium, *Bl. Fil. Jav. t.* 79. Polypodium microsorum, *Metten. Cat. Hort. Herrenh.*—Malayan Archipelago.

†† *Fronds pinnatifid.*

6. D. **Heraclea**, *J. Sm.* Polypodium (§ Drynaria) Heracleum, *Kunze*; *Hook. Gard. Ferns, t.* 1. Drynaria morbillosa, *J. Sm. Cat. Cult. Ferns,* 1857.—Malayan Archipelago.

DIVISION II. Desmobrya.

Fronds in vernation terminal, uniserial or fasciculate, their bases adherent and continuous with the stem, which is either a caudex or sarmentum.

TRIBE IV.—ACROSTICHEÆ.

Sori undefined (amorphous), naked. *Fertile fronds* or segments always more or less contracted; the under side (or rarely both sides) densely sporangiferous. Acrostichum, *Linn.*

§ 1. *Elaphoglosseæ. Fronds always simple. Veins free or rarely combined at the margin or reticulated.*

* *Veins free.*

29. ELAPHOGLOSSUM, *Schott.; J. Sm.*

Vernation uniserial and sarmentose, or subfasciculate and decumbent, squamose. *Stipes* often pseudo-articulate, node

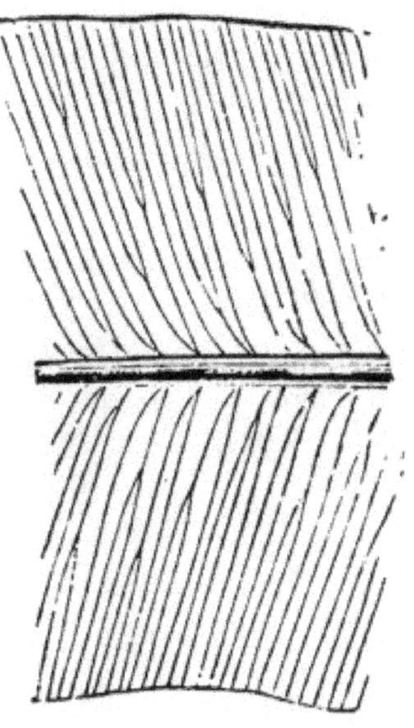

Genus 29.—Portion of barren frond, under side. No. 3.

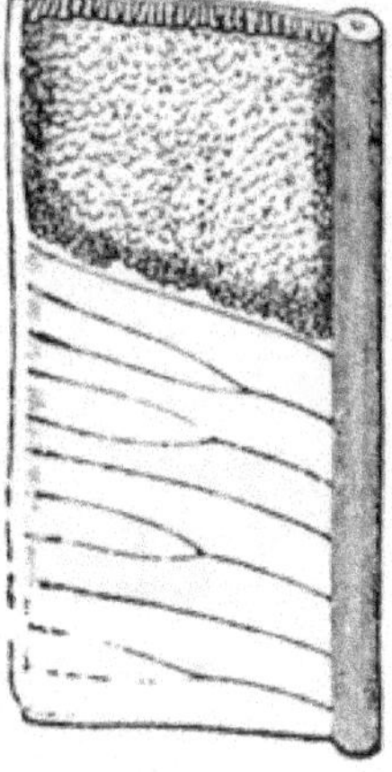

Genus 29.—Portion of fertile frond, under side. No. 8.

elevated. *Fronds* simple entire, from 2 inches to 2½ feet high, smooth or squamose. *Veins* simple or forked, parallel, direct, their apices free and clavate. *Fertile fronds* plain, the under side sporangiferous.

* *Fronds smooth or nearly so.*

† *Vernation sarmentose. Fronds distant.*

1. E. **stigmatolepis,** *J. Sm.* Acrostichum stigmatolepis, *Fée, Acrost. t.* 24, *f.* 2.—Ceylon.

2. E. **Funckii,** *J. Sm.* Acrostichum Funckii, *Fée, Acrost. t.* 6, *f.* 1. Acrostichum (Elaphoglossum) Funckii, *Hook. Sp. Fil.* 5, *p.* 205.—Venezuela and Trinidad.

†† *Vernation fasciculate, decumbent.*

3. E. **conforme,** *Schott.* Acrostichum conforme, *Sw. Syn. Fil. t.* 1, *f.* 1.—South Africa and Java.

4. E. **callæfolium,** *J. Sm.* Acrostichum callæfolium, *Bl. Fil. Jav. t.* 4.—Java.

5. E. **Sieberi,** *J. Sm.* Acrostichum Sieberi, *Hook. et Grev. Ic. Fil. t.* 237.—Mauritius.

6. E. **crassinerve,** *J. Sm.* Acrostichum crassinerve, *Kunze.*—Brazil.

7. E. **latifolium,** *J. Sm.* Acrostichum latifolium, *Sw.; Hook. Fil. Exot. t.* 42.—Tropical America.

8. E. **Herminieri,** *J. Sm.* Acrostichum Herminieri, *Bory, in Fée, Acrost. t.* 11. Acrostichum (Elaphoglossum) Herminieri, *Hook. Sp. Fil.* 5, *p.* 216. — Tropical America and Trinidad.

9. E. **microlepis,** *J. Sm.* Acrostichum microlepis, *Kunze.*—Venezuela.

** *Fronds more or less densely squamiferous.*

10. E. **piloselloides,** *J. Sm.* Acrostichum piloselloides, *Presl. Reliq. Haenk. t.* 2, *f.* 1; *Hook. Fil. Exot. t.* 29.—Tropical America.

11. E. **rubiginosum**, *J. Sm.* Acrostichum rubiginosum, *Fée, Acrost. t.* 5, *f.* 1, *et t.* 13, *f.* 1. E. brachyneuron, *J. Sm.* Acrostichum brachyneuron, *Fée, Acrost. t.* 22, *f.* 1. A. Schiedei, *Kunze.* A. frigida, *Linden.*—Tropical America.

12. E. **cuspidatum**, *J. Sm.* Acrostichum cuspidatum, *Willd.; Fée, Acrost. t.* 14, *f.* 2.—West Indies and Tropical America.

13. E. **Blumeanum**, *J. Sm. En. Fil. Phil.* Acrostichum Blumeanum, *Fée.* A. viscosum, *Bl.* (*not Sw.*)—Malay and Philippine Islands.

14. E. **muscosum**, *J. Sm.* Acrostichum muscosum, *Sw.*—West Indies and Tropical America.

15. E. **squamosum**, *J. Sm.* Acrostichum squamosum, *Sw.* A. hirtum, *Sw.* A. paleaceum, *Hook. et Grev. Ic. Fil. t.* 235.—Madeira, West Indies and Tropical America.

16. E. **vestitum**, *R. T. Lowe in Hook. et Grev. Ic. Fil. t.* 235 (A. paleaceum on plate).—Madeira and West Indies.

*** *Fronds fringed or squamiferous at the margin only.*

17. E. **apodum**, *Schott.* Acrostichum apodum, *Hook. et Grev. Ic. Fil. t.* 99.—West Indies.

18. E. **undulatum**, *J. Sm.* Acrostichum undulatum, *Willd.* (*Plum. Fil. t.* 126).—Dominica.

19. E. **scolopendrifolium**, *J. Sm.* Acrostichum scolopendrifolium, *Radd. Fil. Bras. t.* 16.—Brazil.

*** Veins combined at the margin.*

30. ACONIOPTERIS, *Presl.*

Vernation uniserial; sarmentum short, thick, squamose. *Fronds* contiguous, elliptical, lanceolate, 6–12 inches long, smooth or squamiferous. *Veins* simple or forked, parallel, their apices combined near the margin by a straight or zig-zag vein. *Fertile* frond linear, plane, wholly sporangiferous on the under side.

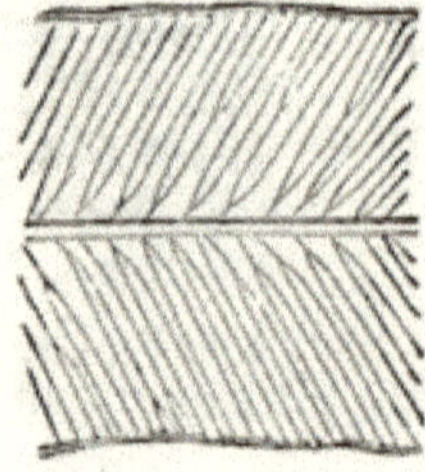

Genus 30.—Portion of mature frond, under side. No. 2.

1. A. **nervosa,** *J. Sm.* Acrostichum nervosum, *Bory.* Aconiopteris subdiaphana, *Presl. Pterid.; Hook. et Bauer. Gen. Fil. t.* 79 *B.* Acrostichum subdiaphanum, *Hook. et Grev. Ic. Fil. t.* 205.—St. Helena and Bourbon.

2. A. **longifolia,** *Fée, Acrost. t.* 41. Acrostichum longifolium, *Jacq.* (*Plum. Fil. t.* 135). Elaphoglossum longifolium, *J. Sm. Cat. Cult. Ferns,* 1857. Olfersia longifolia, *Presl.*—Dominica.

**** Veins reticulated, uniform.*

31. HYMENODIUM, *Fée.*

Vernation fasciculate, decumbent, densely crinite. *Fronds*

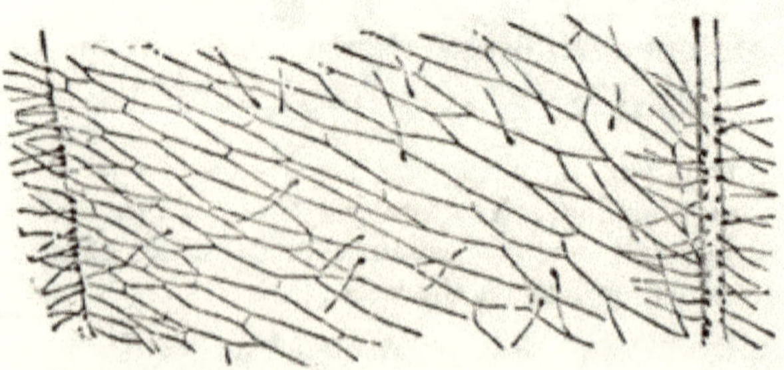

Genus 31.—Portion of frond, under side. No. 1.

simple, entire, squamiferous. 6-8 inches long. *Veins* uniform, reticulated; areoles large, elongated, trapezoid or hexagonoid. *Fertile fronds* broad, densely sporangiferous on the under side.

1. **H. crinitum,** *Fée.* Acrostichum crinitum, *Sw. Plum. Fil. t.* 125; *Hook. et Grev. Ic. Fil. t.* 1; *Hook. Fil. Exot. t.* 6. Dictyoglossum crinitum, *J. Sm. Cat. Kew Ferns*, 1846.—West Indies.

---

32. ANETIUM, *Kunze.*

Vernation uniserial; sarmentum slender, furnished with thin membranous reticulated shining lanceolate scales. *Fronds* distant, simple, oblong-elliptical, acuminate, 6-20 or more inches long, smooth, membraneous. *Veins* uniform, reticulated,

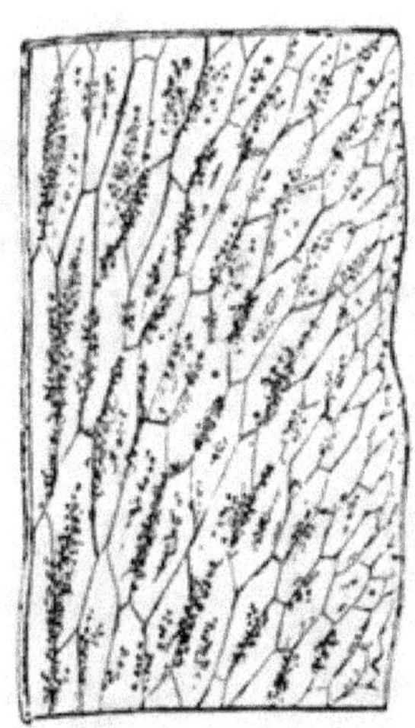

Genus 32.—Portion of mature frond, under side. No. 1.

forming trapezoid or hexagonal transverse elongated areoles. *Receptacles* undefined, the sporangia being thinly scattered or collected in small irregular groups over the whole under surface of the frond, or evident on the veins.

1. **A. citrifolium,** *Splitg.* Acrostichium citrifolium, *Linn. Plum. Fil. t.* 116. Antrophyum citrifolium, *Fée.*

Hemionitis citrifolia, *Hook. Sp. Fil.*—West Indies and Tropical America.

---

2. *Polybotryæ. Vernation generally uniserial, distant or contiguous. Fronds pinnate or bi-tripinnate, rarely flabellate, segments adherent or articulate with the rachis. Veins free or combined at the margin, or anastomosing in various ways.*

* *Veins free.*

† *Segments adherent.*

33. RHIPIDOPTERIS, *Schott.*

Vernation uniserial; sarmentum slender, filiform. *Fronds* distant, 3-6 inches long, the sterile flabelliform, entire, bi-tri-

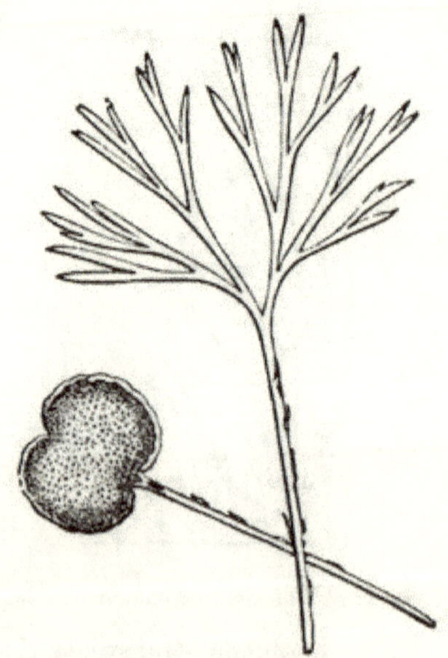

Genus 33.—Fertile and barren fronds. No. 1.

lobed or dichotomously multifid. *Veins* flabellately forked, free. *Fertile frond* subrotund, entire or bilobed, sporangiferous on the under side.

1. **R. peltata,** *Schott.* Acrostichum peltatum, *Schk. Fil. t.* 12 (*Plum. Fil. t.* 50, *f. A*). Acrostichum fœniculaceum, *Hook. et Grev. Ic. Fil. t.* 119.—West Indies and Tropical America.

---

34. MICROSTAPHYLA, *Presl.*

Vernation decumbent, subfasciculate; sarmentum short, squamose. *Fronds* numerous, contiguous, 3-8 inches high, the sterile linear-lanceolate, sub-entire, unequally crenate or laciniately pinnatifid, glandulose, segments and laciniæ cuneiform,

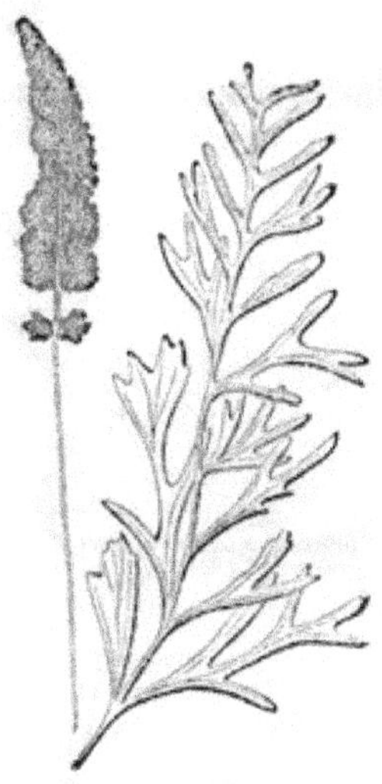

Genus 34.—Portion of fertile and barren fronds, natural size. No. 1.

entire or bi-trilobed. *Veins* simple or forked. *Fertile fronds* contracted, shorter and less divided than the sterile, sporangiferous on the under side.

1. **M. bifurcata,** *Presl. Epim,* Acrostichum bifurcatum, *Sw.; Hook. 2nd Cent. of Ferns, t.* 91; *Schk. Fil. t.* 2.—St. Helena.

---

35. EGENOLFIA, *Schott. Fée.*

Vernation decumbent, uniserial, subhypogeous. *Fronds* contiguous, stipate, pinnate, 1–3 feet high, generally viviparous, sterile pinnæ linear-lanceolate, sub-entire or dentate, laciniated

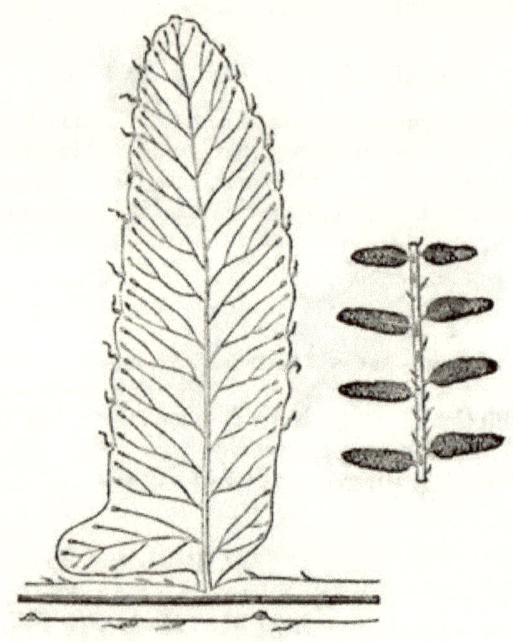

Genus 35.—Fertile and barren fronds. No. 1.

or pinnatifid, sinus mucronate. *Veins* forked or pinnate; venules free. *Fertile segments* more or less contracted; venules evident, contiguous, forming a concrete amorphous receptacle, sometimes forming moniliform spikes.

1. **E. appendiculata,** *J. Sm.* Acrostichum appendiculatum, *Willd.; Hook. Exot. Fl. t.* 108. Acrostichum viviparium, *Sw.* Polybotrya viviparia, *Hook. Exot. Fl. t.* 107. Acrostichum setosum, *Wall.* Acrostichum Hamiltoniana, *Wall.* Egenolfia Hamiltoniana, *Schott. Gen. Fil.* 34.—East Indies and Ceylon.

---

36. PSOMIOCARPA, *Presl. in part.*

Vernation fasciculate, erect. *Fronds* stipate, deltoid, sub-bipinnate, the sterile 6-8 inches high, pilose, with articulated hairs; pinnæ 3-4 inches long; pinnules sessile, decurrent,

Genus 36.—Portion of fertile and barren fronds. No. 1.

oblong elliptical, $\frac{1}{2}$–$\frac{3}{4}$ inch long, unequally dentate or sub-laciniated. *Veins* forked; venules free. *Fertile frond* 14-18 inches high, long, stipate, slender, wholly contracted, forming a sporangiferous panicle.

P. apiifolia, *Presl. Epim. Bot.* Polybotrya apiifolia, *J. Sm. En. Fil. Philipp.; Kunze, in Schk. Fil. t.* 62; *Gard. and Field Sert. t.* 30, 31; *Hook. Sp. Fil.* 5, 248.—Luzon

---

37. POLYBOTRYA, *Humb. et Bonpl.*

Vernation uniserial; sarmentum scandent, squamose. *Fronds* bi-tripinnate, 2-3 feet long. *Veins* pinnate; venules free. *Fertile segments* convolute, pinnatifid or spicæform, wholly sporangiferous.

1. **P. osmundacea,** *Humb. et Bonpl. Nov. Gen.* 1, *t.* 2; *Hook. Gen. Fil. t.* 78 *B.* P. cylindrica, *Kaulf.; Fée, Acrost. t.* 36. Polybotrya speciosa, *Schott. Gen. Fil. t.* 7.—Tropical America.

2. **P. acuminata,** *Link; Metten. Fil. Hort. Lip. t.* 2, *f.* 1–6.—Brazil.

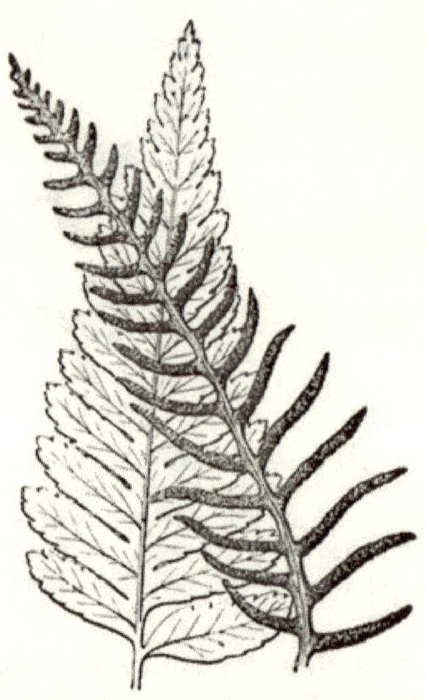

Genus 37.—Portion of fertile and barren fronds. No. 1.

3. **P. incisa,** *Link; Fée, Acrost. t.* 35.—Brazil.

4. **P. caudata,** *Kunze; Fée, Acrost. t.* 34.—West Indies and Tropical America.

---

†† *Segments articulated with the rachis.*

38. LOMARIOPSIS, *Fée.*

Vernation uniserial; sarmentum scandent, squamose. *Fronds* pinnate, 1–3 feet high; pinnæ linear-elliptical, broad, lanceolate, acuminate, 2–10 inches long, articulate with the rachis.

Veins uniform, simple or forked, direct, parallel, free. *Fertile*

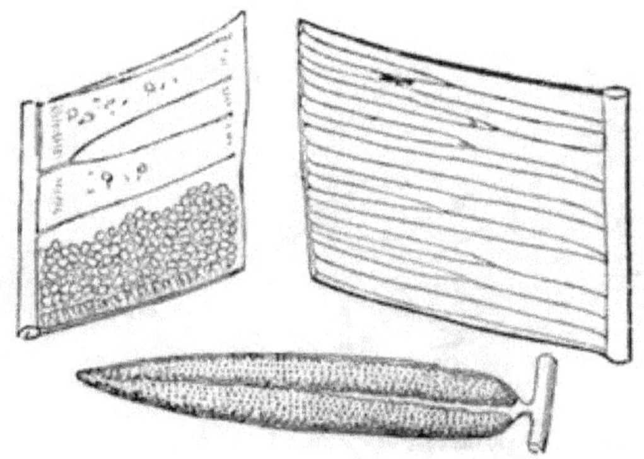

Genus 38.—Portions of fertile and barren fronds, natural size, and portion of fertile, enlarged. No. 2.

pinnæ plane, often broad, sporangiferous on the under side; margin membranous, narrow, subindusiform.

1. **L. sorbifolia,** *Fée.* Acrostichum sorbifolium, *Linn.*; (*Plum. Fil. t.* 117). Stenochlæna sorbifolia, *J. Sm. Gen. Fil.*—West Indies.

2. **L. longifolia,** *J. Sm.* Lomaria longifolia, *Kaulf. Lowe's New Ferns, t.* 37. Acrostichum Yapurense, *Hook. Gard. Ferns, t.* 57. Acrostichum phlebodes, *Kunze*; *Hook. Sp. Fil.* 5, *p.* 24, sub Acrostichum sorbifolium.—West Indies and Tropical America.

3. **L. heteromorpha,** *J. Sm.* Stenochlæna heteromorpha, *J. Sm. Gen. Fil.* 1841. Lomaria filiformis, *A. Cunn. Hook. Sp. Fil.* 3, *t.* 149. Lomaria propinqua, *A. Cunn.*—New Zealand.

** *Veins combined at the margin.*

39. OLFERSIA, *Radd.*; *Presl.*

Vernation uniserial, contiguous; sarmentum scandent, squamose. *Fronds* pinnate, 1–3 feet long. *Veins* uniform, simple or forked, direct, parallel, their apices combined by a transverse

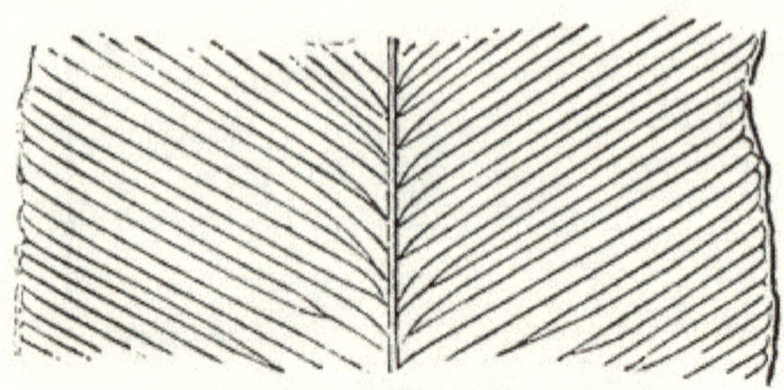

Genus 39.—Portion of the barren pinna, under side. No. 1.

marginal vein. *Fertile pinnæ* linear or pinnatifid, convolute, wholly sporangiferous.

1. **O. cervina,** *Presl; Hook. Fil. Exot. t.* 43; *Lowe's Ferns,* 7, *tt.* 39, 40. Acrostichum cervinum, *Sw.; Plum. Fil. t.* 154; *Hook. et Grev. Ic. Fil. t.* 81. O. Corcovadensis, *Radd. Fil. Bras. t.* 14; *Hook. Gen. Fil. t.* 79 *A.* Acrostichum linearifolium, *Presl.*—Tropical America.

*** *Veins angularly or compoundly anastomosing.*

40. SOROMANES, *Fée.*

Vernation uniserial; sarmentum thick, scandent, squamose.

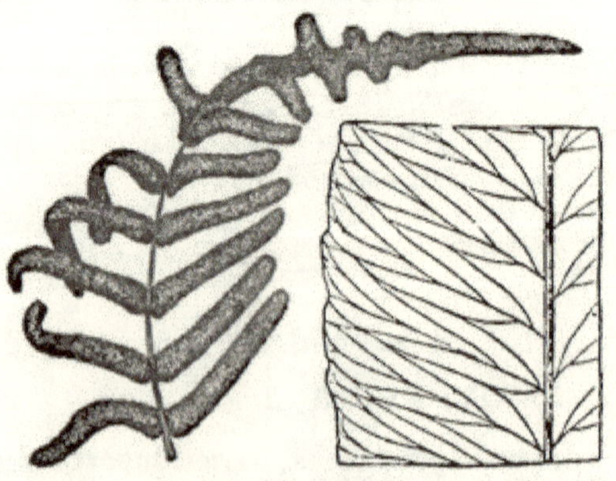

Genus 40.—Portions of fertile and barren fronds. No. 1.

Sterile fronds pinnate, 1–2 feet long. *Veins* pinnate; venules acutely anastomosing, forming oblique elongated areoles; apices next the margin free and clavate. *Fertile fronds* bipinnate; segments convolute, wholly sporangiferous.

1. **S. serratifolium,** *Fée, Acrost. t.* 43. Polybotrya serratifolia, *Klotzsch.*—Venezuela.

---

41. STENOSEMIA, *Presl.*

Vernation fasciculate, erect. *Fronds* ternately pinnate, 6–18 inches high; pinnæ laciniately lobed, bulbiferous. *Veins* pinnate; the lower venules transversely anastomosing, forming

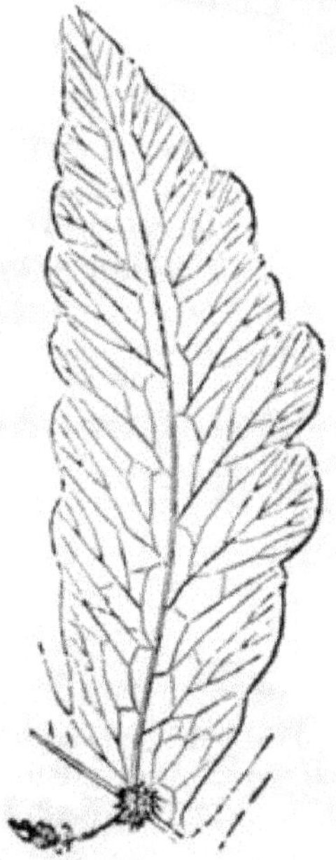

Genus 41.—Portion of mature frond, upper side. No 1.

elongated costal and sub-costal areoles, the superior venules free. *Fertile segments* linear, rachiform, convolute, nearly wholly sporangiferous.

1. **S. aurita,** *Presl.* Acrostichum auritum, *Sw.*; *Lowe's Ferns,* 7, *tt.* 52, 53. Polybotrya aurita, *Bl. Fl. Jav. t.* 1; *Hook. Fil. Exot. t.* 81.—Java.

42. PŒCILOPTERIS, *Eschw.; Presl.*

Vernation uniserial, distant or contiguous, subfasciculate and decumbent. *Fronds* pinnate, 1–3 feet long, often bulbiferous. *Primary veins* costæform, pinnate; venules arcuately or angularly anastomosing, producing on their exterior sides or angles one or more free or anastomosing veinlets, forming unequal areoles. *Sporangia* amorphous, or sometimes in defined lines on the venules (*Jenkinsia, Hook.*).

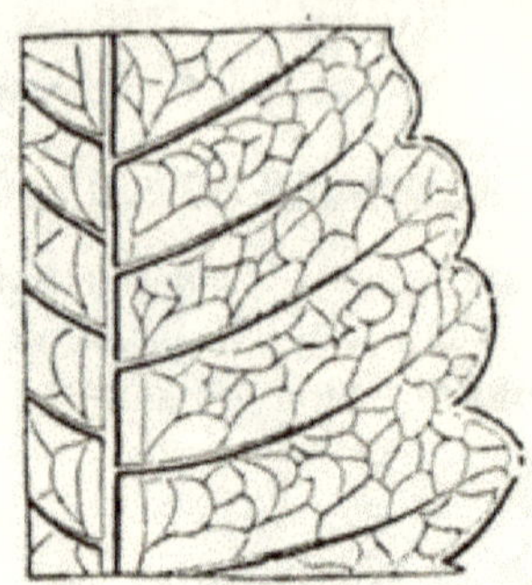

Genus 42.—Portion of barren frond. No. 3.

1. **P. flagellifera,** *J. Sm.* Acrostichum flagelliferum, *Wall.; Hook. et Grev. Ic. Fil. t.* 23; *Blume, Fl. Jav. t.* 13.—East Indies.

2. **P. crispatula,** *J. Sm.* Acrostichum crispatulum, *Wall.*—East Indies.

3. **P. prolifera,** *J. Sm.* Acrostichum proliferum, *Blume; Hook. Ic. Pl. t.* 681, 2. Heteroneuron proliferum, *Fée, Acrost. t.* 55. Acrostichum virens, *Wall.; Hook. et Grev. Ic. Fil. t.* 221.—East Indies.

4. **P. punctulata,** *Presl.* Acrostichum punctulatum, *Linn.* Heteroneuron punctulatum, *Fée, Acrost. t.* 54.—Mauritius and West Tropical Africa.

---

43. GYMNOPTERIS, *Bernh.; Presl.*

Vernation uniserial and sarmentose, or contiguous subfasciculate and decumbent. *Fronds* simple, lobed or pinnate, from 6 inches to 2–3 feet high. *Primary veins* costæform; venules compound anastomosing, with free variously directed veinlets terminating in the areoles. *Sporangia* amorphous.

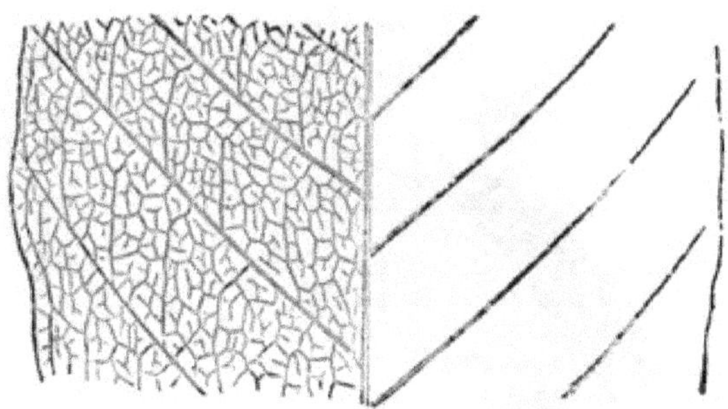

Genus 43.—Portion of sterile frond, under side. No. 2.

1. G. **quercifolia,** *Bernh.; Presl; Hook. Ic. Pl. t.* 905; *Hook. Fil. Exot. t.* 80. Acrostichum quercifolium, *Retz.; Sw.; Schk. Fil. t.* 3. Gymnopteris Nicnerii, *Hort.*—Ceylon.

2. G. **nicotianæfolia,** *Presl; Fée, Acrost. t.* 46. Acrostichum nicotianæfolium, *Sw.; Hook. Gard. Ferns, t.* 26.—West Indies.

3. G. **acuminata,** *Presl.* Acrostichum acuminatum, *Willd.;* (*Plum. Fil. t.* 115).—West Indies.

4. G. **aliena,** *Presl; Hook. Gen. Fil. t.* 85. Acrostichum alienum, *Sw.; Plum. Fil. t.* 10.—Tropical America.

5. G. **Gaboonense,** *J. Sm.* Acrostichum (Gymnopteris) Gaboonense, *Hook. Sp. Fil.* 5, *p.* 270.—Tropical West Africa.

---

§ 3. *Acrostichæ. Vernation fasciculate. Fronds pinnate, 4–8 feet high; pinnæ adherent. Veins uniform, reticulated; areoles small subquadrangular, or large hexagonoid.*

44. NEUROCALLIS, *Fée.*

Vernation fasciculate, decumbent. *Fronds* pinnate, 3–4 feet high, smooth; sterile pinnæ elliptical-lanceolate, acuminate, entire, 8–10 inches long, 2 inches wide, sessile, adherent with

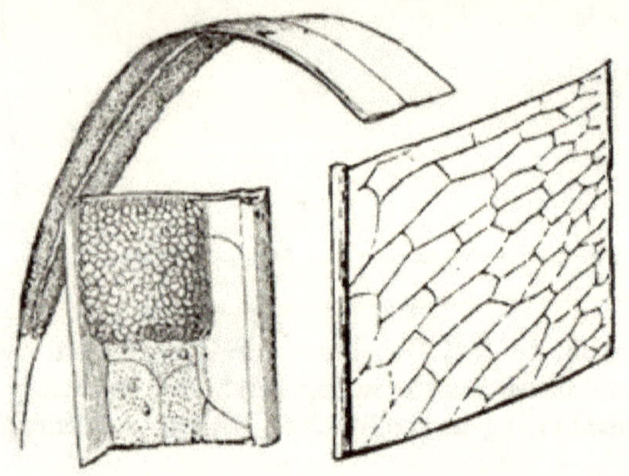

Genus 44.—Portions of fertile and barren fronds. No 1.

the rachis. *Veins* uniform, reticulated; areoles oblong, hexagonoid. *Fertile fronds* contracted; pinnæ linear, acuminate, plane, wholly sporangiferous on the under side; sporangia destitute of indusoid scales.

1. **N. præstantissima,** *Fée, Acrost. t.* 52; *Fée, Gen. Fil. t.* 4 *A.* Acrostichum præstantissimum, *Bory, Hb.; Hook. Gard. Ferns, t.* 58.—Dominica and Guadeloupe.

---

45. ACROSTICHUM, *Linn.* (*in part*); *J. Sm.*

Vernation fasciculate, erect, caudiciform. *Fronds* pinnate, smooth, 2–8 feet high; pinnæ entire, broad, the upper densely sporangiferous on their under side. *Veins* uniform, reticulated, forming numerous elongated subquadrangular parallel areoles.

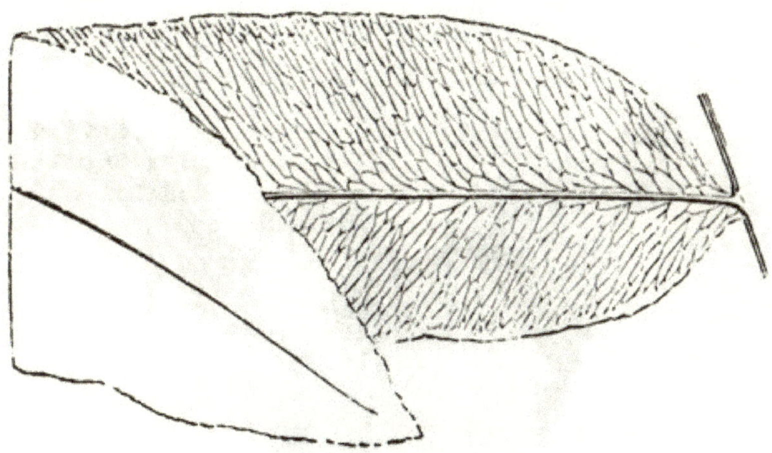

Genus 45.—Pinna of barren frond. No. 1.

1. A. aureum, *Linn.; Sw.; Plum. Fil. t.* 104; *Schk. Fil. t.* 1; *Hook. Gen. Fil. t.* 81 *A; Lowe's Ferns*, 7, *t.* 42. Chrysodium aureum, *Fée.* Acrostichum fraxinifolium, *R. Br.* Acrostichum marginatum, *Schk. Fil. t.* 3 *B.*—Tropics and sub-Tropics of both spheres, generally in swamps.

---

§ 4. *Platycereæ. Rhizome obsolete; sterile frond sessile, depressed, conchiform; fertile fronds stipate, repeatedly forked; segments broad. Veins compound anastomosing.*

46. PLATYCERIUM, *Desv.; Bl.*

Vernation articulate, rhizome obsolete. *Sterile fronds* sessile, oblique reniform, depressed or elongated and subascending, alternately overlapping each other, forming an epiphytal spongy conchiform mass, often 1–2 feet in diameter. *Fertile fronds* stipitate, rising from the sinus of the sterile, once or many times dichotomously forked, 2–6 feet in length; segments broad, obtuse, densely covered with stellated scales, coriaceous. *Veins* internal, compound anastomosing. *Receptacle* amorphous, occupying more or less of the under side of the segments, or on a sessile or petiolate lobe.

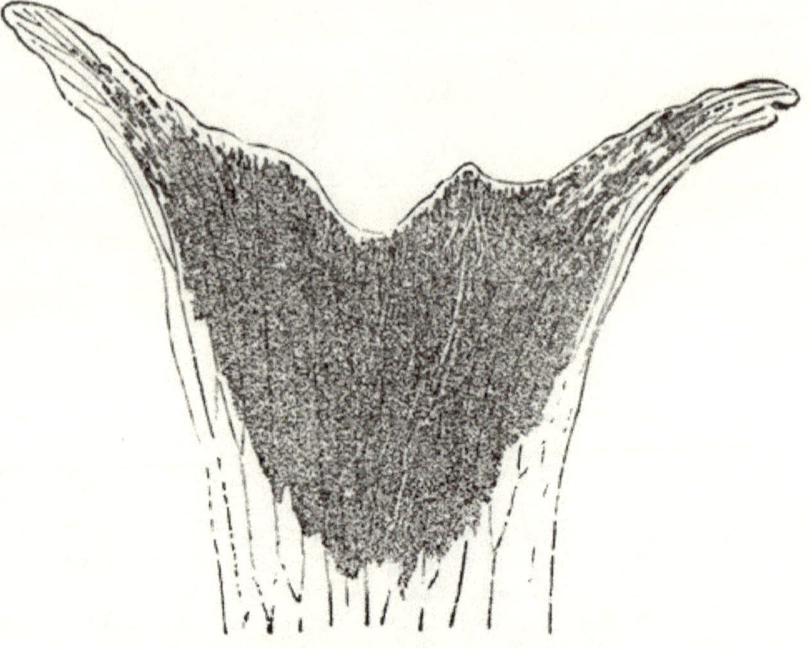

Genus 46.—Portion of mature frond, under side. No. 1.

1. **P. alcicorne,** *Gaud.; Lowe's Ferns,* 7, *t.* 63. Acrostichum alcicorne, *Sw.; Bot. Reg. t.* 262-3.—East Indies, Malayan Archipelago, and Australia.

2. **P. Stemaria,** *Desv.* Acrostichum Stemaria, *Beauv.* Platycerium Æthiopicum, *Hook. Gard. Ferns, t.* 9.—West Africa.

3. **P. grande,** *J. Sm.; Hook. Fil. Exot. t.* 86. Acrostichum grande, *A. Cunn.; Hook. et Bauer, Gen. Fil. t.* 80 *P.*—Malayan Archipelago and Australia.

4. **P. biforme,** *Blume, Fl. Jav. t.* 18. Acrostichum fuciforme, *Wall.*—Malacca and Java.

5. **P. Wallichii,** *Hook. Fil. Exot. t.* 97.—Malacca.

Tribe V.—GRAMMITIDEÆ.

Sori linear, sometimes only oval or oblong, oblique or trans verse, marginal or costal, or more or less complete, reticulated naked.

* *Veins free.*

§ 1. *Grammiteæ. Fronds linear, entire or rarely forked, generally smooth.*

47. GRAMMITIS, *Sw. in part.*

Vernation fasciculate, or uniserial and sarmentose, becoming cæspitose. *Fronds* linear-lanceolate, entire, rarely subpinnatifid, plane, opaque, smooth or pilose, 6–10 inches high. *Veins* simple

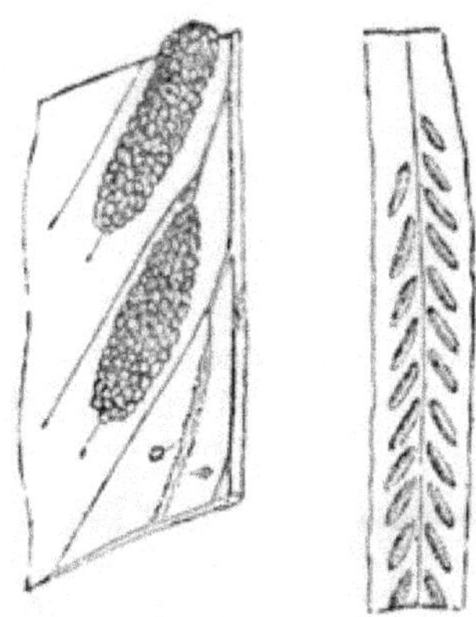

Genus 47.—Portion of frond, natural size; ditto, enlarged.

or forked, generally clavate, free; the anterior venule fertile. *Receptacles* elongated, medial-terminal. *Sori* ovate, oblong or linear oblique, sometimes punctiform transverse-uniserial.

1. **G. marginella,** *Sw. Syn. Fil. Schk. Fil. t.* 7. Polypodium marginellum, *Sw. Fl. Ind. Occ.*—St. Helena.
2. **G. Australis,** *R. Br.* Grammitis Billardieri, *Willd.; Kunze, Anal. t.* 9, *f.* 2.—New South Wales.

---

48. XIPHOPTERIS, *Kaulf.*

Vernation contiguous, sub-fasciculate; sarmentum slender, sub-erect. *Fronds* 2–6 inches high, linear, dentate-serrate or pinnatifid below, sub-entire, and plicate or nearly plane above. *Veins* simple, free, very short. *Receptacles* costal or medial,

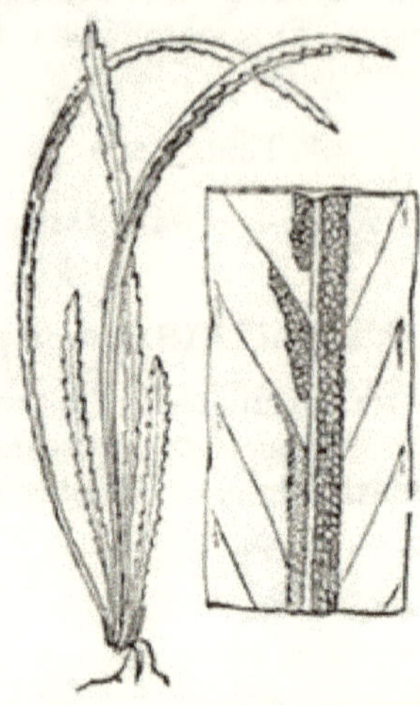

Genus 48.—Plant natural size, and portion of frond enlarged. No. 1.

elongated. *Sori* oblong, confluent, mostly contiguous to and parallel with the midrib, confined to the upper part of the frond.

1. **X. serrulata,** *Kaulf.; Fée, Gen. Fil. t.* 10 *B; Hook. Gard. Ferns, t.* 44; *Lowe's New Ferns, t.* 42 *A.* Grammitis serrulata, *Sw.; Schk. Fil. t.* 7; *Hook. Exot. Fil. t.* 78. Polypodium serrulatum, *Metten.*—West Indies and Tropical America.

---

§ 2. *Gymnogrammeæ. Fronds pinnate or bi-tripinnatifid or decompound, smooth, or generally pilose, tomentose, or farinose.*

49. LEPTOGRAMMA, *J. Sm.*

Vernation fasciculate, erect or decumbent. *Fronds* bipinnatifid, 1–3 feet high. *Veins* of laciniæ pinnate; venules free. *Receptacles* medial, elongated. *Sori* oblong or linear, naked. *Sporangia* in some species pilose.

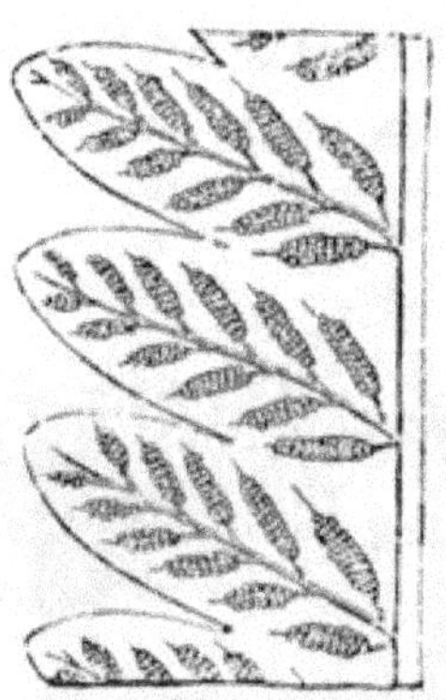

Genus 49.—Portion of fertile frond.

1. **L. totta,** *J. Sm. Gen. Fil.* Polypodium tottum, *Willd.* Gymnogramma totta, *Schlecht.; Bl. Fl. Jav. t.* 38. Grammitis totta, *Presl.* Gymnogramma Lowei, *Hook. et Grev. Ic. Fil. t.* 89.—South Africa and Madeira.

2. **L. asplenioides,** *J. Sm.* Gymnogramma asplenioides, *Sw.; Kaulf.* Gymnogramma aspidioides, *Kaulf.* Ceterach aspidioides, *Willd.; Radd. Fil. Bras. t.* 21, *f.* 1. Phegopteris aspidioides, *Metten. Fil. Hort. Lip. t.* 17, *f.* 1.—Tropical America.

3. **L. Linkiana,** *J. Sm.* Gymnogramma Linkiana, *Kunze; Fée.* Grammitis Linkiana, *Presl.*—Brazil.

4. **L. rupestris,** *J. Sm.* Gymnogramma rupestris, *Kunze.* Phegopteris rupestris, *Metten.*—Tropical America.

5. **L. gracile,** *J. Sm.* Gymnogramma gracilis, *Hew. in Mag. Nat. Hist.* (1838). Grammitis Hewardii, *Moore.* Leptogramma attenuata, *J. Sm. En. Fil. Hort. Kew.* (1856).—Jamaica.

6. **L. villosa,** *J. Sm.* Gymnogramma villosa, *Link; Lowe's Ferns,* 1, *t.* 11.—Tropical America.

7. **L. polypodioides,** *J. Sm.* Ceterach polypodioides, *Radd. Fil. Bras. t.* 22. Gymnogramma polypodioides, *Spreng.* Gymnogramma Raddiana, *Link.*—Brazil.

———

50. GYMNOGRAMMA, *Desv.*

Vernation fasciculate, erect. *Fronds* pinnate, bipinnatifid, or multifid, rarely simple, smooth, villose, or farinose, from a

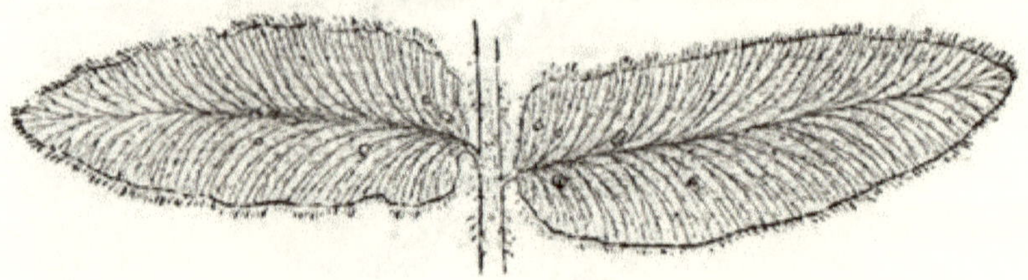

Genus 50.—Portion of mature frond, upper side. No. 1.

few inches to 2–3 feet high. *Veins* forked; venules free. *Receptacles* medial, elongated. *Sori* linear, simple, or forked, often becoming confluent, naked.

§ 1. *Neurogramma. Fronds pinnate or bipinnate, villose.*

1. G. **rufa,** *Desv.; Lowe's Ferns,* 1, *t.* 6 *A.* Hemonitis rufa, *Sw.; Schk. Fil. tt.* 17, 21.—Tropical America.

2. G. **tomentosa,** *Desv.; Lowe's Ferns,* 1, *t.* 6 *B; Hook. Fil. Exot. t.* 13. Hemionitis tomentosa, *Radd. Fil. Bras. t.* 19.—Tropical America.

§ 2. *Trismeria. Fronds pinnate; pinnæ bi-trifoliate; segments linear, covered with white or yellow farina.*

3. G. **trifoliata,** *Desv.; Hook. Gard. Ferns, t.* 4; *Lowe's New Ferns, t.* 31. Acrostichum trifoliatum, *Linn.;* (*Plum. Fil. t.* 144;) *Schk. Fil. tt.* 3 *et* 22. Trismeria argentea et aurea, *Fée, Gen. Fil. t.* 14 *A.*—West Indies and Tropical America.

§ 3. *Ceropteris. Fronds bi-tripinnatifid or multifid, covered with waxy farina on the under side.*

4. G. **Calomelanos,** *Kaulf.; Hook. Gen. Fil. t.* 37; *Hook. Gard. Ferns, t.* 50. Acrostichum Calomelanos, *Linn.; Plum. Fil. t.* 40; *Schk. Fil. t.* 5; *Lang. et Fisch. Ic. Fil. t.* 3.—Tropical America.

5. G. **Tartarea,** *Desv.* Acrostichum Tartareum, *Sw.*—Tropical America.

6. **G. ochracea,** *Presl.*—Tropical America.

7. **G. L'Herminieri,** *Bory (accord. to Link).*—Guadeloupe. (*Link.*)

8. **G. chrysophylla,** *Kaulf.* Acrostichum chrysophyllum, *Sw.; Plum. Fil. t.* 41.—West Indies.

9. **G. Martensii,** *Bory (accord. to Link).* (Hybrid, *J. Sm.*)

10. **G. sulphurea,** *Desv.* Acrostichum sulphureum, *Sw. Schk. Fil. t.* 4. *Var.* Wettenhalliana, *Moore, in Gard. Chron.* 1861, *p.* 934.—West Indies.

11. **G. pulchella,** *Linden's Cat.; Moore, in Gard. Chron.* 1856; *Hook. Fil. Exot. t.* 74; *Lowe's New Ferns, t.* 5.—Venezuela.

12. **G. Peruviana,** *Desv.; Kunze, Fil. t.* 32. *Var.* Argyrophylla, *Moore, in Gard. Chron.* 1856; *Lowe's New Ferns, t.* 6. *Var.* dealbata, *Moore.* *Var.* laciniata, *Moore, Gard. Chron.* 1863.—Tropical America.

§ 4. *Anogramme. Fronds bi-tripinnatifid, smooth. (Annuals.)*

13. **G. leptophylla,** *Desv.; Hook. et Grev. Ic. Fil. t.* 25; *Hook. Brit. Ferns, t.* 1; *Lowe's Ferns,* 1, *t.* 7. Grammitis leptophylla, *Sw.* Polypodium leptophyllum, *Linn.; Schk. Fil. t.* 26.—South of Europe, &c.

14. **G. chærophylla,** *Desv.; Hook. et Grev. Ic. Fil. t.* 45; *Lowe's Ferns,* 1, *t.* 8.—Tropical America.

15. **G. Pearcii,** *Moore, in Gard. Chron.* 1864, *p.* 340.—Peru.

§ 4. *Pleurosorus. Fronds pinnatifid or pinnate, piloso-glandulose.*

16. **G. rutæfolia,** *Hook. et Grev. Ic. Fil. t.* 90; *Hook. Fil. Exot. t.* 5; *Hook. Ic. Pl. t.* 935; *Lowe's New Ferns, t.* 45 *A.* Gymnogramma subglandulosa, *Hook. et Grev. Ic. Fil. t.* 91. Grammitis Hispanica, *Goss.* Grammitis rutæfolia, *R. Br.*—Australia and South of Spain.

§ 5. *Eriosorus. Fronds bipinnatifid, lanose-tomentose.*

17. **G. ferruginea,** *Kunze.* G. lanata, *Klotzsch.* *Var.* monstrosa, *Hort.*—Tropical America.

51. CONIOGRAMMA, *Fée.*

Vernation contiguous, decumbent, subsarmentose. *Fronds* pinnate or bipinnate, 2-5 feet high, smooth; pinna and pin-

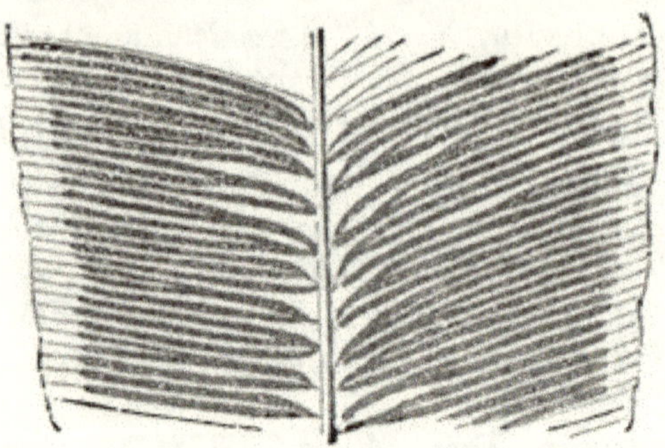

Genus 51.—Portion of fertile pinna—under side. No. 1.

nules broad elliptical-lanceolate, distant, smooth. *Veins* forked, parallel, free. *Receptacles* medial, elongated, occupying nearly the whole length of the venules. *Sori* linear, forked, contiguous, naked.

1. **C. Javanica,** *Fée.* Gymnogramma Javanica, *Bl. Fl. Jav. t.* 41; *Lowe's New Ferns, t.* 7.—Malayan Archipelago.

---

52. LLAVEA, *Lag.*

Vernation fasciculate, erect. *Fronds* tri-quadripinnate,

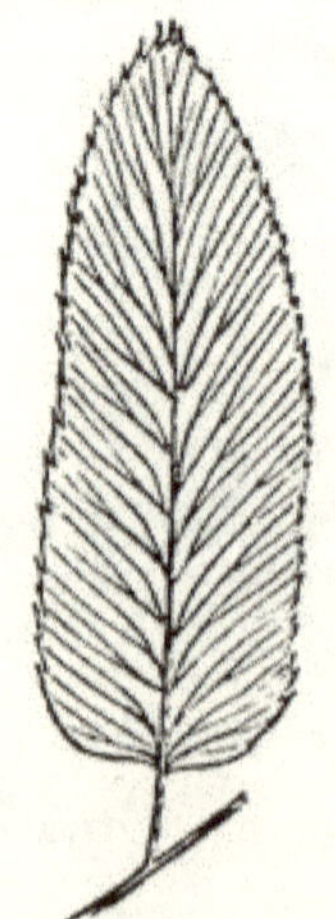

Genus 52.—Barren pinna. No. 1.

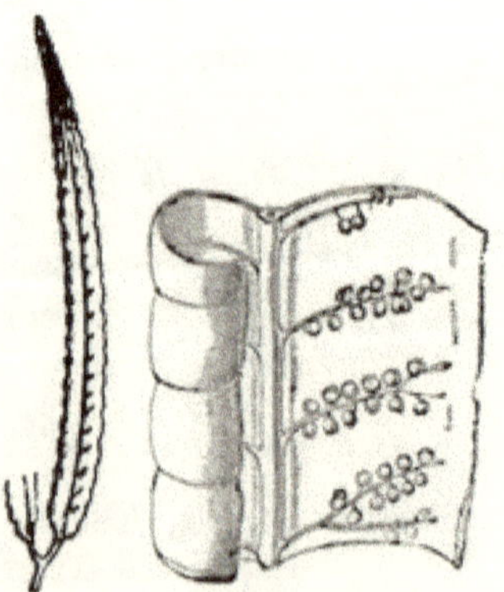

Fertile pinna, natural size; ditto, unfolded. No. 1.

1–2½ feet long, smooth, contracted and fertile above, sterile below; *Sterile* pinnules oblong, elliptical, oblique sub-cordate, serrulate, 1–1¼ inches in length. *Veins* forked; venules free. *Fertile* pinnules linear, 2–3 inches long, revolute, margin conniving and forming an universal indusium. *Sporangia* occupying nearly the whole length of the contracted venules, forming linear forked confluent sori.

1. **L. cordifolia,** *Lag.; Hook. Bot. Mag. t.* 5159. Ceratodactylis osmundioides, *J. Sm. in Hook. et Bauer, Gen. Fil. t.* 36; *Lowe's New Ferns, t.* 30. Botryogramma Karwinskii, *Fée, Gen. Fil. t.* 15 *C.* Allosorus Karwinskii, *Kunze, Fil. t.* 4; *Hook. Ic. Pl. t.* 387–8.—Mexico. **Tr.**

** *Veins anastomosing.*

§ 3. *Hemioniteæ. Fronds simple, pinnate or rarely bipinnate. Sori more or less complete reticulated.*

53. DICTYOGRAMMA, *Fée.*

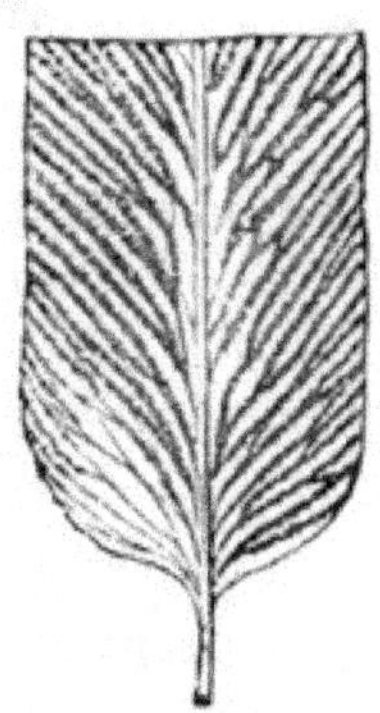
Genus 53.—Portion of fertile frond. No. 1.

Vernation uniserial, contiguous; sarmentum short. *Fronds* pinnate or bipinnate, 1–3 feet high, smooth; pinnæ elliptical-lanceolate, 6–10 inches long. *Venation* subuniform, reticulated; areoles unequal, generally elongated, oblique. *Receptacles* superficial. *Sori* linear, reticulated, naked.

1. **D. Japonica,** *Fée, Gen. Fil. t.* 15 *A.* Hemionitis Japonica, *Thunb.* Gymnogramma Japonica, *Hook. Sp. Fil.*—Japan, Formosa.

---

54. HEMIONITIS, *Linn.*

Vernation fasciculate, erect, short. *Fronds* simple, cordate, palmate orpinnate, smooth or villose. *Veins* uniform reticulated.

Sporangia occupying the whole of the vernation, forming reticulate, often confluent sori. *Receptacles* medial, elongated. *Sori* reticulated.

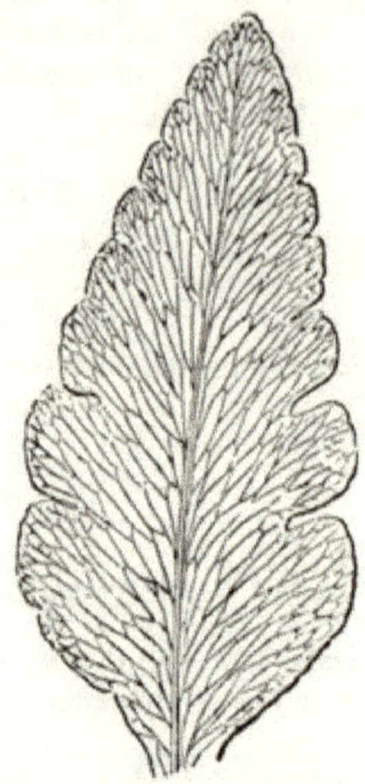

Genus 54.—Segment of barren frond, under side. No. 2.

1. **H. cordifolia,** *Roxb.; Hook. Fil. Exot. t.* 35; *Hook. et Grev. Ic. Fil. t.* 64; *Hook. et Bauer, Gen. Fil. t.* 74. H. sagittata, *Fée.*—East Indies.

2. **H. palmata,** *Linn.; Plum. Fil. t.* 151; *Hook. Ex. Fl. t.* 33; *Schott. Gen. Fil. t.* 9; *Lowe's Ferns,* 7, *t.* 37.—West Indies.

3. **H. pedata,** *Sw. Syn. Fil. t.* 1, *f.* 3. Gymnogramma pedata, *Kaulf.*—Mexico. **Tr.**

---

55. ANTROPHYUM, *Kaulf.*

Vernation uniserial, contiguous; sarmentum short (undefined), squamose. *Fronds* simple, linear-lanceolate or oblong-elliptical or subrotund, smooth, coriaceous, with or without a defined midrib. *Veins* uniform, reticulated. *Receptacles* medial, elongated, immersed, rarely superficial, forming linear, continuous or interrupted reticulated sori.

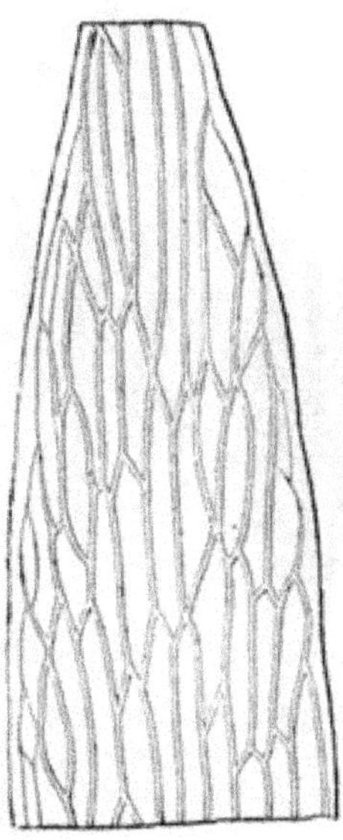

Genus 55.—Outline portion of fertile frond, under side. No. 4.

1. **A. lineatum**, *Kaulf.* Polytænium lineatum, *Desv.; J. Sm. Gen. Fil.; Hook. Gen. Fil. t.* 107. Vittaria lanceolata, *Sw.; Schk. Fil. t.* 101 *B.*—West Indies.

2. **A. lanceolatum**, *Kaulf.* Hemionitis lanceolata, *Linn.;* (*Plum. Fil. t.* 127, *f.* c); *Schk. Fil. t.* 6.—West Indies.

3. **A. Cayennense**, *Kaulf.; Kunze, Anal. t.* 19, *f.* 2. Hemionitis Cayennensis, *Desv.; Presl.* — Tropical America.

4. **A. reticulatum**, *Kaulf.* Hemionitis reticulata, *Forst. Schk. Fil. t.* 6.—Indian, Malayan, and Pacific Islands.

---

§ 4. *Vittariæ. Fronds simple, linear. Sori transverse, continuous, marginal or anti-marginal.*

56. VITTARIA, *Sm.*

Vernation uniserial, contiguous; sarmentum short, furnished with hyaline squamæ. *Fronds* simple, linear, smooth, rigid or flaccid and pendulous, from a few inches to 2–3 feet in length. *Veins* simple, forming an acute angle with the midrib, their

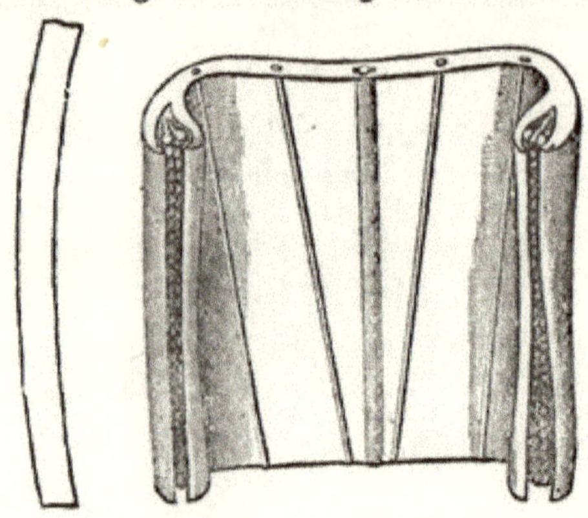

Genus 56.—Outline portion of frond, natural size; ditto enlarged. No. 1.

apices prolonged into a transverse marginal vein, which becomes the receptacle. *Sporangia* seated in an extrorse slit of the margin. *Sori* marginal, linear, continuous.

1. V. zosteræfolia, *Bory; Fée, Mem. Fil. t.* 2, *f.* 2; *Lowe's Ferns,* 2, *t.* 65 *B.*—Mauritius.

---

57. HAPLOPTERIS, *Presl.*

Vernation uniserial, contiguous; sarmentum short, becoming

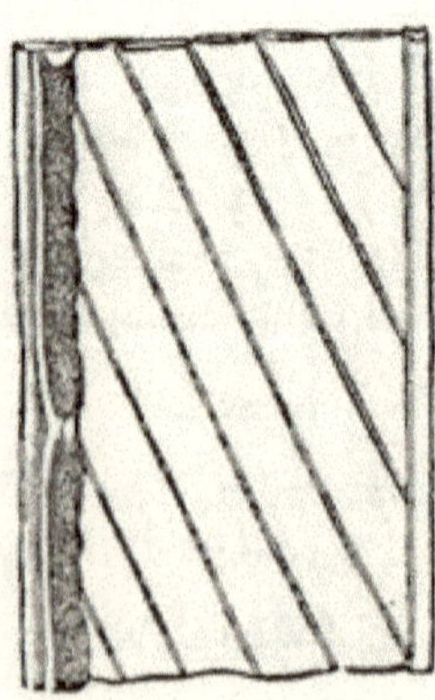

Genus 57.—Portion of frond, slightly enlarged. No. 1.

cæspitose. *Fronds* simple, narrow, linear or lanceolate, 1–2 feet long, smooth. *Veins* simple, parallel, their apices combined by a transverse intra-marginal vein, which is immersed in a groove, and becomes the receptacle, forming a linear, continuous, inter-marginal, naked sorus.

1. **H. scolopendrina,** *Presl, Tent. Pterid. t.* 8, *f.* 21. Pteris scolopendrina, *Bory; Sw.* Tæniopsis scolopendrina, *J. Sm. Gen. Fil.* 1841. Tæniopteris Forbesii, *Hook. et Bauer. Gen. Fil. t.* 76 *B.* Vittaria Zeylanica, *Fée, Vittar. t.* 1, *f.* 3.—Ceylon and Mauritius.

2. **H. lineata,** *J. Sm.* Vittaria lineata, *Sw.; Schk. t.* 101 *B; J. Sm. Cat.* 1857; *Lowe's Ferns,* 2, *t.* 65 *A.* Tæniopsis lineata, *J. Sm. Gen. Fil.* 1841.—Tropical America.

---

58. PTEROPSIS, *Desv.*

Vernation uniserial, contiguous; sarmentum short, cæspitose. *Fronds* simple, linear, acuminate, 6–18 inches long, rigid, smooth. *Veins* uniform, reticulated, forming transverse elongated, hexagonoid areoles. *Receptacles* compital, elongated on the exterior, transverse anastomose, forming a linear, continuous, marginal sorus.

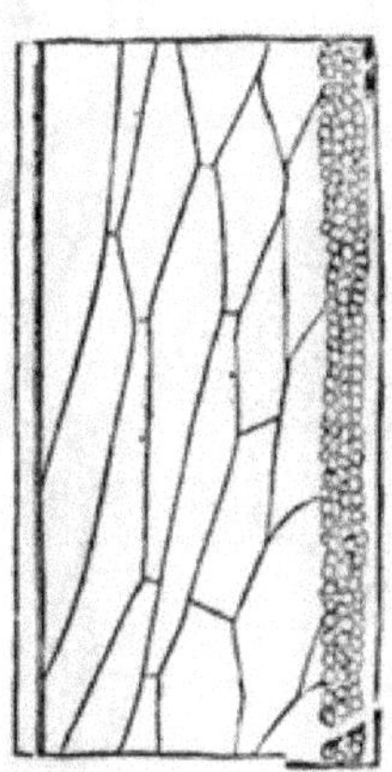

Genus 58.—Portion of frond, slightly enlarged. No. 1.

1. P. angustifolia, *Desv.; Hook. et Bauer. Gen. Fil. t.* 77 *B.* Tænitis angustifolia, *R. Br.* Pteris angustifolia, *Sw.; Willd.* Pteris tricuspidata, *Linn.; Plum. Fil. t.* 140, *var.* comosa, *J. Sm.*—West Indies.

---

59. DICTYOXIPHIUM, *Hook.*

Vernation fasciculate, erect. *Fronds* simple, linear-lanceolate, attenuated and decurrent on the stipes, 1–3 feet long. *Veins* compound anastomosing. *Receptacles* compital, elongated, immersed in an extrorse marginal groove, which is indusiform. *Sori* linear, continuous.

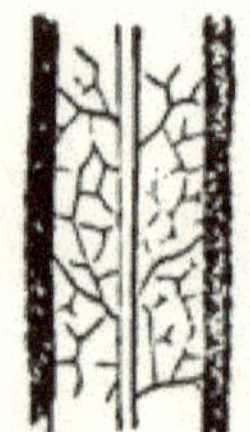

Genus 59.—Portion of fertile frond, under side. No. 1.

1. D. Panamense, *Hook. Gen. Fil. t.* 62; *J. Sm. Gen. Fil.; Lowe's Ferns,* 8, *t.* 69. Lindsæa Panamensis, *Mettn. Hook. Fil. Exot. t.* 54.—Panama.

---

§ 5. *Ceratopteriæ. Fertile fronds contracted; segments revolute, forming an universal indusium, enclosing the sporangia.*

60. CERATOPTERIS, *Brongn.*

Vernation fasciculate, erect (annual). *Fronds* fragile; the

Genus 60.—Portions of fertile and barren fronds, natural size; portion of fertile enlarged. No. 1.

fertile decompound; segments forked, linear; margins revolute, membranaceous, conniving, indusiform. *Veins* transversely elongated, distantly anastomosing. *Sporangia* occupying the transverse venules, superficial, large, disposed in a simple series, constituting two linear sub-parallel sori.

1. **C. thalictroides,** *Brongn.; Hook. Gen. Fil. t.* 12; *Lowe's Ferns,* 2, *t.* 66. Ellobocarpus oleraceus, *Kaulf.* Parkeria pteridioides, *Hook. Ex. Fl. t.* 147; *Hook. et Grev. Ic. Fil. t.* 97. Ceratopteris Parkeri, *J. Sm. Gen. Fil.* 1841.—Tropics.

Tribe VI.—PHEGOPTERIDEÆ.

Sori punctiform, intra-marginal or rarely on marginal dents, naked or each furnished with a special indusium, which is either peltate or lateral and interiorly attached, rarely calyciform; or the margin of contracted fronds revolute, forming an universal indusium; or the dents of the margin reflexed and indusiform.

* *Veins anastomosing in various ways.*

† *Sori naked.*

§ 1. *Dictyopteriæ. Primary veins costæform, generally well defined. Sori punctiform or linear, in oblique or transverse rows or lines, or rarely reticulated between the primary veins.*

61. DRYOMENIS, *Fée; J. Sm.*

Vernation uniserial, contiguous or subfasciculate, subhypogeous. *Fronds* simple, pinnatifid or pinnate, smooth,

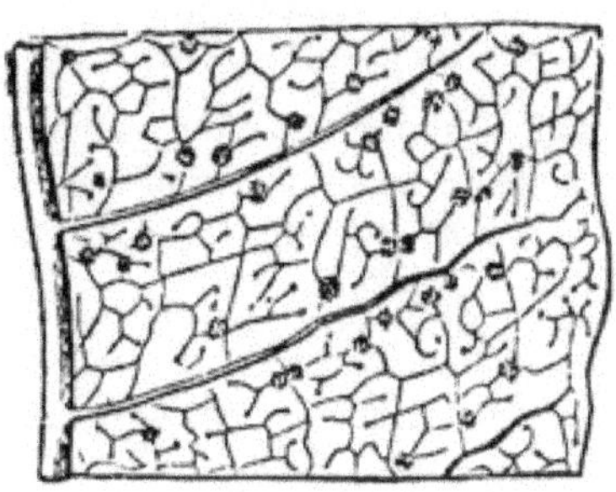

Genus 61.—Portion of frond. No. 1.

1–2½ feet high, submembraneous. *Primary veins* costæform, transversely combined and compound anastomosing, with free veinlets terminating in areoles. *Receptacles* punctiform, compital. *Sori* round or by confluence unequal oblong, oblique biserial or irregular, naked.

1. **D. plantaginea,** *J. Sm. in Seemann's Bot. Voy. Herald.* Polypodium plantagineum, *Linn.; Jacq. Coll. t.* 3, *f.* 1; (*Plum. Fil. t.* 128). Aspidium plantagineum, *Grisb.; Hook. Sp. Fil.* (*in part*). Pleopeltis plantaginea, *Moore, Ind.*—West Indies.

---

62. DICTYOPTERIS, *Presl* (*in part*).

Vernation fasciculate, decumbent or sub-erect. *Fronds* coriaceous, deltoid, bipinnatifid or bipinnate, 3–4 feet high; ultimate segments or pinnules sub-entire or sinuous-pinnatifid. *Veins* costæform; venules and veinlets anastomosing (rarely

Genus 62.—Portion of fertile pinna; ditto barren and fertile. No. 1.

few free, excurrent), forming oblique, somewhat elongated areoles, the costal ones transversely elongated. *Receptacles* medial or compital. *Sori* round, large, irregular, sometimes crowded near the margin, naked.

1. **D. irregularis,** *Presl.* Polypodium irregulare, *Presl. Rel. Hænk. t.* 4, *f.* 3.—East Indies, Malayan and Philippine Islands.

---

63. MENISCIUM, *Schreb.*

Vernation fasciculate and decumbent, or uniserial and sarmentose. *Fronds* pinnate, rarely simple, 1–8 feet high. *Primary veins* costæform, pinnate; each opposite pair of venules

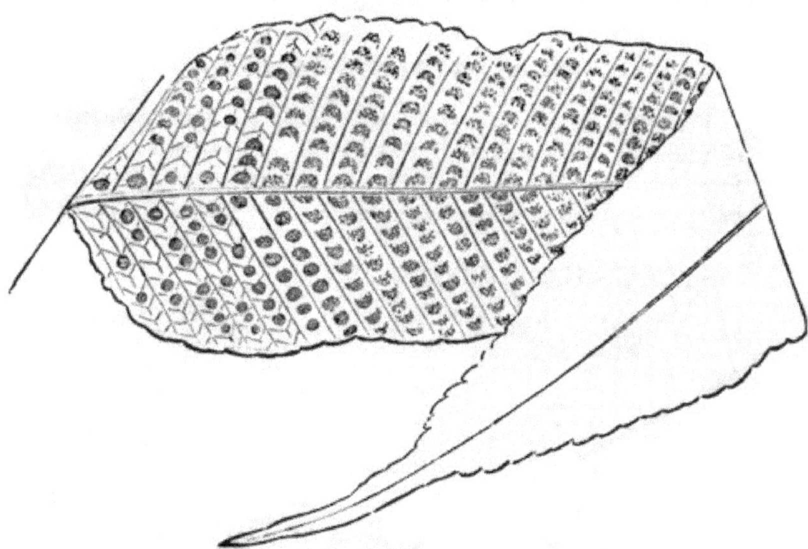

Genus 63.—Portion of mature frond. No. 4.

angularly or arcuately anastomosing and sporangiferous, producing from their junction an excurrent, free, sterile veinlet. *Receptacles* medial, linear, continued across the junction of the venules, forming arcuate, transverse sori. *Sporangia* in some species pilose.

* *Fronds simple.*

1. **M. simplex,** *Hook. Lond. Journ. Bot. v.* 1, *t.* 11; *Hook. Fil. Exot. t.* 83.—Hong-kong.

2. **M. giganteum,** *Metten.; Hook. Sp. Fil.* 5, *p.* 163.—Tropical America.

** *Fronds pinnate.*

3. **M. triphyllum,** *Sw.; Hook. et Grev. Ic. Fil. t.* 120; *Kunze, Fil. t.* 52.—India, Ceylon.

4. **M. palustre,** *Radd. Fil. Bras. t.* 20; *Hook. Gen. Fil. t.* 40; *Lowe's Ferns,* 2, *t.* 45.—Brazil.

5. **M. dentatum**, *Presl.*—Brazil.

6. **M. reticulatum**, *Sw.; Schk. Fil. t.* 5. Polypodium reticulatum, *L.*—Tropical America.

---

64. GONIOPTERIS, *Presl.*

Vernation fasciculate, erect or decumbent. *Fronds* pinnatifid or pinnate, rarely simple, 1–4 feet high. *Primary veins* costæform, pinnate; venules opposite, the whole or only the lower pair, or more, angularly anastomosing, producing from their

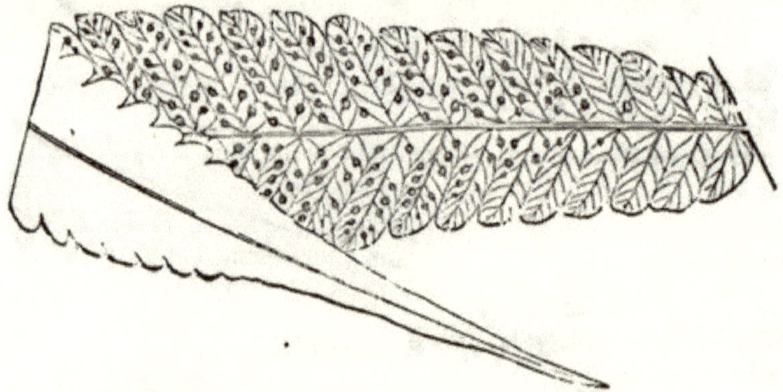

Genus 64.—Portion of mature frond. No. 8.

junction an excurrent sterile veinlet, which is either free or anastomoses in the angle next above it. *Sporangiferous receptacles* punctiform, medial (between the base and angular junction of the venules). *Sori* round, obliquely biserial. *Sporangia* pilose in some species.

1. **G. scolopendroides**, *Presl.* Polypodium scolopendroides, *Sw.;* (*Plum. Fil. t.* 91); *Hook. Fil. Exot. t.* 18. Goniopteris subpinnata, *Hort.*—Jamaica.

2. **G. gracilis**, *Moore, in Gard. Chron.* 1856; *Lowe's Ferns*, 1, *t.* 9 *A.*—Jamaica.

3. **G. reptans**, *Presl.* Polypodium reptans, *Sw.; Sloane's Jam.* 2, *t.* 30. Polypodium compositum, *Link.*—Jamaica.

4. **G. asplenoides,** *Presl.* Polypodium asplenoides, *Sw.; Sloane's Jam.* 1, *t.* 43, *f.* 2; *Lowe's Ferns,* 1, *t.* 34 *B.*—Jamaica.

5. **G. crenata,** *Presl; Hook. Gen. Fil. t.* 38. Polypodium crenatum, *Sw.;* (*Plum. Fil. t.* 111); *Lowe's Ferns,* 1, *t.* 26 *b.*—West Indies.

6. **G. megalodes,** *Presl.* Polypodium megalodes, *Schk. Fil. t.* 19 *b.*—West Indies.

7. **G. Gheisbeghtii,** *J. Sm.* Polypodium Gheisbeghtii, *Lind. Cat.* 1858. Meniscium pubescens, *Linn. Cat.* 1858. Polypodium crenatum, *Hook. Fil. Exot. t.* 84 (*non Sw.*).—Tropical America.

8. **G. tetragona,** *Presl.* Polypodium tetragonum, *Sw.; Schk. Fil. t.* 18 *b.*—West Indies.

9. **G. serrulata,** *J. Sm.* Polypodium serrulatum, *Sw.; Presl; Sloane's Jam. t.* 43, *f.* 1.—Jamaica.

10. **G. prolifera,** *Presl.* Meniscium proliferum, *Sw.; Hook. 2nd Cent. Ferns, t.* 15.—East Indies.

11. **G. vivipara,** *J. Sm.* Polypodium viviparum, *Radd. Fil. Bras. t.* 32. Polypodium proliferum, *Lowe's Ferns, t.* 31. Goniopteris fraxinifolia, *Presl* (*non* Polypodium fraxinifolium, *Jacq.*). Polypodium fraxinifolium, *Lowe's Ferns, t.* 31.—Brazil.

12. **G. pennigera,** *J. Sm.* Polypodium pennigerum, *Forst.*—New Zealand.

13. **G. Fosteri,** *Moore.*—New Zealand.

†† *Sori indusiate.*

§ 2. *Aspidiæ. Sori punctiform. Indusium orbicular, reniform or rarely calyciform.*

a. *Indusium orbicular or reniform.*

65. NEPHRODIUM, *Schott.*

Vernation fasciculate, decumbent or erect, rarely uniserial and sarmentose. *Fronds* 1–6 feet high, simple or pinnate; pinnæ entire, sinuose or pinnatifid. *Veins* costæform, pinnate; the lower pair of venules only, or more, or the whole, angularly anastomosing, producing from their junction an excurrent, anastomosing, sterile veinlet. *Receptacles* medial or subterminal. *Sori* round. *Indusium* reniform, rarely nearly orbicular.

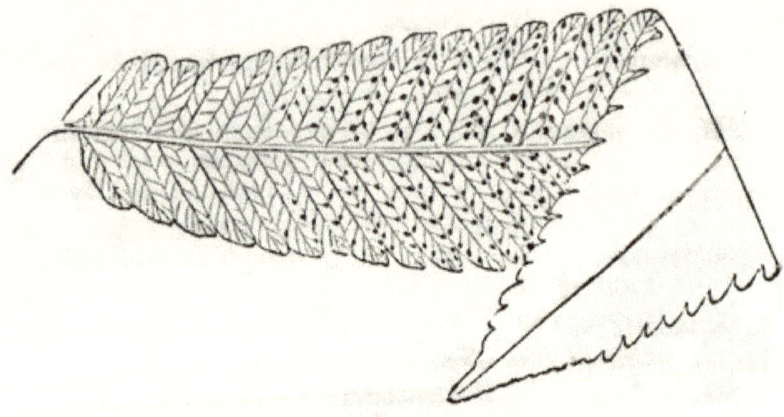

Genus 65.—Portion of mature frond. No. 5.

* *Vernation uniserial, distant.*

1. **N. unitum,** *R. Br.; Hook. Gen. Fil. t.* 48 *B.* Polypodium unitum, *Linn.* Aspidium unitum, *Schk. Fil. t.* 33 *B, f.* 1. Aspidium serra, *Schk. Fil. t.* 33, *f.* 2.—Tropics.

2. **N. pteroides,** *J. Sm.* Polypodium pteroides, *Retz.* Aspidium pteroides, *Sw.* Aspidium terminans, *Wall.* Nephrodium terminans, *J. Sm. Cat. Fil. Hort. Kew.* 1846.—East Indies.

3. **N. venulosum,** *Hook. Sp. Fil.* 5, *p.* 17.—Fernando Po.

** *Vernation fasciculate, erect or decumbent.*

4. **N. Hookeri,** *J. Sm.* Aspidium Hookeri, *Wall.; Hook. Ic. Pl. t.* 922.—East Indies.

5. **N. granulosum,** *J. Sm.* Polypodium granulosum, *Presl. Reliq. Hænk. t.* 4, *f.* 2. Aspidium glandulosum, *Blume.; Lowe's Ferns,* 7, *t.* 9. Nephrodium multilineatum, *Moore and Houlst.* (*non Presl*).—Philippine Islands, Java.

6. **N. articulatum,** *Moore and Houlst. in Gard. Mag. of Bot.* 1851; *J. Sm. Cat. Cult. Ferns,* 1857; *Lowe's Ferns,* 6, *t.* 29.—Ceylon.

7. **N. refractum**, *J. Sm.* Polypodium refractum, *Fisch. et Mey.*; *Lowe's Ferns*, 2, *t.* 48. Goniopteris refracta, *J. Sm. Cat. Cult. Ferns*, 1857.—Brazil.

8. **N. truncatum**, *J. Sm.* Aspidium truncatum, *Gaud. in Freycinet's Voy. t.* 10; *Lowe's Ferns*, 6, *t.* 12.—Sandwich Islands.

9. **N. abortivum**, *J. Sm.* Aspidium abortivum, *Blume.* Aspidium decurtatum, *Kunze.*—Java.

10. **N. venustum**, *J. Sm.* Aspidium venustum, *R. Hew. in Mag. Nat. Hist.* 1838, *p.* 464.—Jamaica.

11. **N. molle**, *R. Br.*; *Schott, Gen. Fil. t.* 14; *Hook. Gen. Fil. t.* 48 *B.* Aspidium molle, *Sw.*; *Schk. Fil. t.* 34 *B.* A. violascens, *Link.* *Var.* corymbiferum, *Moore, in Gard. Chron.* 1856; *Lowe's Ferns*, 7, *t.* 13.—Tropics, very general. T.

12. **N. patens**, *J. Sm.*—Demerara.

---

66. MESOCHLÆNA, *R. Br.*

Vernation fasciculate, erect. *Fronds* 2–4 feet high, bipinnatifid. *Veins* costæform, pinnate; the lower pair of venules anastomosing, the others free, parallel. *Receptacles* medial,

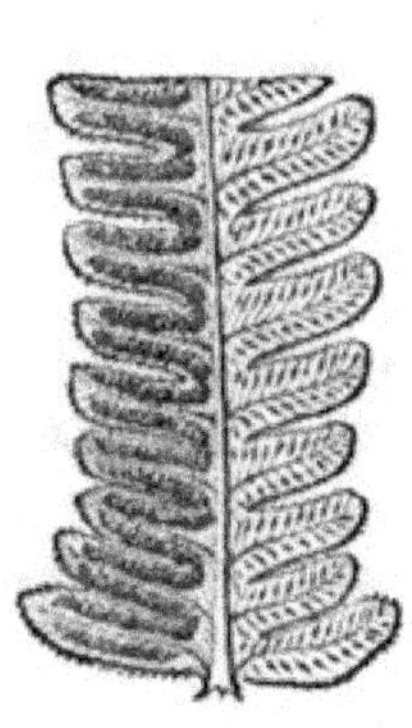

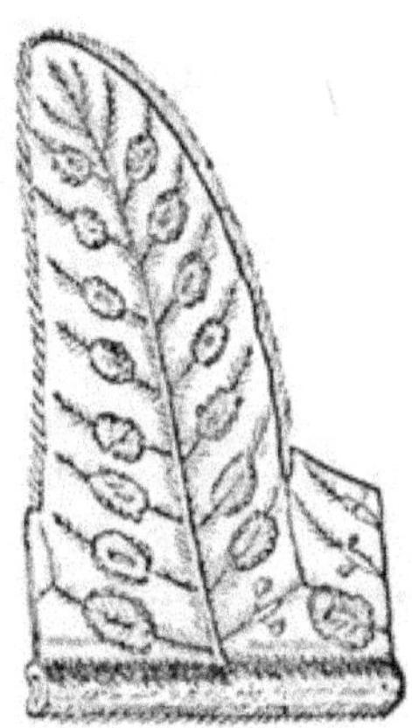

Genus 66.—Portions of barren and fertile frond, natural size; ditto enlarged. No. 1.

elongated. *Sori* oblong-linear. *Indusium* hippocrepiform, attached longitudinally on the centre of the receptacle, having sporangia in its axis on each side, its margin free.

1. **M. Javanica,** *R. Br. in Horsf. Fl. Jav.; Lowe's Ferns,* 7, *t.* 15. Nephrodium Javanica, *Hook. Fil. Exot. t.* 62. Sphærostephanos asplenioides, *J. Sm. in Hook. Gen. Fil. t.* 24; *Kunze, Fil. t.* 10, 11.—Singapore and Java.

---

67. CYCLODIUM, *Presl.*

Vernation fasciculate, subdecumbent. *Fronds* pinnate, 2–3 feet high; sterile pinnæ broad-elliptical, linear-lanceolate. *Veins* pinnately forked; venules acutely anastomosing, producing from their angular junctions an excurrent free or anastomosing

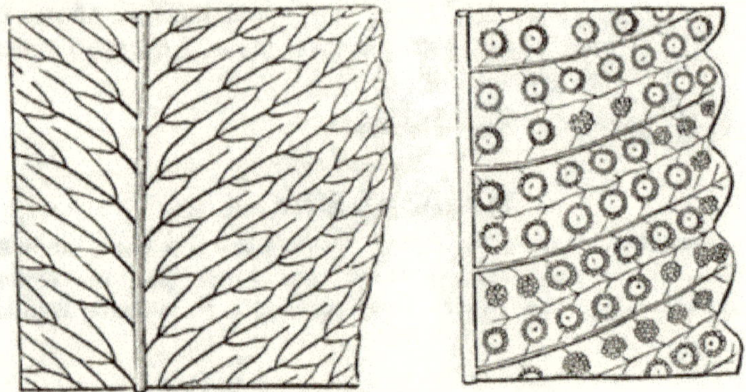

Genus 67.—Portion of barren and fertile frond.

veinlet; fertile pinnæ contracted, linear-lanceolate. *Veins* uniform, reticulated. *Receptacles* medial, punctiform. *Sori* round, confluent, and occupying the whole under surface. *Indusium* orbicular.

1. **C. confertum,** *Presl.* Aspidium confertum, *Kaulf.; Hook. et Grev. Ic. Fil. t.* 121; *Hook. Gen. Fil. t.* 49 *B; J. Sm. Cat. Cult. Ferns,* 1857. Aspidium Hookeri, *Kl.*— Guiana, Bahia.

---

68. CYRTOMIUM, *Presl.*

Vernation fasciculate, erect. *Fronds* 1–3 feet high, pinnate; pinnæ elliptical-lanceolate, 6–8 inches long, 1–4 wide, falcate more or less, auriculate at the base, the margin sub-entire or spinulose. *Veins* two or three times forked, or pinnate; venules alternate, the lower exterior branch free, the others acutely anastomosing, producing from their junctions free or anastomosing veinlets. *Receptacles* punctiform on or below their

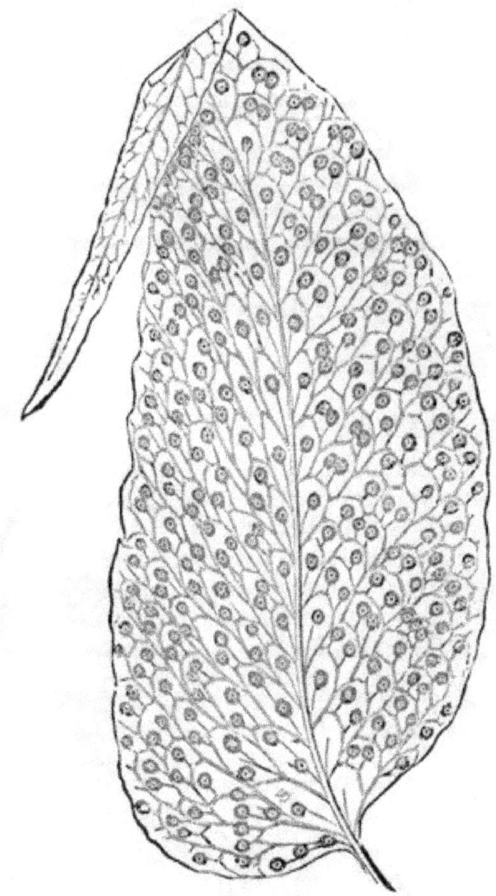

Genus 68.—Portion of mature frond, under side.

apices or points of junction of the venules. *Sori* round, transversely uniserial, or numerous and oblique-serial. *Indusium* orbicular.

1. C. falcatum, *Presl.* Polypodium falcatum, *Linn.; Thunb. Fl. Jap. t.* 36. Aspidium falcatum, *Sw.; Lang. et Fisch. t.* 15; *Lowe's Ferns,* 6, *t.* 9; *Hook. Fil. Exot. t.* 92.—Japan, China.

2. **C. caryotideum,** *Presl; Hook. Gen. Fil. t.* 49 *C.* Aspidium caryotideum, *Wall.; Hook. et Grev. Ic. Fil. t.* 69; *Hook. Gard. Ferns, t.* 13. Aspidium anomophyllum, *Zenk. Pl. Nilgh. t.* 1.—East Indies and Natal. **T.**

3. **C. juglandifolium,** *Moore.* Polypodium juglandifolium, *Humb.* Amblia juglandifolia, *Presl; Fée, Gen. Fil. t.* 22 *B, f.* 1. Phanerophlebia juglandifolia, *J. Sm.; Hook. Gen. Fil. t.* 49 *A.* Aspidium juglandifolium, *Kunze; Metten. Fil. Hort. Lips. t.* 22, *f.* 6–7.—Tropical America.

---

69. FADYENIA, *Hook.*

Vernation fasciculate, erect, caudex undefined. *Fronds* simple, entire, 5–6 inches long; the sterile lanceolate, attenuated and proliferous at the apex; the fertile linear, ligulate, obtuse, erect. *Veins* forked; venules acutely anastomosing; the lower exterior venule of each fascicle free, and sporangiferous on its apex. *Receptacles* punctiform. *Sori* round, transversely uniserial. *Indusium* reniform, suboblong, hippocrepiform.

1. **F. prolifera,** *Hook. Gen. Fil. t.* 53 *B; Lowe's Ferns,* 6, *t.* 2. Aspidium proliferum, *Hook. et Grev. Ic. Fil. t.* 96; *Hook. Fil. Exot. t.* 36.—Jamaica.

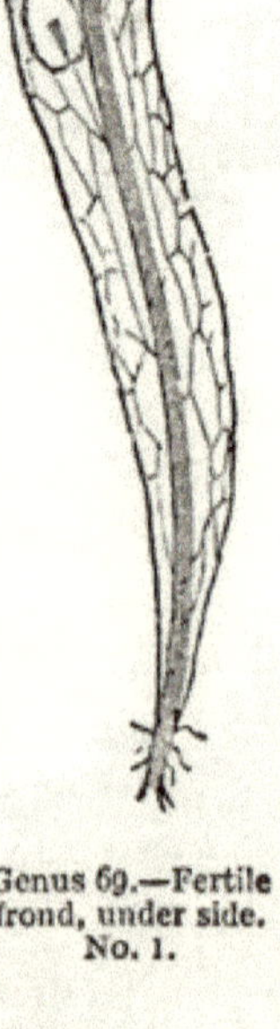

Genus 69.—Fertile frond, under side. No. 1.

---

70. ASPIDIUM, *Sw.* (*in part*); *Schott.*

Vernation fasciculate, erect. *Fronds* 1–4 feet high, entire, lobed, pinnate, bipinnatifid or bipinnate; ultimate segments generally broad. *Primary veins* costæform; venules simply or compoundly anastomosing. *Receptacles* compital, or on the apex of free veinlets terminating in the areoles. *Sori* round. *Indusium* orbicular or reniform.

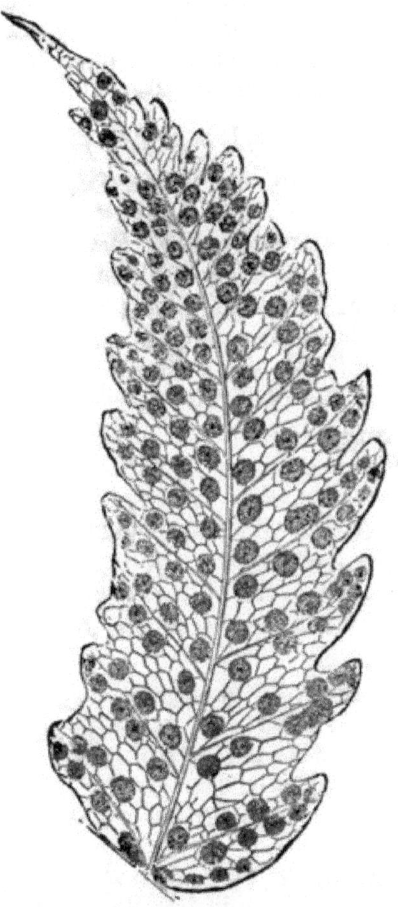

Genus 70.—Pinna of mature frond, under side. No. 3.

* *Fronds cordate, lobed, or trifoliate.*

1. A. **Plumieri,** *Presl, Rel. Hœnk.* (*excl. syn.* Polypodium angulatum, *Willd.*). Polypodium trifoliatum, *Linn. Sp. Pl.* (*not of Linn. Herb.*); *Plum. Fil. t.* 148.—Martinique and Dominica.

2. A. **Pica,** *Desv.* Polypodium Pica, *Linn.* Sagenia Pica, *Moore.* Aspidium ebenum, *J. Sm. Cat. Cult. Ferns,* 1857. Bathmium ebenum, *Fée.*—Mauritius.

3. **A. trifoliatum**, *Sw.; Schk. Fil. t.* 28; *Hook. Gen. Fil. t.* 33; *Schott, Gen. Fil. cum Ic.; Lowe's Ferns*, 6, *t.* 29. Polypodium trifoliatum, *Linn. fide specimen in Linn. Herb.; Jacq. Ic. Rar. t.* 638. Bathmium trifoliatum, *Link.* Aspidium heracleifolium, *Willd.* (*Plum. Fil. t.* 147).—Tropical America.

** *Fronds pinnate or subpinnatifid; pinnæ entire or lobed.*

4. **A. macrophyllum**, *Sw.* (*Plum. Fil. t.* 145); *Metten. Fil. Hort. Lips. t.* 22, *f.* 13; *Lowe's Ferns*, 6, *t.* 46. Cardiochlæna macrophylla, *Fée.* Bathmium macrophyllum, *Link.*—Tropical America.

5. **A. repandum**, *Willd.* Bathmium repandum, *Fée.* Sagenia repanda, *Moore.* Sagenia platyphylla, *J. Sm. En. Fil. Phil.* Aspidium platyphyllum, *Metten. Fil. Hort. Lips. t.* 21.—Malayan Islands.

6. **A. latifolium**, *J. Sm. Enum. Fil. Phil.* Polypodium latifolium, *Forst.; Schk. Fil. t.* 24. Aspidium melanocaulon, *Blume; Hook. Sp. Fil.* 4, *p.* 53. Aspidium nigripes, *Hort.*—Malayan and Pacific Islands.

7. **A. coadunatum**, *Wall.; Hook. et Grev. Ic. Fil. t.* 202 *Metten. Fil. Hort. Lips. t.* 22, *f.* 3–4; *Lowe's Ferns* 6, *t.* 50. Sagenia coadunata, *J. Sm. Gen. Fil.*—East Indies.

8. **A. cicutarium**, *Sw.* Polypodium cicutarium, *Linn. fide specimen Linn. Herb.* Sagenia cicutaria, *Moore, Ind.* Aspidium Hippocrepis, *Sw.* (*Plum. Fil. t.* 150). Polypodium Hippocrepis, *Jacq. Ic. rar. t.* 641. Sagenia Hippocrepis, *Presl; Hook. et Bauer, Gen. Fil. t.* 53 *A.* —Jamaica.

9. **A. apiifolium**, *Schk. Fil. t.* 56 *B*. Sagenia apiifolia, *J. Sm* Microbrochis apiifolia, *Presl.* Aspidium sinuatum *Gaud.; Labill. Sert. Aust. Caled. t.* 1. Bathmium Billardieri, *Fée.* — Sandwich Islands and New Caledonia.

10. **A. dilaceratum**, *Kunze, in part; Metten. Fil. Hort. Lips. t.* 22, *f.* 14, 16.—Jamaica.

11. **A. subtriphyllum**, *Hook. Sp. Fil.* 4, *p.* 52. Polypodium subtriphyllum, *Hook. et Arn. Bot. of Beech. Voy. t.* 50.—China, Hong-kong, Ceylon.

12. **A. variolosum**, *Wall.; Hook. Sp. Fil.* 4, *p.* 51.—India.

13. **A. giganteum,** *Blume; Hook. Sp. Fil.* 4, *p.* 50. *Var.* β minor, *Thwaites, Enum. Pl. Zeyln. p.* 390.—Ceylon.

*** *Fronds pinnatifid or pinnate; the segments sessile, decurrent.*

14. **A. Pteropus,** *Kunze; Hook. Sp. Fil.* 4, *p.* 47. Aspidium decurrens, *J. Sm. Cat.* 1857.—Ceylon.

---

71. PLEOCNEMIA, *Presl.*

Vernation fasciculate, erect, caudex arborescent. *Fronds* bi-tripinnatifid, 4–6 feet long. *Veins* of laciniæ costæform; venules forked, the lower ones arcuately and angularly anastomosing, forming unequal areoles next the costa, the upper ones free. *Receptacles* medial on the free or anastomosed venules, punctiform. *Sori* round. *Indusium* reniform.

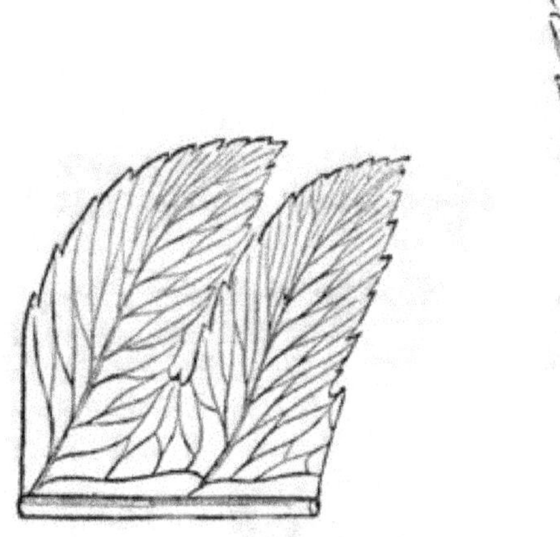
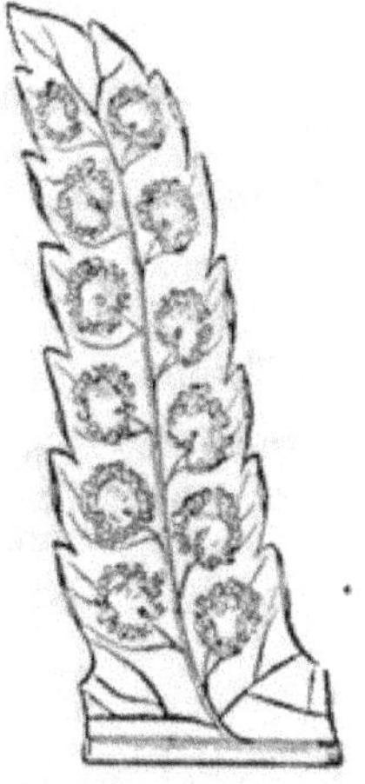

Genus 71.—Portions of barren and fertile pinnæ. No. 1.

1. **P. Leuzeana,** *Presl; Hook. Gen. Fil. t.* 97. Polypodium Leuzeanum, *Gaud. in Frey. Voy. t.* 6. Nephrodium Leuzeanum, *Hook.*—Philippine and Fiji Islands.

---

b. Indusium calyciform.

72. HYPODERRIS, *R. Br.*

Vernation uniserial, subsarmentose. *Fronds* simple, entire or trilobed, 1–2 feet long. *Primary veins* costæform; venules compound anastomosing. *Receptacles* punctiform, compital,

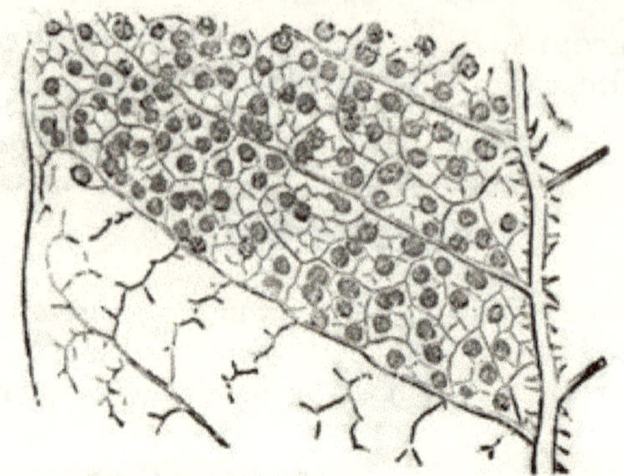

Genus 72.—Portion of mature frond, under side. No. 1.

included within a calyciform, obscure, membranous indusium. *Sori* round, irregular or oblique, biserial between the primary veins.

1. H. Brownii, *J. Sm.; Hook. Gen. Fil. t.* 1; *Hook. Gard. Ferns, t.* 24; *Lowe's Ferns,* 7, *t.* 14. Woodsia Brownii, *Metten.*—Trinidad and Guiana.

---

73. TRICHIOCARPA, *Hook.*

Vernation fasciculate, decumbent. *Fronds* bi-tripinnatifid,

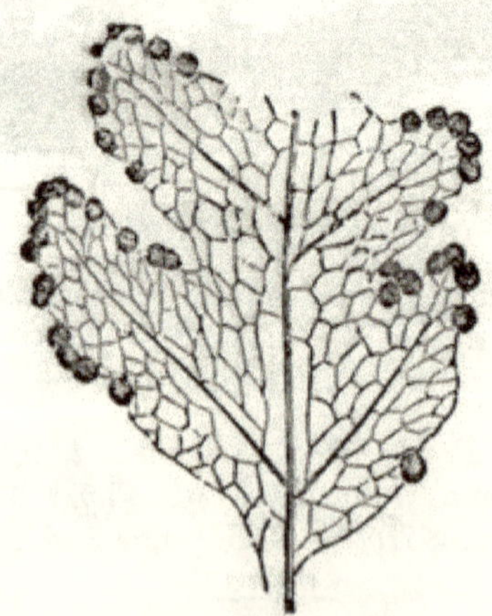

Genus 73.—Portion of pinna, fertile frond, under side. No. 1.

deltoid, 6–18 inches high; pinnæ distant, pinnate below, pinnatifid decurrent above; pinnules lanceolate, pinnatifid or sinuose lobed. *Veins* uniform, reticulated; areoles transverse oblong; marginal veinlets free, exserted beyond the margin, bearing a globose pediculate sorus. *Indusium* calyciform, spreading, entire.

1 **T. Moorei,** *J. Sm.; Lowe's Ferns,* 8, *t.* 37. Deparia Moorei, *Hook. Journ. Bot. and Kew Gard. Misc. v.* 4, *t.* 3; *Hook. Fil. Exot. t.* 28. Cionidium Moorei, *Moore, Ind.*—New Caledonia.

** *Veins free.*

† *Sori indusiate.*

§ 3. *Oreopteriæ. Sori punctiform. Indusium orbicular or reniform, plane or cucullate, rarely calyciform.*

a. Indusium orbicular or reniform.

74. POLYSTICHUM, *Roth (in part); Schott; Presl.*

Vernation fasciculate and erect, or uniserial and subsarmentose. *Fronds* pinnate, bi-tripinnate, or decompound; pinnæ

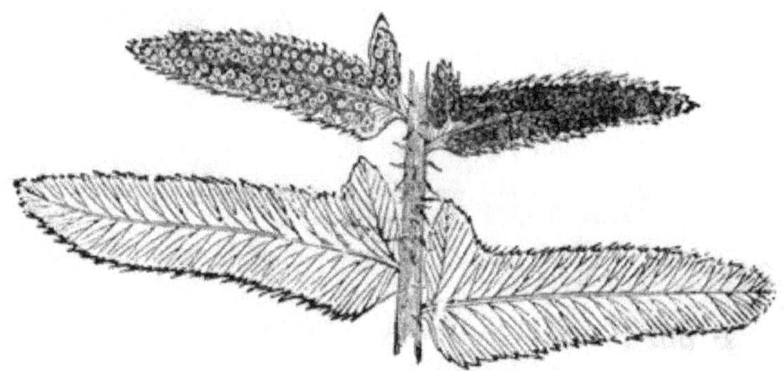

Genus 74.—Portion of mature frond, under side. No. 1.

and pinnules auriculated and lobed, dentate, rigid, spinulose, mucronate. *Veins* simply or pinnately forked; venules free, the lower exterior branch or more (of each fascicle) soriferous on, or generally below its apex. *Receptacles* punctiform. *Sori* round. *Indusium* orbicular or subreniform.

§ 1. *Vernation fasciculate, caudex erect.* (*Polystichum verum.*)

* *Fronds pinnate.*

1. **P. acrostichoides,** *Schott.* Aspidium acrostichoides, *Sw.; Lowe's Ferns,* 6, *t.* 19. Aspidium auriculatum, *Schk. Fil. t.* 30.—North America.

2. **P. falcinellum,** *Presl.* Aspidium falcinellum, *Sw.; Lowe's Ferns,* 6, *t.* 7. *Hook. Fil. Exot. t.* 53.—Madeira.

3. **P. mucronatum,** *Presl.* Aspidium mucronatum, *Sw.; Schk. Fil. t.* 29 *B, C; Hook. Sp. Fil.* 4, *t.* 216.—Jamaica.

4. **P. Lonchitis,** *Roth; Schott, Gen. Fil. t.* 9; *Lindl. and Moore's Brit. Ferns, t.* 9; *Sowerby's Ferns, t.* 15. Polypodium Lonchitis, *Linn.; Eng. Bot. t.* 797. Aspidium Lonchitis, *Sw.; Schk. Fil. t.* 29; *Hook. Brit. Ferns, t.* 9. — Europe, Britain.

5. **P. triangulum,** *Fée.* Polypodium triangulum, *Linn.* (*Plum. Fil. t.* 72). Aspidium triangulum, *Sw.; Hook. Fil. Exot. t.* 33. Aspidium trapezoides, *Sw.* Aspidium mucronatum, *Lowe's Ferns,* 6, *t.* 31 *B* (*non Sw.*).

Genus 74.—Portion of mature frond, under side. No. 11.

Var. laxum, *Lowe's New Ferns, p.* 143. — West Indies.

6. **P. obliquum,** *J. Sm.* Aspidium obliquum, *Don.* Aspidium cæspitosum, *Wall.; Hook. Sp. Fil.* 4, *t.* 213.—Nepal, Japan.

** *Fronds bipinnate.*

7. **P. aculeatum,** *Roth; Lindl. and Moore's Brit. Ferns, t.* 10; *Sowerby's Ferns, t.* 17. Polystichum aculeatum,

β intermedium, *Hook. Brit. Ferns, t.* 11. Polypodium aculeatum, *Linn.* Aspidium aculeatum, *Sw.; Schk. Fil. t.* 39; *Eng. Bot.* 1562.

Var. lobatum, *Lindl. and Moore's Brit. Ferns, t.* 11. Polypodium lobatum, *Huds.* Aspidium lobatum, *Sw.; Schk. Fil. t.* 40. Polystichum lobatum, *Presl; Hook. Gen. Fil. t.* 48 *C; Sowerby's Ferns, t.* 16. Aspidium aculeatum, et A. lobatum, *Hook. Brit. Ferns, t.* 10.

Var. angulare. Aspidium angulare, *Willd.; Eng. Bot. t.* 2776. Polystichum angulare, *Presl; Lindl. and Moore's Brit. Ferns,* 12 *A; Sowerby's Ferns, t.* 18. Aspidium aculeatum, *Hook.*

Var. angulare, *Hook. Brit. Ferns, t.* 12. Aspidium Braunii, *Spenn.* Polystichum Braunii, *Fée.*

Var. argutum, *Moore; Lindl. and Moore's Brit. Ferns, t.* 10 *B.*

Var. alatum, *Moore; Lindl. and Moore's Brit. Ferns, t.* 10 *C.*

Var. hastulatum, *Moore; Lindl. and Moore's Brit. Ferns, t.* 12 *B.*

Var. irregulare, *Moore; Lindl. and Moore's Brit. Ferns, t.* 12 *C.*

Var. biserratum, *Moore; Lindl. and Moore's Brit. Ferns, t.* 12 *D.*

Var. imbricatum, *Moore; Lindl. and Moore's Brit. Ferns, t.* 12 *E.*

Var. sub-tripinnatum, *Moore; Lindl. and Moore's Brit. Ferns, t.* 13 *A.*

Var. tripinnatum, *Moore; Lindl. and Moore's Brit. Ferns, t.* 13 *B; Lowe's Ferns,* 6, *t.* 24.

Var. proliferum, *Wollast; Lindl. and Moore's Brit. Ferns, t.* 13 *C.*

Var. cristatum, *Moore, Lowe's New Ferns,* 1, *t.* 27.

Var. aristatum, *Wollast; Lowe's New Ferns, t.* 56.

Var. acro-cladon, *Moore, Proc. Hort. Soc.* 4, *p.* 136. —Temperate Regions of the Northern Hemisphere.

8. **P. squarrosum**, *Fée* Aspidium squarrosum, *Don.* Aspidium rufo-barbatum, *Wall.*—East Indies.

9. **P. anomalum**, *J. Sm.* Polypodium anomalum, *Hook. et Arn.*; *Hook. Kew Gard. Misc.* 8, *t.* 9.—Ceylon.

10. **P. obtusum**, *J. Sm.* Aspidium obtusum, *Kunze*; *Hook. Sp. Fil.* 4, *t.* 221.—Philippine Islands.

11. **P. proliferum**, *Presl.* Aspidium proliferum, *R. Br.*—Tasmania.

12. **P. vestitum**, *Presl.* Polypodium vestitum, *Forst.* Aspidium vestitum, *Sw.*; *Schk. Fil. t.* 43; *Lowe's Ferns*, 6, *t.* 38.—New Zealand.

13. **P. pungens**, *Presl.* Aspidium pungens, *Kaulf.*; *Schlecht. Fil. t.* 10.—South Africa.

§ 2. *Vernation uniserial, sarmentose. Fronds deltoid tri-quadripinnate* (Tectaria, *Cav.*).

14. **P. coriaceum**, *Schott.* Aspidium coriaceum, *Sw.* (*excl. syn. Forst.*); *Schk. Fil. t.* 50.—West Indies.

15. **P. Capense**, *J. Sm.* Aspidium Capense, *Willd.* (*in part*).—South Africa.

16. **P. flexum**, *Remy.* Aspidium flexum, *Kunze.* Aspidium coriaceum, *Lowe's Ferns*, 6, *t.* 26. Aspidium Berteroanum, *Col. Pl. Chil. t.* 70; *Hook. Sp. Fil.* 4, *t.* 229.—Chili, Juan Fernandez.

17. **P. amplissimum**, *Presl.* Aspidium amplissimum, *Metten.* Aspidium fallax, *Fisch. MSS.* Lastrea fallax, *Moore.*—Brazil.

18. **P. frondosum**, *J. Sm.* Aspidium frondosum, *R. T. Lowe.* Nephrodium læte-virens, *R. T. Lowe.*—Madeira.

19. **P. aristatum**, *Presl.* Polypodium aristatum, *Forst.* Aspidium aristatum, *Sw.*; *Schk. Fil. t.* 42. A. curvifolium, *Kunze.* Polystichum curvifolium, *Hort.*—Norfolk Island.

20. **P. coniifolium**, *Presl.* Aspidium coniifolium, *Wall.*—East Indies and Ceylon.

21. **P. denticulatum**, *J. Sm.* Aspidium denticulatum, *Sw.*; *Lowe's New Ferns*, *t.* 59.—Jamaica.

22. P. **amabile**, *J. Sm.* Aspidium amabile, *Blume; Hook. Sp. Fil.* 4, *t.* 225. Aspidium rhomboideum, *Wall.* Polystichum rhomboideum, *Schott.*—East Indies, Java.

23. P. **setosum**, *Presl.* Aspidium setosum, *Sw.; Lang. et Fisch. Fil. t.* 17.—Japan.

---

75. LASTREA, *Presl; J. Sm.*

Vernation uniserial and sarmentose, or fasciculate and erect

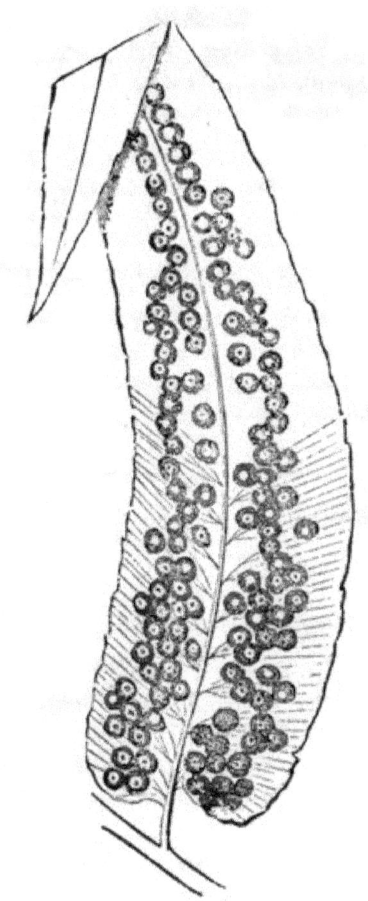

Genus 75.—Pinna of fertile frond. No. 30.

or decumbent. *Fronds* varying from pinnatifid, bipinnatifid, to decompound-multifid, 1–6 feet high. *Veins* simple, or once or several times forked, or costæform and pinnate; venules free. *Receptacles* punctiform, terminal or medial. *Sori* round. *Indusium* reniform or subrotund (as in figure), attached by its sinus on the interior side.

* *Vernation uniserial, sarmentose.* (Thelypteris.)

† *Fronds lanceolate, bipinnatifid.*

1. L. **palustris**, *J. Sm.* Thelypteris palustris, *Schott.* Lastrea Thelypteris, *Presl; Lindl. and Moore's Brit. Ferns, t.* 29; *Hook. Gen. Fil. t.* 45 *A* 2; *Sowerby's Ferns, t.* 7. Polypodium Thelypteris, *Linn.* Aspidium Thelypteris, *Sw.; Schk. Fil. t.* 52; *Eng. Bot. t.* 1018. Nephrodium Thelypteris, *Desv.; Hook. Brit. Ferns, t.* 13.—Europe.

2. L. **Noveboracensis**, *Presl.* Polypodium Noveboracense, *Linn.; Schk. Fil. t.* 46. Aspidium Thelypteroides, *Sw.*—North America.

3. L. **invisa**, *Presl.* Aspidium invisum, *Sw.; Schk. Fil. t.* 18.—West Indies.

4. L. **augescens**, *J. Sm.* Aspidium augescens, *Link; Kunze, Fil. t.* 59; *Lowe's Ferns,* 6, *t.* 10. Nephrodium Ottonianum, *Kunze.*—Tropical America.

†† *Fronds deltoid, decompound.*

5. L. **decomposita**, *J. Sm.* Nephrodium decompositum, *R. Br.; Hook. Fil. Fl. Nov. Zel. t.* 79 (*exclus. name,* glabellum). Aspidium decompositum, *Spreng.; Lowe's Ferns,* 6, *t.* 33.—Australia, Tasmania, and New Zealand.

6. L. **velutina**, *J. Sm.* Nephrodium (Lastrea) velutinum, *Hook. Sp. Fil.* 4, *p.* 145; *Hook. Fil. Nov. Zel. t.* 80. Aspidium velutinum, *A. Rich.*—New Zealand.

7. L. **pubescens**, *Presl.* Polypodium pubescens, *Linn.* Nephrodium pubescens, *Hook. et Grev. Ic. Fil. t.* 62.—Jamaica.

8. L. **quinquangularis**, *J. Sm.* Aspidium quinquangulare, *Kunze.* Aspidium pubescens, *Lowe's Ferns,* 6, *t.* 25.—Native country unknown.

9. **L. recedens**, *J. Sm.* Polypodium recedens, *J. Sm. En. Fil. Phil.* Aspidium recedens, *Lowe's Ferns*, 7, *t.* 1.—Ceylon and Philippine Islands.

10. **L. elegans**, *Moore and Houlst.*—Ceylon.

11. **L. pilosissima**, *J. Sm.* Aspidium pilosissima, *G. Don, in Herb.* 1822; *J. Sm.*—Sierra Leone.

12. **L. subquinquifida**, *J. Sm.* Aspidium subquinquifidum, *Beauv. Fl. Owar. t.* 19. Aspidium protensum, *Sw.*—West Tropical Africa.

13. **L. funesta**, *Moore.* Aspidium funestum, *Kunze.*—Tropical West Africa, Brazil.

** *Vernation fasciculate, caudex erect or subdecumbent.*

† *Fronds lanceolate, bipinnatifid, rarely pinnatifid. Veins generally simple, rarely forked.* (Oreopteris.)

14. **L. decursive-pinnata**, *J. Sm.* Polypodium decursive-pinnatum, *Hook. 2nd Cent. t.* 49. Lastrea decurrens, *J. Sm. Cat.* 1846 *and* 1857.—China, Japan.

15. **L. montana**, *Moore.* Polypodium montanum, *Vogler.* Polystichum montanum, *Roth.* Polypodium Oreopteris, *Ehrhart; Eng. Bot. t.* 1019. Aspidium Oreopteris, *Sw.; Schk. Fil. t.* 35, 36. Lastrea Oreopteris, *Presl; Lindl. and Moore's Brit Ferns, t.* 28; *Sowerby's Ferns, t.* 8. Nephrodium Oreopteris, *Hook. Brit. Ferns, t.* 14.—*Var.* Nowelliana, *Moore; Lowe's New Ferns, p.* 99.—Europe.

16. **L. patens**, *Presl; Hook. Gen. Fil. t.* 45 *A* 1. Aspidium patens, *Sw.; Radd. Fil. Bras. t.* 40; *Lowe's Ferns*, 7, *t.* 3, 4.—Tropical America.

17. **L. concinna**, *J. Sm.* Polypodium concinnum, *Willd.* Phegopteris concinna, *Fée.* Polypodium molliculum, *Kunze.* Phegopteris mollicula, *J. Sm. Cat. Cult. Ferns*, 1857.—Tropical America.

18. **L. contermina**, *Presl.* Aspidium conterminum, *Willd.* (*Plum. Fil. t.* 47). Aspidium polyphyllum, *Kaulf.* A. rivulorum, *Link.*—Tropical America, West Indies.

19. **L. immersa**, *J. Sm.* Aspidium immersum, *Blume; Metten. Fil. Hort. Lips. t.* 18, *f.* 1–3. Lastrea verrucosa, *J. Sm. En. Fil. Phil.* Aspidium impressum, *Kunze.*—Malayan Islands.

20. L. **cana**, *J. Sm.* Aspidium canum, *Wall.* Nephrodium pubescens, *D. Don*, (*non Sw.*).—East Indies.

21. L. **strigosa**, *Presl.* Aspidium strigosum, *Willd.*; *Lowe's Ferns*, 7, *t.* 10. Polypodium crinitum, *Poir.* Lastrea crinita, *Moore*; *Hook. et Grev. Ic. Fil. t.* 66.—Mauritius.

22. L. **similis**, *J. Sm. En. Fil. Phil.* (*n.* 390, *Cuming*). Aspidium submarginale, *Hort. Berol.*—Malacca.

23. L. **Kaulfussii**, *Presl.* Aspidium Kaulfussii, *Link*; *Lowe's Ferns*, 7, *t.* 5.—Brazil.

24. L. **chrysoloba**, *Presl.* Aspidium chrysolobum, *Link.*—Brazil.

25. L. **Caripense**, *J. Sm.* Polypodium Caripense, *H. et B.* Polypodium submarginale. *Lang. et Fisch. Fil. t.* 13; *Lowe's Ferns*, 2, *t.* 49 (*without indusiæ*). Phegopteris submarginalis, *J. Sm. Cat.* 1857.—Tropical America.

26. L. **vestita**, *J. Sm.* Polypodium vestitum, *Radd. Fil. Bras. t.* 36.—Brazil, West Indies.

27. L. **falciculata**, *Presl.* Aspidium falciculatum, *Radd. Fil Bras. t.* 47.—Brazil.

28. L. **Sprengelii**, *J. Sm.* Aspidium Sprengelii, *Kaulf.* Aspidium glandiferum, *Karst.*—Tropical America and West Indies.

29. L. **deltoidea**, *Moore.* Aspidium deltoideum, *Sw.* Nephrodium deltoideum, *Desv.*; *Hook. Sp. Fil.* 4, *p.* 103.—West Indies.

†† *Fronds lanceolate or deltoid, bi-tripinnatifid or bipinnate, rarely pinnate, usually firm and subcoriaceous. Veins forked, generally immersed.* (Dryopteris.)

30. L. **podophylla**, *J. Sm.* Aspidium (Lastrea) podophyllum, *Hook. in Journ. Bot. and Kew Misc. v.* 5, *t.* 1. Aspidium Sieboldi, *Van Houtte, Cat.*; *Metten. Fil. Hort. Lips. t.* 20, *f.* 1–4. Pycnopteris Sieboldi, *Moore.*—Japan and Hong-kong. **T.**

31. **L. Filix-mas,** *Presl; Lindl. and Moore's Brit. Ferns, t.* 14; *Sowerby's Ferns, t.* 9. Polypodium Filix-mas, *Linn.* Aspidium Filix-mas, *Sw.; Schk. Fil. t.* 44; *Eng. Bot.* 1458. Nephrodium Filix-mas, *Michx; Hook. Brit. Ferns, t.* 15.

Var. paleacea, *Moore; Lindl. and Moore's Brit. Ferns, t.* 17 *B.* Aspidium paleaceum, *Don.* Nephrodium Filix-mas, *var.* paleaceum, *Hook. Fil. Exot. t.* 98. Lastrea Pseudo-mas, *Wollast.* L. Filix-mas, *var.* Borreri, *Johns;* Nephrodium affine, *R. T. Lowe.*

Var. pumila, *Moore; Lindl. and Moore's Brit. Ferns, t.* 17 *A.* Aspidium pumilum, *Lowe's Ferns,* 6, *t.* 15.

Var. cristata, *Moore; Lindl. and Moore's Brit. Ferns, t.* 16 *A.*

Var. incisa, *Moore; Lindl. and Moore's Brit. Ferns, t.* 15. Aspidium affine, *Fisch. et Mey.* Lastrea affinis, *Moore.*

Var. polydactyla, *Moore; Lindl. and Moore's Brit. Ferns, t.* 16 *B.*

Var. ramosissima, *Moore, Gard. Chron.* 1864.

Temperate regions of the earth generally.

32. **L. remota,** *Moore.* Aspidium remotum, *A. Braun; Lowe's New Ferns, t.* 22. Nephrodium remotum, *Hook. Brit. Ferns, t.* 22. Aspidium Boottii, *Tuckerman.* A. dilatatum, *var.* Boottii, *A. Gray.*—Europe and North America.

33. **L. lacera,** *J. Sm.* Polypodium lacerum, *Thunb.* Aspidium lacerum, *Eaton.*—Japan.

34. **L. hirtipes,** *J. Sm.* Aspidium hirtipes, *Blume.* Nephrodium (Lastrea) hirtipes, *Hook. Sp. Fil.* 4, *p.* 115. Aspidium atratum, *Wall.*—India and Ceylon.

35. **L. rigida,** *Presl; Lindl. and Moore's Brit. Ferns, t.* 18; *Sowerby's Ferns, t.* 11. Aspidium rigidum, *Sw.; Schk. Fil. t.* 38; *Eng. Bot. t.* 2724; *Lowe's Ferns,* 6, *t.* 21. Nephrodium rigidum, *Desv.; Hook. Brit. Ferns, t.* 16.—Europe.

36. L. **elongata**, *Presl.* Polypodium elongatum, *Ait.* Aspidium elongatum, *Sw.; Hook. et Grev. Ic. Fil. t.* 234.—Madeira.

37. L. **varia**, *Moore.* Polypodium varium, *Linn.* Aspidium varium, *Sw.; Hook. Sp. Fil.* 4, *t.* 226. Lastrea opaca, *Hook.*—China, Japan.

38. L. **Napoleonis**, *J. Sm.* Aspidium Napoleonis, *Bory, Hook. Sp. Fil.* 4, *t.* 255.—St. Helena.

39. L. **marginalis**, *Presl.* Polypodium marginale, *Linn. Lowe's Ferns*, 6, *t.* 26. Aspidium marginalis, *Sw. Schk. Fil. t.* 45 *B; Lowe's Ferns*, 6, *t.* 6.—North America.

40. L. **Goldiana**, *Presl.* Nephrodium Goldianum, *Hook. et Grev. Ic. Fil. t.* 102.—North America.

41. L. **erythrosora**, *J. Sm.* Nephrodium erythrosorum, *Eat.; Hook. Sp. Fil.* 6, *t.* 253.—Japan.

42. L. **Mexicana**, *Lieb.* Nephrodium Mexicanum, *Presl.* Aspidium Mexicanum, *Kunze.*—Mexico.

††† *Fronds tripinnate,* 6–10 *feet long; stipes thick, paleaceous; pinnæ* 2–2½ *feet long, more or less villose; pinnules* 4–8 *inches long, lanceolate-acuminate, deeply pinnatifid; segments entire or pinnatifidly lobed; ultimate lobes unisorous. Indusium almost peltate.* (Megopteris.)

43. L. **villosa**, *Presl.* Polypodium villosum, *Sw.* (*Plum. Fil. t.* 27). Aspidium villosum, *Sw.; Schk. Fil. t.* 46. Nephrodium (Lastrea) villosum, *Hook. Sp. Fil. t.* 264.—West Indies.

†††† *Fronds lanceolate or deltoid, bi-tripinnate, generally fragile, crenate, dentate, or unequally laciniated, often spinulose Veins forked.* (Lophodium.)

44. L. **dilatata**, *Presl; Lindl. and Moore's Brit. Ferns, t.* 22; *Sowerby's Ferns, t.* 13. Aspidium dilatatum, *Sm.; Eng. Bot. t.* 1461. Nephrodium spinulosum, *var.* dilatatum, *Hook. Brit. Ferns, t.* 19.

Var. tanacetifolia, *Moore.* Polypodium tanacetifolium, *Hoffm.* Aspidium depastum, *Schk. Fil. t.* 51. Aspidium erosum, *Schk. Fil. t.* 45.

Var. nana, *Moore; Lindl. and Moore's Brit. Ferns, t.* 26 *C, D.*

Var. dumetorum, *Moore; Lindl. and Moore's Brit. Ferns, t.* 25. Aspidium dumetorum, *Sm.* Nephrodium spinulosum, δ dumetorum, *Hook. Brit. Ferns, t.* 21.

Var. collina, *Moore; Lindl. and Moore's Brit. Ferns, t.* 26 *A, B.* Lastrea collina, *Newm.*

Var. Chanteriæ, *Moore; Lindl. and Moore's Brit. Ferns, t.* 24.

Var. glandulosa, *Moore; Lindl. and Moore's Brit. Ferns, t.* 23.

Europe, North America, and North-East Asia.

45. **L. cristata,** *Presl; Lindl. and Moore's Brit. Ferns, t.* 19; *Sowerby's Ferns, t.* 10. Polypodium cristatum, *Linn.* Aspidium cristatum, *Sw.; Schk. Fil. t.* 37; *Eng. Bot. t.* 2125. Nephrodium cristatum, *Mich.; Hook. Brit. Ferns, t.* 17.—β, Lancastriense, *J. Sm.* Aspidium Lancastriense, *Spreng.; Schk. Fil. t.* 41.—Europe and North America.

46. **L. intermedia,** *Presl.* Aspidium intermedium, *Willd.*—North America.

47. **L. spinulosa,** *Presl; Lindl. and Moore's Brit. Ferns, t.* 21; *Sowerby's Ferns, t.* 12. Aspidium spinulosum, *Sw.; Schk. Fil. t.* 48. Nephrodium spinulosum, *a*, bipinnatum, *Hook. Brit. Ferns, t.* 18.

Var. uliginosa, *J. Sm.* Lastrea uliginosa, *Newm.* Lastrea cristata, *var.* uliginosa, *Moore; Lindl. and Moore's Brit. Ferns, t.* 20.—Europe.

48. **L. æmula,** *J. Sm.* Polypodium æmulum, *Ait.* Aspidium æmulum, *Sw.* Nephrodium spinulosum, γ æmulum, *Hook. Brit. Ferns, t.* 20. Nephrodium Fœnisecii, *R. T. Lowe.* Lastrea Fœnisecii, *Watson; Lindl. and Moore's Brit. Ferns, t.* 27; *Sowerby's Ferns, t.* 14. Lastrea recurva, *Newm.* Lastrea concava, *Newm.*—Europe and Madeira.

49. **L. glabella,** *J. Sm.* Nephrodium glabellum, *A. Cunn.* Aspidium glabellum, *Lowe's Ferns,* 6, *t.* 36.—New Zealand.

50. **L. Shepherdi,** *J. Sm.* Aspidium Shepherdi, *Kunze.* Lastrea acuminata, *Houlst. et Moore.* Aspidium acuminatum, *Hort. Ang.; Lowe's Ferns,* 6, *t.* 11. Lastrea atro-virens, *J. Sm. Cat. Cult. Ferns,* 1857.—Native country unknown. T.

51. **L. hirta,** *Presl.* Aspidium hirtum, *Sw.; Schk. Fil. t.* 46 *B. Lowe's Ferns,* 7, *t.* 11. Polypodium crystallinum, *Kunze, Fil. t.* 135.—Jamaica.

52. **L. sancta,** *J. Sm.* Polypodium sanctum, *Sw.* Phegopteris sancta, *Fée.*—West Indies.

53. **L. hispida,** *Moore and Houlst.* Aspidium hispidum, *Sw.; Schk. Fil. t.* 49; *Lowe's Ferns,* 7, *t.* 8. Polystichum hispidum, *J. Sm. Cat.* 1857. Polypodium setosum, *Forst.*—New Zealand.

Fronds bi-tripinnatifid, 6–18 *inches high, flaccid, ultimate dents unisorous. Indusium equal with the dent, and forming with it a bilabiate cyst.* (Diclisodon, *Moore.*)

54. **L. deparioides,** *J. Sm.* Nephrodium (Lastrea) deparioides, *Hook. Sp. Fil.* 4, *p.* 139. Aspidium deparioides, *Hook. Fil. Exot. t.* 3. Diclisodon deparioides, *Moore.*—Ceylon.

---

b. Indusium inflated, cucullate.

76. CYSTOPTERIS, *Bernh.*

Vernation sub-fasciculate and decumbent or sub-erect, or uniserial and short sarmentose. *Fronds* slender, bi-tripinnatifid, 4–15 inches high. *Veins* forked; venules free. *Receptacles* punctiform, medial. *Sori* round. *Indusium* lateral, oblong or reniform, cucullate, dentate or fimbriate.

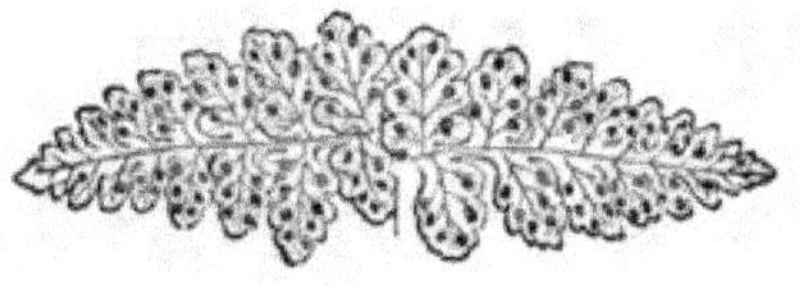

Genus 76.—Portion of fertile frond, under side. No. 3, var.

1. **C. tenuis,** *Schott; Lowe's Ferns,* 7, *t.* 35. Aspidium tenue, *Sw.; Schk. Fil. t.* 53 *B.* Aspidium atomarium, *Muhl.; Willd.* Cystopteris atomaria, *Presl.*—North America.

2. **C. bulbifera,** *Bernh.; Lowe's Ferns,* 7, *t.* 36. Polypodium bulbiferum, *Linn.* Aspidium bulbiferum, *Sw.; Schk. Fil. t.* 57.—North America.

3. **C. fragilis,** *Bernh.; Hook. Gen. Fil. t.* 52 *B; Lindl. and Moore's Brit. Ferns, t.* 46 *A, f.* 1; *Sowerby's Ferns, t.* 19. Polypodium fragile, *Linn.* Aspidium fragile, *Sw.; Schk. Fil. t.* 54. Cyathea fragilis, *Sm.; Eng. Bot. t.* 1587.

Var. dentata, *Hook.; Lindl. and Moore's Brit. Ferns, t.* 46 *A, f.* 4; *Lowe's Ferns,* 7, *t.* 32. Cystopteris dentata, *Hook.; Sowerby's Ferns, t.* 21. Cyathea dentata, *Sm. Eng. Bot. t.* 1588. Polypodium dentatum, *Dicks.*

Var. Dickieana, *Moore; Lindl. and Moore's Brit. Ferns, t.* 46 *A, f.* 5, 6; *Hook. Brit. Ferns, t.* 23, *f.* 4, 5. Cystopteris dentata, *var.* Dickieana, *Bab.; Sowerby's Ferns, t.* 22. C. Dickieana, *Sim.; Lowe's Ferns,* 7, *t.* 33.

Var. sempervirens, *Moore; Lindl. and Moore's Brit. Ferns, t.* 46 *A, f.* 2, 3.

Var. angustata, *Link; Moore's Nat. Print. Ferns, Oct. Ed.* 2, *t.* 102 *C; Sowerby's Ferns, t.* 20.—Temperate Regions of the Northern Hemisphere.

4. C. **regia,** *Presl; Lindl. and Moore's Brit. Ferns, t.* 46 *B.* Polypodium regium, *Linn.* Cystea regia, *Sm.* Aspidium regium, *Sw.* Cyathea incisa, *Sm. Eng. Bot. t.* 163. Polypodium alpinum, *Jacq. Ic. Rar. t.* 642. Aspidium alpinum, *Sw.; Schk. Fil. t.* 62. Cystopteris alpina, *Desv.; Hook. Brit. Ferns, t.* 24; *Sowerby's Ferns, t.* 23.—Europe.

5. C. **montana,** *Bernh.; Lindl. and Moore's Brit. Ferns, t.* 46 *C, f.* 1–3; *Hook. Brit. Ferns, t.* 25; *Sowerby's Ferns, t.* 24. Aspidium montanum, *Sw.; Schk. Fil. t.* 63.—Europe.

---

c. *Indusium calyciform.*

77. WOODSIA, *R. Br.*

Vernation fasciculate, erect, cæspitose. *Fronds* bi-tripinnatifid, rarely pinnate, 1–12 inches high, smooth or squamiferous. *Veins* simple or forked, free, the lower exterior branch sporangiferous on or below its apex. *Receptacles* punctiform. *Sori* round. *Indusium* calyciform, its margin nearly entire or deeply laciniated, laciniæ usually terminating in long hairs, which involve the sporangia.

Genus 77.—Frond of No. 1, and pinna of No. 5.

§ 1. *Woodsia vera. Stipes with a special articulation. Membrane of indusium nearly obsolete, fringed with articulated hairs, which involve the sporangia.*

1. **W. Ilvensis,** *R. Br.; Hook. Brit. Ferns, t.* 8; *Eng. Bot. t.* 2616; *Sowerby's Ferns, t.* 5; *Lindl. and Moore's Brit. Ferns, t.* 47 *A.* Acrostichum Ilvense, *Linn.* Polypodium Ilvense, *Sw.; Schk. Fil. t.* 19.—Europe, Britain. **T.**

2. **W. hyperborea,** *R. Br. in Trans. Linn. Soc.* 11, *t.* 11; *Hook. Gen. Fil. t.* 119; *Hook. Brit. Ferns, t.* 7; *Sowerby's Ferns, t.* 6. Polypodium hyperboreum, *Sw.; Eng. Bot. t.* 2023; *Schk. Fil. t.* 17 *B.*—Europe, Britain. **T.**

§ 2. *Physematium. Stipes not articulated. Membrane of indusium complete, at length somewhat sinuose-laciniated.*

3. **W. polystichoides,** *Eaton; Hook. 2nd Cent. Ferns, t.* 2. *β* Veitchii, *Hook. Gard. Ferns, t.* 32. Woodsia Veitchii, *Hance, MSS.*—Japan, Manchuria.

4. **W. obtusa,** *Hook.; Hook. Gard. Ferns, t.* 43; *Lowe's Ferns,* 7, *t.* 29. Polypodium obtusum, *Sw.; Schk. Fil. t.* 21. Woodsia Perriniana, *Hook. et Grev. Ic. Fil. t.* 68.—North America.

5. **W. mollis,** *J. Sm.; Lowe's Ferns,* 7, *t.* 26. Physematium molle, *Kunze, Anal Pterid. t.* 27. Woodsia Mexicana, *R. Br.*—Mexico. **H.**

---

§ 4. *Arthropteræ. Fronds always pinnate; pinnæ entire or subpinnatifid, always articulated with the rachis. Sori punctiform, terminal. Indusium reniform, rarely obsolete or wanting.*

78 ARTHROPTERIS, *J. Sm.*

Vernation uniserial, distant; sarmentum slender, scandent; stipes pseudo-articulated; the node of articulation basal or more or less elevated. *Fronds* pinnate, 1–1½ foot long;

pinnæ entire, dentate or pinnatifid. *Veins* forked or pinnate; venules free, their apices clavate, the lower exterior one sporangiferous. *Receptacles* punctiform. *Sori* terminal, round. *Indusium* reniform or absent.

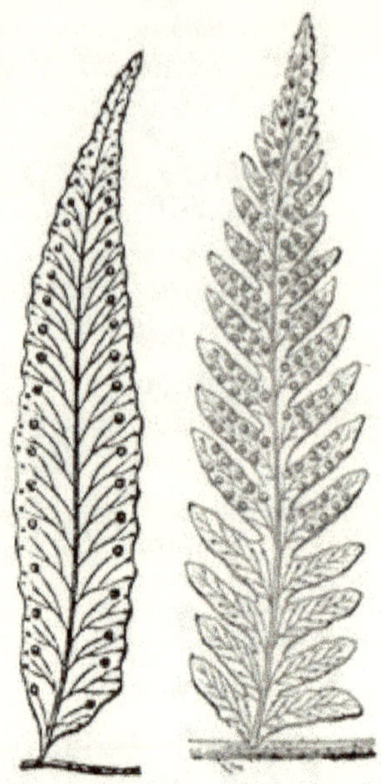

Genus 78.—Pinna of No. 1, and ditto No. 3, natural size.

a. Indusium absent.

1. **A. tenella,** *J. Sm. in Hook. Fil. Fl. Nov. Zeal. t.* 82. Polypodium tenellum, *Forst. Schk. Fil. t.* 16. Polypodium filipes, *Moore, in Gard. Chron.* (1855), *p.* 368; *Lowe's Ferns,* 2, *t.* 38. Arthropteris filipes, *J. Sm. Cat. Cult. Ferns* (1857).—New Zealand.

b. Indusium small, soon obliterated.

2. **A. obliterata,** *J. Sm.* Nephrodium obliteratum, *R. Br.* Nephrolepis obliterata, *Hook. Sp. Fil.* 4, *p.* 154. Lindsæa Lowei, *Hort.*—Australia, Malayan and Polynesian Islands.

c. Indusium evident, pinnæ pinnatifid.

3. **A. albo-punctata,** *J. Sm.* Aspidium albo-punctatum, *Willd.* Nephrodium albo-punctatum, *Desv.; Hook. Fil. Exot. t.* 89. Aspidium leucosticton, *Kunze.* Aspidium (Lastrea) Boutonianum, *Hook. Ic. Pl. t.* 93.—Mauritius, W. Tropical Africa.

———

79. NEPHROLEPIS, *Schott.*

Vernation fascicuiate, erect, stoloniferous. *Fronds* pinnate, linear, 1–6 feet long; pinnæ numerous, oblong or linear-lanceolate and falcate, entire, dentate or deeply crenate, upper side of the base auriculated, sessile, articulated to the rachis. *Veins* forked; venules free, clavate, the lower exterior one fertile. *Sori* terminal, round, sub-marginal, transverse uniserial. *Indusium* reniform or nearly orbicular (as in figure), sometimes equal and conniving with the soriferous crenule, forming a marginal bilabiate cyst.

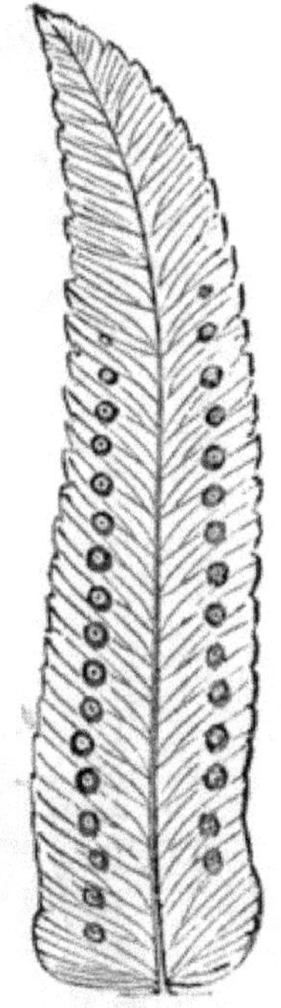

Genus 79.—Pinna of fertile frond, under side. No. 5.

1. **N. pectinata,** *Schott; Lowe's Ferns,* 7, *t.* 18. Aspidium pectinatum, *Willd.* Aspidium trapezoides, *Schk. Fil. t.* 29 *B.* Aspidium Schkuhrii, *Link.* — Tropical America.

2. **N. undulata,** *J. Sm.; Lowe's Ferns,* 7, *t.* 20. Aspidium undulatum, *Sw.*—West Africa.

3. **N. tuberosa,** *Presl; Lowe's Ferns,* 7, *t.* 25. Aspidium tuberosum, *Bory.* Nephrodium edule, *D. Don.*—East Indies.

4. **N. exaltata,** *Schott, Gen. Fil. t.* 3; *Hook. Gen. Fil. t.* 35; *Lowe's Ferns,* 7, *t.* 19. Polypodium exaltatm, *L inn.;* (*Plum. Fil. t.* 63). Aspidium exaltatum, *uw.; Schk. Fil. t.* 32 *B; Radd. Fil. Bras. t.* 46. Nephrodium exaltatum, *R. Br.*—Tropical America.

5. **N. ensifolia,** *Presl; Lowe's Ferns,* 7, *t.* 22. Aspidium ensifolium, *Sw.; Schk. Fil. t.* 32. Aspidium acutum, *Sw.; Schk. Fil. t.* 31. Nephrolepis platyotis, *Kunze; Metten. Fil. Hort. Lips. t.* 26, *f.* 1.—Tropical America and Java.

6. **N. hirsutula,** *Presl; Lowe's Ferns,* 7, *t.* 21. Aspidium hirsutulum, *Sw.; Schk. Fil. t.* 33. Lepidoneuron hirsutulum, *Fée.* Aspidium pilosum, *Lang. et Fisch. Fil. t.* 16.—East Indies.

7. **N bisserata,** *Schott.* Aspidium bisseratum, *Sw.; Schk. Fil. t.* 33. Nephrodium bisseratum, *Presl.*—Philippine Islands.

8. **N. davallioidos,** *Moore; Lowe's Ferns,* 7, *t.* 23; *Hook. Fil. Exot. t.* 60. Aspidium davallioides, *Sw.; Hook. Ic. Plant. t.* 395–6.—Malayan Archipelago.

---

80. CYCLOPELTIS, *J. Sm.*

Vernation fasciculate, decumbent. *Fronds* pinnate, 1–3 feet high; pinnæ entire, falcate, lanceolate, 4–9 inches long, sessile, auriculated at the base, articulated with the rachis. *Veins* two or three times forked; venules free, the lower interior and exterior ones sporangiferous on or below their apices. *Receptacles* punctiform. *Sori* round, transverse, biserial. *Indusium* orbicular.

1. **C. semicordata,** *J. Sm. En. Fil. Hort. Kew.* (1846); *Lowe's Ferns,* 6, *t.* 3. Aspidium semicordatum, *Sw.;* (*Plum. Fil. t.* 113). Lastrea semicordata, *Presl.* Hemicardium Nephrolepis, *Fée.* Polystichum semicordatum, *Moore.*—West Indies.

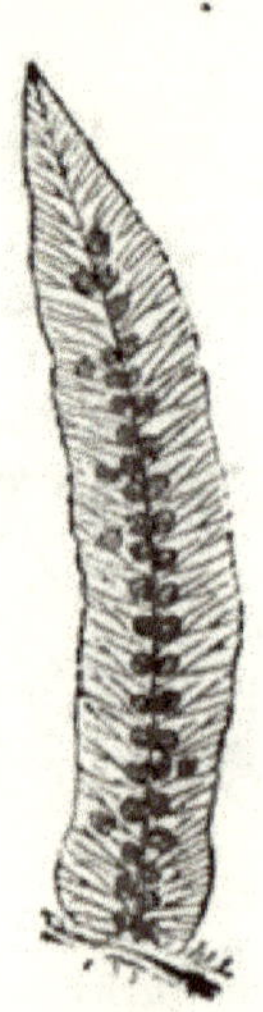

Genus 80.—Pinna of mature frond, under side. No. 1.

---

§ 5. *Didymochlænæ. Fronds always bipinnate; pinnæ and pinnules articulate with the rachis. Sori punctiform, oblong, terminal. Indusium oblong hippocrepiform.*

81. DIDYMOCHLÆNA, *Desv.*

Vernation fasciculate and erect, subarboreous. *Fronds* bipinnate, 2–6 feet long; pinnules oblong-elliptical, oblique, base truncate, subsessile, articulated with the rachis. *Veins* radiating, forked; costa excentric; venules direct, free, their apices clavate, the anterior one sporangiferous. *Receptacles* oblong. *Sori*

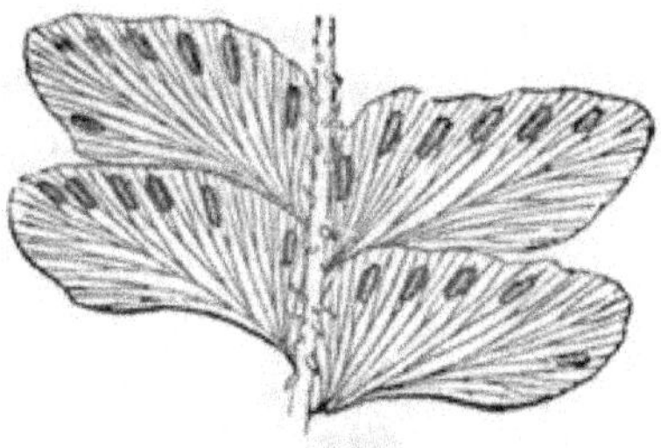

Genus 81.—Portion of mature frond, under side. No. 1.

terminal, elliptical, marginal. *Indusium* oblong, attached longitudinally, hippocrepiform.

1. **D. lunulata**, *Desv.; Hook. Gard. Ferns, t.* 17. Didymochlæna truncatula, *J. Sm. Cat. Cult. Ferns* (1857). Aspidium truncatulum, *Sw.* Aspidium squamatum, *Willd.; (Plum. Fil. t.* 56). Didymochlæna sinuosa, *Desv.; Hook. Gen. Fil. t.* 8. Diplazium pulcherrimum, *Radd. Fil. Bras. t.* 59.—Malayan Archipelago and Tropical America.

---

†† *Sori destitute of special indusium.*

§ 6. *Struthioptereæ. Fertile fronds contracted; segments revolute, forming a universal indusium, enclosing crowded punctiform sori.*

82. STRUTHIOPTERIS, *Willd.*

Vernation fasciculate, erect. *Fronds* pinnate or bipinnate 1–3 feet high. *Veins* pinnate; venules free, the fertile pinnæ contracted, linear, with membranous, revolute, conniving margins, forming a universal indusium. *Receptacles* medial, base of the pedicels of the sporangia concrete, forming thickened receptacles. *Sori* round, confluent.

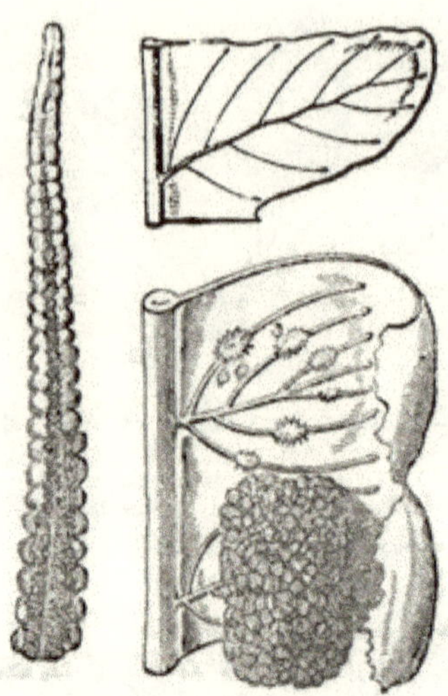

Genus 82.—Portions of sterile and fertile pinna, natural size, and ditto fertile, enlarged. No. 1.

1. S. **Germanica,** *Willd.; Lowe's Ferns*, 2, *t.* 63; *Hook. Gen. Fil. t.* 69. Onoclea Struthiopteris, *Sw.; Schk. Fil. t.* 105.—Germany.

2. S. **Pennsylvanica,** *Willd.*—North America.

---

§ 7. *Phegopterieæ. Sori punctiform, rarely oblong, naked, or seated in the axis of reflexed indusiform dents.*

83. AMPHIDESMIUM, *Schott.*

Vernation fasciculate, decumbent and criniferous. *Fronds* pinnate, 4–6 feet long, smooth; pinnæ linear-lanceolate, adherent. *Veins* simple or rarely forked, parallel. *Receptacles* punctiform. *Sori* medial, round, criniferous, irregular, often more than one on the same vein.

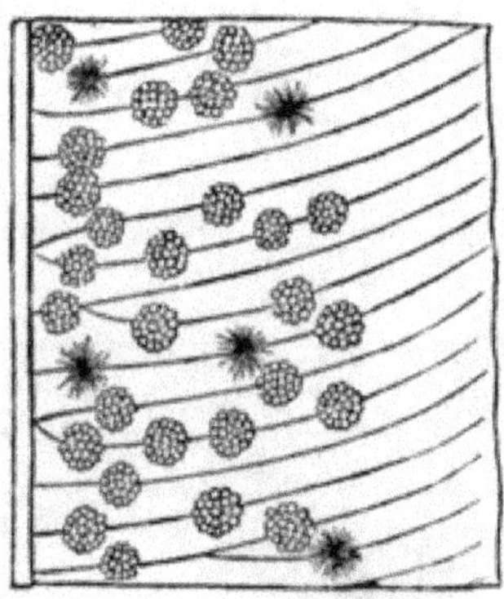

Genus 83.—Portion of pinnæ, slightly enlarged. No. 1.

1. **A. blechnoides**, *Klotzsch.* Polypodium blechnoides, *Rich.* Alsophila blechnoides, *Hook. Sp. Fil.* Amphidesmium rostratum, *J. Sm. Gen. Fil.* Polypodium rostratum, *Humb.* Metaxya rostrata, *Presl; Hook. Gen. Fil. t.* 42*B.* Amphidesmium Parkeri, *Schott.* Polypodium Parkeri, *Hook. et Grev. Ic. Fil. t.* 232.—West Indies and Tropical America.

---

84. PHEGOPTERIS, *Fée; J. Sm.*

Vernation uniserial and sarmentose, or fasciculate and erect, or decumbent. *Fronds* varying from pinnate to decompound-

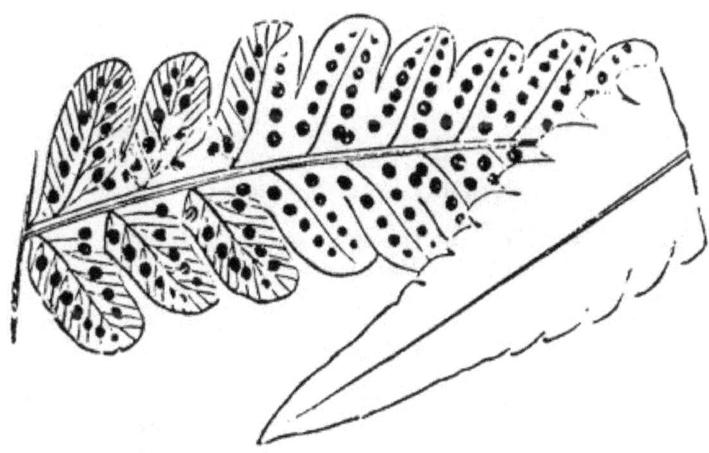

Genus 84.—Portion of fertile frond, under side. No. 7.

multifid, and from 1–6 feet high. *Veins* forked or pinnate; venules free. *Receptacles* punctiform, terminal, medial, or basal. *Sori* round, rarely oblong, naked.

§ 1. *Vernation fasciculate, erect, becoming cæspitose.* (Anopodium.)

1. **P. alpestris**, *J. Sm.* Polypodium alpestre, *Hoppe; Lindl. and Moore's Brit. Ferns, t.* 7, *A, B, C; Sowerby's Ferns,* 2, *t.* 49; *Lowe's Ferns,* 1, *t.* 39. Aspidium alpestre, *Hoppe; Sw.; Schk. Fil. t.* 60, *exclus. syn. Linn.*—Europe, Britain.

2. **P. flexilis**, *J. Sm.* Polypodium flexile, *Moore.* Polypodium alpestre, *var.* flexile, *Lindl. and Moore's Brit. Ferns, t.* 7, *D, E; Hook. Brit. Ferns, t.* 6.—Scotland.

§ 2. *Vernation fasciculate, erect; caudex becoming elevated, solitary.* (Desmopodium.)

3. **P. hastæfolia**, *J. Sm.* Polypodium hastæfolium, *Sw.; Hook. et Grev. Ic. Fil. t.* 203; *Lowe's Ferns,* 2, *t.* 55.—Jamaica.

4. **P. Walkeræ**, *Hook. Sp. Fil.* 4, *p.* 233.—Ceylon.

5. **P. Sieberianum**, *Fée.* Polypodium Sieberianum, *Kaulf.; Hook. Sp. Fil.* 4, *p.* 235.—Mauritius.

6. **P. decussata**, *J. Sm.* Polypodium decussatum, *Linn.* (*Plum. Fil. t.* 24); *Lowe's Ferns,* 2, *t.* 54. Polypodium grammicum, *Spr.*—West Indies.

7. **P. macroptera**, *Fée.* Polypodium macropterum, *Kaulf.* Polypodium formosum, *Lowe's Ferns,* 2, *t.* 53. Alsophila Fischeriana, *Regel, Hort. Petrop.*—Brazil.

8. **P. ampla**, *Fée.* Polypodium amplum, *Humb.; Lowe's Ferns,* 9, *t.* 52.—Martinique.

9. **P. spectabilis**, *Fée.* Polypodium spectabile, *Kaulf.; Lowe's Ferns,* 2, *t.* 43.—Tropical America.

10. **P. lachnopoda**, *J. Sm.* Polypodium lachnopodium, *J. Sm. En. Fil. Hort. Kew* (1846); *Lowe's Ferns,* 1, *t.* 33.—Jamaica.

11. **P. drepana,** *J. Sm.* Aspidium drepanum, *Sw.; Schk. Fil. t.* 43 *C.* Polystichum drepanum, *Presl.* Polypodium drepanum, *Lowe's Ferns*, 2, *t.* 34.—Madeira.

12. **P. rufescens,** *Metten.* Polypodium rufescens, *Blume, Fil. Jav. t.* 91; *Thwait. Enum. Pl. Zeyl.* 394.—Java, Ceylon.

§ 3. *Vernation fasciculate, decumbent.* (Catapodium, *J. Sm.*)

13. **P. divergens,** *Fée.* Polypodium divergens, *Willd.; Schk. Fil. t.* 26 *B; Lowe's Ferns*, 2, *t.* 23. Polypodium multifidum, *Jacq. Ic. Rar. t.* 643.—West Indies.

14. **P. effusa,** *Fée.* Polypodium effusum, *Sw.; Sloane, Hist. Jam. t.* 57, *f.* 3; *Schk. Fil. t.* 26 *C.*—West Indies.

15. **P. trichodes,** *J. Sm.* Polypodium trichodes, *Reinw.; J. Sm. En. Fil. Phil.* Polypodium tenericaule, *Wall. Cat.; Hook. Sp. Fil. t.* 269. Aspidium uliginosum, *Kunze.*—Malayan Archipelago.

16. **P. unidentata,** *J. Sm.* Polypodium unidentatum, *Hook. Sp. Fil.* 4, *p.* 247.—Sandwich Islands.

§ 4. *Vernation uniserial, distant.* (Phegopteris vera.)

17. **P. aurita,** *J. Sm.* Gymnogramma aurita, *Hook. Ic. Pl t.* 974 *and* 989. Grammitis aurita, *Moore.* Leptogramma aurita, *Hort.*—East Indies.

18. **P. hexagonoptera,** *Fée.* Polypodium hexagonopterum, *Michx.; Lowe's Ferns*, 1, *t.* 49.—North America.

19. **P. Robertiana,** *J. Sm.* Polypodium Robertianum, *Hoff.; Lindl. and Moore's Brit. Ferns, t.* 6; *Hook. Brit. Ferns, t.* 5. Phegopteris calcarea, *Fée; J. Sm. Cat. Cult. Ferns* (1857). Polypodium calcareum, *Sm. Eng. Bot. t.* (1525).—Temperate Zone of the Northern Hemisphere, Britain.

20. **P. Dryopteris,** *Fée.* Polypodium Dryopteris, *Linn.; Schk. Fil. t.* 25; *Eng. Bot. t.* 616; *Lindl. and Moore's Brit. Ferns, t.* 6; *Hook. Brit. Ferns, t.* 4.—Temperate Zone of the Northern Hemisphere, Britain.

21. **P. vulgaris,** *Metten.* Phegopteris polypodioides, *Fée.* Polypodium Phegopteris, *Linn.; Eng. Bot. t.* 2224; *Schk. Fil. t.* 20; *Lindl. and Moore's Brit. Ferns, t.* 4; *Hook. Brit. Ferns, t.* 3.—Temperate Zone of the Northern Hemisphere, Britain.

22. **P. rugulosa,** *Fée.* Polypodium rugulosum, *Labill. Nov. Holl. t.* 241.—Tasmania and New Zealand.

---

85. HYPOLEPIS, *Bernh.*

Vernation uniserial, sarmentose. *Fronds* bi-tripinnate, 1–6 feet high, smooth, pilo-glandulose or aculeate. *Veins* forked or pinnate; venules free, the lower exterior branch sporangiferous on its apex. *Receptacles* punctiform. *Sori* round, marginal, each seated in the axis of a reflexed indusiform crenule.

Genus 85.—Fertile pinna, under side. No. 3.

1. **H. repens,** *Presl; Hook. Sp. Fil.* 2, *t.* 90 *B*; *Hook. et Bauer, Gen. Fil. t.* 67 *B.* Lonchites repens, *Linn.*; (*Plum. Fil. t.* 12).—West Indies.

2. **H. tenuifolia,** *Bernh.; Hook. Sp. Fil.* 2, *t.* 89 *C.* Lonchites tenuifolia, *Forst.* Cheilanthes arborescens, *Sw.*—Malayan Archipelago and Polynesia.

3. **H. amaurorachis,** *Hook. Sp. Fil.; Metten. Fil. Hort. Lips. t.* 16, *f.* 1; *Lowe's New Ferns, t.* 2. Cheilanthes amaurorachis, *Kunze.*—Australia..

4. **H. distans,** *Hook. Sp. Fil.* 2, *t.* 95 *C.*—New Zealand.

Tribe VII.—PTERIDEÆ.

Sori marginal, round, oblong or linear, interrupted or continuous. *Indusium* lateral, exteriorly attached on the margin of the frond, special to each sorus, or sometimes universal to two or more sori.

§ 1. *Cheilantheæ. Sori marginal, round or oblong, distinct or laterally contiguous and confluent forming a compound linear sorus. Indusium special to each receptacle, or more or less linearly continued and common to two or more receptacles.*

86. NOTHOLÆNA, *R. Br.; J. Sm.*

Vernation fasciculate, generally erect, cæspitose. *Fronds* pinnate or bi-pinnate, 6–18 inches high, pilo-tomentose, squamose or farinose. *Veins* forked, free. *Receptacles* terminal.

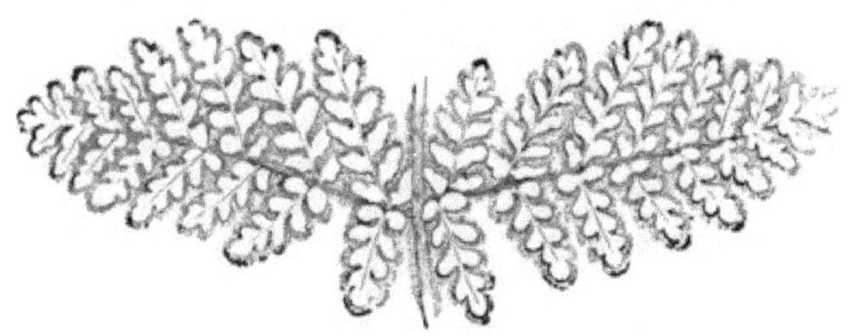

Genus 86.—Mature frond, upper side. No. 10.

Sporangia few to each receptacle, contiguous, forming a marginal row in the axis of the reflexed sub-indusiform margin.

1. **N. trichomanoides,** *R. Br.; Lowe's Ferns*, 1, *t.* 14 *B.* Pteris trichomanoides, *Linn.; (Plum. Fil. t.* 75); *Schk. Fil. t.* 99.—Jamaica.
2. **N. rufa,** *Presl.* Notholæna ferruginea, *Hook. 2nd Cent. of Ferns, t.* 52.—Mexico, Peru.
3. **N. brachypus,** *J. Sm.* Cheilanthes brachypus, *Kunze.* Notholæna squamata, *Hort.* N. squamosa, *Lowe's Ferns*, 1, *t.* 17 *B.*—Mexico.
4. **N. distans,** *R. Br.; Labill. Nov. Cald. t.* 7; *Hook. Ic. Pl. t.* 980; *Lowe's Ferns*, 1, *t.* 19.—Australia, New Zealand, and New Caledonia.
5. **N. mollis,** *Kunze, Fil. t.* 53, *f.* 2.—Chili.

6. **N. Marantæ,** *R. Br.* Acrostichum Marantæ, *Linn*; *Schk. Fil. t.* 4; *Sibth. Fl. Gr. t.* 964.—South of Europe and North Asia.

7. **N. Canariense,** *J. Sm.* Acrostichum Canariense, *Willd.*—Teneriffe and Cape de Verd Islands.

8. **N. sinuata,** *Kaulf.*; *Kunze, Fil. t.* 45; *Bot. Mag. t.* 4699. Acrostichum sinuatum, *Sw.*—Mexico.

9. **N. lævis,** *Mart. et Gal.* N. crassifolia, *Moore et Houlst.*; *Lowe's Ferns*, 1, *t.* 14 *A.*—Mexico.

10. **N. Eckloniana,** *Kunze*; *Lowe's Ferns*, 1, *t.* 17 *A.*—South Africa.

11. **N. lanuginosa,** *Desv.* Acrostichum lanuginosum, *Desf.*; *Fl. Atlan.* 2, *t.* 256; *Schk. Fil. t.* 1. Acrostichum velleum, *Ait.*; *Sibth. Fl. Gr. t.* 656.—South Europe and Madeira.

12. **N. sulphurea,** *J. Sm. in Seemann's Bot. Voy. Herald*, *p.* 233. Pteris sulphurea, *Cav.* Cheilanthes Borsigiana, *Richenb. fil. in Hort. Berol.*; *Lowe's New Ferns*, *t.* 16 *A.*—Peru.

---

87. MYRIOPTERIS, *Fée.*

Vernation uniserial, sarmentose, or subfasciculate and cæspi-

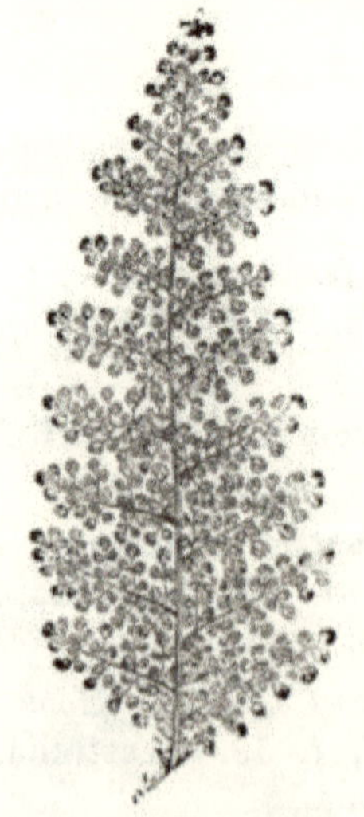

Genus 87.—Mature frond, under side. No. 3.

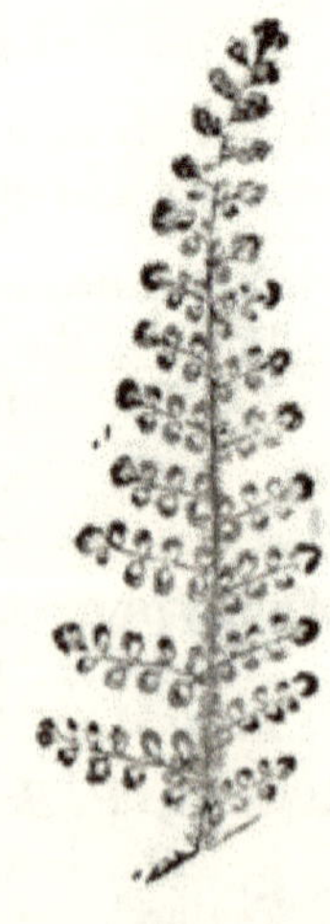

Genus 87.—Pinna of mature frond, under side. No. 1.

tose. *Fronds* 4–18 inches high, slender, decompound, pinnatifid, pilose or piloso-squamiferous; ultimate segments minute, generally orbicular or cuneiform, concave; the margin membranous, revolute, conniving, forming an universal cucullate indusium. *Veins* forked, free. *Receptacles* terminal. *Sporangia* few to each receptacle, confluent under the uniserial indusium.

1. **M. lendigera,** *Fée.* Cheilanthes lendigera, *Sw.; Hook. Sp. Fil, t.* 106 *A.* Notholæna lendigera, *J. Sm. Gen. Fil.* Cheilanthes tenuis, *Hort.; Lowe's Ferns,* 4, *t.* 23.—Tropical America. T.

2. **M. myriophylla,** *J. Sm.* Cheilanthes myriophylla, *Desv.; Hook. Sp. Fil. t.* 105 *A.*—Tropical America. T.

3. **M. elegans,** *J. Sm.* Cheilanthes elegans, *Desv.; Hook. Sp. Fil. t.* 105 *B; Lowe's Ferns,* 4, *t.* 20. Myriopteris Marsupianthus, *Fée, Gen. Fil. t.* 12 *A, f.* 1.—Tropical America. T.

4. **M. tomentosa,** *Fée.* Cheilanthes tomentosa, *Link; Hook. Sp. Fil. t.* 109 *A.*—Mexico and Southern United States.

5. **M. frigida,** *J. Sm.* Cheilanthes frigida, *Linden.* Cheilanthes lendigera, *Lowe's Ferns,* 4, *t.* 24.—Tropical America. T.

6. **M. vestita,** *J. Sm.* Cheilanthes vestita, *Sw.; Schk. Fil. t.* 124; *Hook. Sp. Fil. t.* 108 *B.* Notholæna vestita, *Desv.; J. Sm. Gen. Fil.; Lowe's Ferns,* 1, *t.* 16 *B.*—North America. T.

7. **M. hirta,** *J. Sm.* Cheilanthes hirta, *Sw.; Hook. Sp. Fil.* 2, *t.* 101 *B; Lowe's Ferns,* 4, *t.* 18. Cheilanthes Ellisiana, *Hort.*—South Africa.

---

88. CHEILANTHES, *Sw.* (*in part*).

Vernation fasciculate, erect or decumbent. *Fronds* bi-tripinnate, rarely simple pinnate, 4–18 inches or more in height, smooth, pilose, glandulose, squamose or farinose; ultimate segments often small. *Veins* forked, free. *Receptacles* terminal.

Sori round, marginal, distinct or laterally confluent. *Indusium* reniform or subrotund, and special to each sorus, or more or

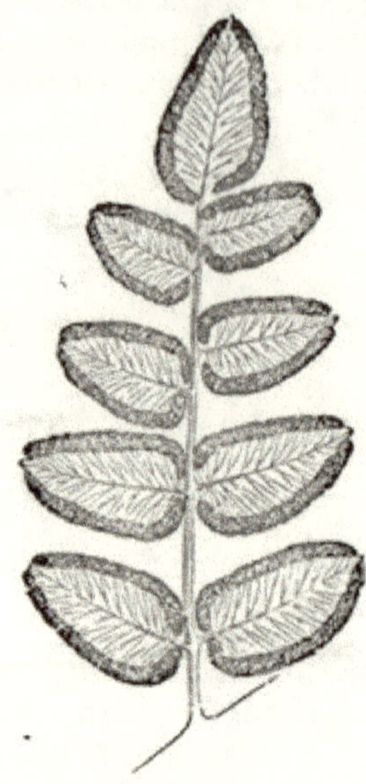

Genus 88.—Pinna of mature frond, under side. No. 13.

Genus 88.—Fertile pinna, under side. No. 14.

less elongated, plain or undulate, and including two or more sori.

§ 1. *Indusium oblong or linear, common to two or more clusters of sporangia. Fronds smooth, pilose, tomentose, squamose, or glandulose.* (Eucheilanthes, *Hook.*)

1. **C. micropteris,** *Sw. Syn. Fil. t.* 3, *f.* 5.—Quito and Brazil.

2. **C. viscosa,** *Link; Hook. Sp. Fil. t.* 93 *B; Lowe's Ferns,* 4, *t.* 25.—Tropical America. **T.**

3. **C. fragrans,** *Sw.* (*in part*); *Hook.; Lowe's Ferns*, 4, *t.* 17 *B.* Polypodium fragrans, *Linn.* Cheilanthes odora, *Sw.; Schk. Fl. t.* 123. C. suaveolens, *Sw.; Schk. Fil. t.* 19; *Sibth. Fil. Gr. t.* 966; *Hook. et Bauer. Gen. Fil. t.* 106 *B.* C. Maderensis, *R. T. Lowe.* — South Europe and Madeira.

4. **C. tenuifolia**, *Sw.; Schk. Fil. t.* 125; *Hook. Sp. Fil. t.* 87 *C.*—East Indies, Malayan Archipelago, Australia, and New Zealand.

5. **C. microphylla**, *Sw.*; (*Plum. Fil. t.* 58); *Hook. Sp. Fil.* 2, *t.* 98 *A.* C. micromera, *Link; Lowe's Ferns*, 4, *t.* 16.—Tropical America. **T.**

6. **C. Sieberi**, *Kunze; Hook. Sp. Fil.* 2, *t.* 97 *B.* Cheilanthes Preissiana, *Kunze Lowe's Ferns*, 4, *t.* 29.—Australia and New Zealand.

7. **C. Alabamensis**, *Kunze; Hook. Sp. Fil. t.* 103 *B; Hook. Fil. Exot. t.* 90.—South United States. **T.**

8. **C. spectabilis**, *Kaulf.; Lowe's Ferns*, 4, *t.* 15. Hypolepis spectabilis, *Link; Hook. Sp. Fil. t.* 88 *B.* Adiantopsis spectabilis, *Fée.* Cheilanthus Brasiliensis, *Radd. Fil. Bras. t.* 75, *f.* 2.—Brazil. **T.**

9. **C. multifida**, *Sw.; Hook. Sp. Fil.* 2, *t.* 100 *B; Hook. Gard. Ferns, t.* 39.—South Africa, St. Helena, Java.

§ 2. *Indusium linear, continuous, sinuose-undulate. Fronds farinose.* (Aleuritopteris, *Fée.*)

10. **C. argentea**, *Kunze; Lang. et Fisch. Ic. Fil. t.* 22. Pteris argentea, *Gmel.; Sw.*—Siberia. **T.**

11. **C. farinosa**, *Kaulf.; Hook. et Grev. Ic. Fil. t.* 134; *Hook. Bot. Mag. t.* 4765. Cassebeera farinosa, *J. Sm. olim.* Pteris farinosa, *Forsk.* Cheilanthes dealbata, *Don.* Allosorus dealbatus, *Presl.* Pteris Argyrophylla, *Sw.*—India and Arabia.

12. **C. pulveracea**, *Presl.; Lowe's Ferns*, 4, *t.* 28. Aleuritopteris Mexicanum, *Fée.*—Mexico. **T.**

§ 3. *Indusium subrotund, special to each cluster of sporangia. Fronds smooth.* (Adiantopsis, *Fée.*)

13. **C. pteroides**, *Sw.; Hook. Sp. Fil. t.* 101 *A; Lowe's Ferns*, 4, *t.* 21. Adiantopsis pteroides, *Moore.*—South Africa.

14. **C. Capensis**, *Sw.; Lowe's New Ferns, t.* 26 *A.* Hypolepis Capensis, *Hook. Sp. Fil. t.* 77. Adiantopsis Capensis, *Moore.*—South Africa.

15. C. **radiata,** *J Sm; Lowe's Ferns,* 4, *t.* 18. Adiantum radiatum, *Linn.; Sw.;* (*Plum. Fil. t.* 100). Hypolepis radiata, *Hook. Sp. Fil. t.* 91 *A.* Adiantopsis radiata, *Fée.*—Tropical America.

16. C. **pedata,** *A. Br.* Hypolepis pedata, *Hook. Sp. Fil.* 2, *t.* 92 *A.* Adiantopsis pedata, *Moore.*—Jamaica.

---

89. CRYPTOGRAMME, *R. Br.*

Vernation fasciculate, erect, cæspitose. *Fronds* bi-tripinnate, 6–10 inches high, the fertile contracted; segments oblong, linear; margins membranous, revolute, oppositely conniving,

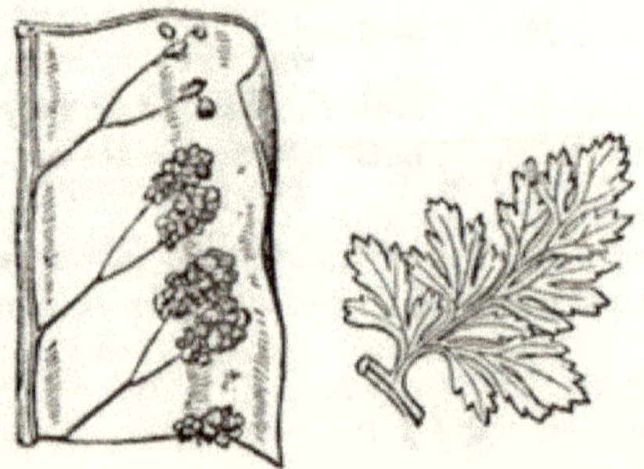

Genus 89.—Portion of barren pinna, natural size, ditto fertile enlarged and unfolded. No. 1.

forming an universal indusium. *Veins* forked, free. *Receptacles* terminal, subelongated. *Sori* defined, oblong, becoming laterally confluent, forming a compound, broad, intramarginal, linear sorus, included under the universal indusium.

1. C. **crispa,** *R. Br.; Hook. et Bauer. Gen. Fil. t.* 115 *B; Hook. Brit. Ferns,* 3, *t.* 34. Pteris crispa, *Linn.; Sw.; Schk. Fil. t.* 98; *Eng. Bot. t.* 1160. Allosorus crispus, *Bernh.; Presl; J. Sm. Gen. Fil.; Lindl. and Moore's Brit. Ferns, t.* 8; *Lowe's Ferns,* 3, *t.* 34. Phorolobus crispus, *Desv.*—Temperate Regions of the Northern Hemisphere.

---

90. CINCINALIS, *Desv. (in part)*; *Fée.*

Vernation fasciculate, erect. *Fronds* tripinnate, 4-12 inches high, slender; pinnæ distant, spreading; pinnules ovate, hastate-lobed or trifoliate, plane, smooth, glaucous or farinose. *Veins*

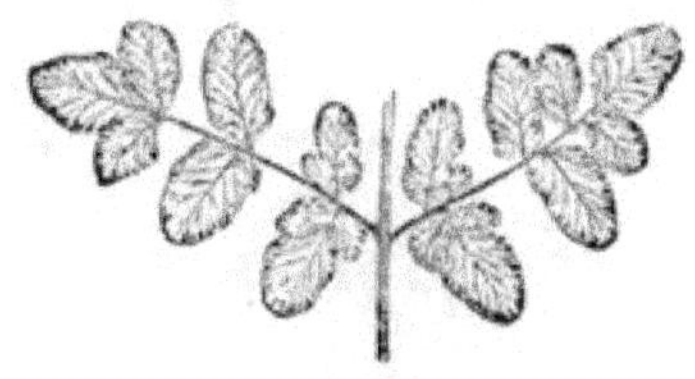

Genus 90.—Portion of mature frond, under side. No. 1.

forked, free. *Receptacles* terminal. *Sporangia* large, globose, sessile, definite, in a short series on the upper portion of the venules, becoming laterally confluent, forming a broad sub-intramarginal sorus. *Indusium* obsolete or very narrow.

* *Fronds smooth, glaucous.*

1. C. **tenera,** *Fée.* Notholæna tenera, *Gill.; Hook. Bot. Mag. t.* 3055; *Kunze, Fil. t.* 22, *f.* 2; *Hook. et Bauer. Gen. Fil. t.* 76 *A*; *Lowe's Ferns*, 1, *t.* 15.—Chili.

** *Fronds farinose.*

2. C. **nivea,** *Desv.* Pteris nivea, *Lam.; Sw.* Notholæna nivea, *Desv.*; *Kunze, Fil. t.* 22, *f.* 1; *Lowe's Ferns*, 1, *t.* 19 *C.* Acrostichum albidulum, *Sw. Syn. Fil. t.* 1, *f.* 2. Notholæna incana, *Presl, Rel. Hænk. t.* 1, *f.* 2.—Tropical America.

3. C. **pulchella,** *J. Sm.* Allosorus pulchellus, *Mart. et Gal. Fil. Mex. t.* 10, *f.* 1. Pellæa pulchella, *Fée; Hook.*—Mexico. **Tr.**

4. C. **Hookeri,** *J. Sm.* Notholæna Hookeri, *Lowe's Ferns*, 1, *obs. sub t.* 19 *C et t.* 13.—Tropical America.

5. C. **flavens,** *Desv.; Fée, Gen. Fil.* 5, *t.* 30; *Lowe's New Ferns, t.* 8. Acrostichum flavens, *Sw.* Gymnogramme flavens, *Kaulf.; Hook. Fil. Exot. t.* 47. Notholæna chrysophylla, *Hort.*—Tropical America.

91. PELLÆA, *Link.*

Vernation fasciculate, erect or decumbent, squamose. *Fronds* palmate, pinnate, or bi-tripinnate, 5 inches to 6 feet high; pinnules articulated to the rachis or to a short petiole. *Veins* forked; venules free. *Receptacles* terminal, generally con-

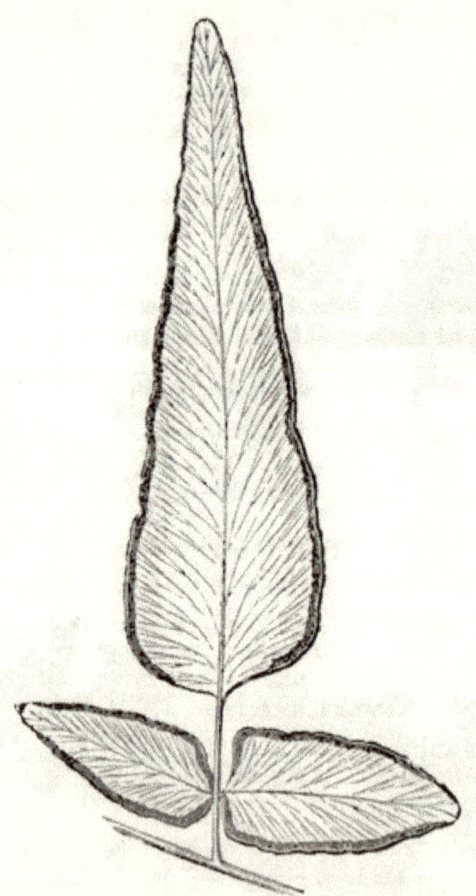

Genus 91.—Fertile pinna, under side. No. 6.

tiguous, forming a more or less broad, compound, continuous marginal sorus, or rarely distant, and each receptacle forming definite round sori. *Indusium* linearly continuous or subrotund and special to each receptacle.

1. **P. geraniifolia**, *Fée.* Pteris geraniifolia, *Radd. Fil. Bras. t.* 67; *Hook. Ic. Pl. t.* 915. Pteris concolor, *Lang. et Fisch. Ic. Fil. t.* 21. Platyloma geraniifolia, *Lowe's Ferns*, 3, *t.* 27.—East Indies, Polynesia, Tropical America.

2. **P. cuneata**, *J. Sm.* Cheilanthes cuneata, *Link; Kunze, Fil. t.* 36.—Cheilanthes rufescens, *Link.*—Mexico.

3. **P. profusa**, *J. Sm.* Cheilanthes profusa, *Kunze, Fil. t.* 17. Notholæna profusa, *Presl.*—South Africa.

4. **P. intramarginalis**, *J. Sm.* Pteris intramarginalis, *Kaulf.*; *Kunze, Anal. Pterid. t.* 17, *f.* 1. Pteris fallax, *Mart. et Gal. Fil. Mex. t.* 14, *f.* 2. Cheilanthes intramarginalis, *Hook.* Platyloma intramarginalis, *Lowe's Ferns*, 3, *t.* 31.—Mexico.

5. **P. glauca**, *J. Sm.* Pteris glauca, *Cav.* Cheilanthes glauca, *Metten. Cheil. t.* 31, *f.* 18, 19.—Mexico.

6. **P. hastata**, *Link; Fée; Hook. Sp. Fil. t.* 116 *B*; *Hook. Fil. Exot. t.* 50. Platyloma hastatum et adiantoides, *Lowe's Ferns*, 3, *t.* 32, 33. Pteris hastata, *Sw.* Allosorus hastatus, *Presl; Hook. Gen. Fil. t.* 5.—South Africa.

7. **P. consobrina**, *Hook. Sp. Fil.* 2, *t.* 117 *A.* Pteris consobrina, *Kunze.*—South Africa.

8. **P. atropurpurea**, *Link; Fée.* Pteris atropurpurea, *Linn.*; *Schk. Fil. t.* 99.—North America. **T.**

9. **P. Calomelanos**, *Link; Fée.* Pteris Calomelanos, *Sw.*; *Schlecht. Adumb. t.* 24. Allosorus Calomelanos, *Presl; Hook. in Bot. Mag. t.* 4769. Platyloma Calomelanos, *J. Sm.*; *Lowe's Ferns*, 3, *t.* 26.—South Africa. **Tr.**

10. **P. ternifolia**, *Link; Fée; Hook. Fil. Exot. t.* 15. Pteris ternifolia, *Cav.*; *Hook. et Grev. Ic. Fil. t.* 126. Platyloma ternifolium, *J. Sm.*; *Lowe's Ferns*, 3, *t.* 24 *B.*—Tropical America. **T.**

11. **P. Wrightiana**, *Hook. Sp. Fil.* 2, *t.* 115 *B.*—New Mexico.

12. **P. sagittata**, *Link.* Pteris sagittata, *Cav.* Allosorus sagittatus, *Presl*: *Kunze, Fil. t.* 24.—Peru. **T.**

13. **P. flexuosa,** *Link ; Fée.* Pteris flexuosa, *Kaulf. ; Hook. Ic. Pl. t.* 119. Allosorus flexuosus, *Kunze, Fil. t.* 23; *Hook. in Bot. Mag. t.* 4762. Platyloma flexuosum, *J. Sm.; Lowe's Ferns,* 3, *t.* 25.—Tropical America. T.

14. **P. cordata,** *J. Sm.* (*non Fée*). Pteris cordata, *Cav.* Allosorus cordatus, *Presl ; Hook. in Bot. Mag. t.* 4698.—Tropical America. T.

---

92. PLATYLOMA, *J. Sm.* (*in part*)

Vernation uniserial; sarmentum subhypogæous. *Fronds* pinnate, 1–2 feet high; pinnæ entire, opaque. *Veins* forked; venules free. *Receptacles* terminal, oblong, contiguous, forming a broad, compound, continuous, marginal sorus. *Indusium* linear, continuous, narrow, subobsolete.

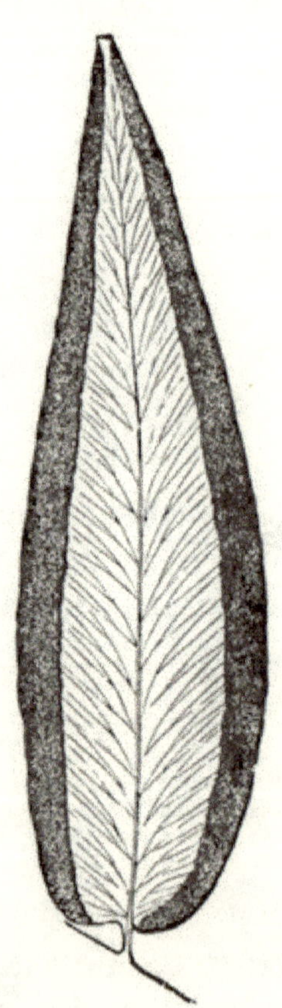

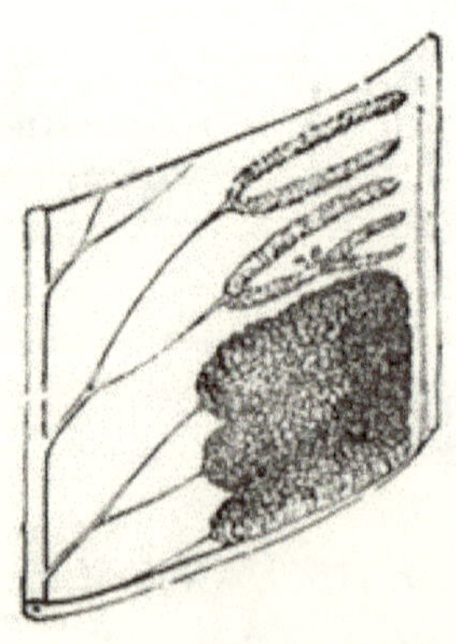

Genus 92.—Fertile pinnæ, under side, No. 1; ditto enlarged, No. 3.

1. **P. Brownii**, *J. Sm.; Lowe's Ferns*, 3, *t.* 29. Adiantum paradoxum, *R. Br.* Pellæa paradoxa, *Hook. Fil. Exot. t.* 21; *Sp. Fil.* 2, *t.* 3 *A.* Pellæa cordata, *Fée* (*non J. Sm.*).—Australia.

2. **P. falcatum**, *J. Sm.; Lowe's Ferns*, 3, *t.* 30 *A B*; *Hook. Gen. Fil. t.* 115 *A* (*excl. name* P. Brownii). Pteris falcata, *R. Br.* Pteris seticaulis, *Hook. Ic. Pl. t.* 207. Pellæa falcata, *Fée; Hook. Sp. Fil.* 2, *t.* 111 *B.*—East Indies, Australia.

3. **P. rotundifolium**, *J. Sm.; Lowe's Ferns*, 3, *t.* 24 *A.* Pteris rotundifolia, *Forst.; Schk. Fil. t.* 99; *Hook. Ic. Pl. t.* 422. Pellæa rotundifolia, *Hook. Fil. Exot. t.* 48.—New Zealand.

---

93. ADIANTUM, *Linn.*

Vernation distant, sarmentose or subfasciculate, decumbent and cæspitose. *Fronds* simple-reniform, pinnate or bi-tripinnate; pinnæ and pinnules articulated with the petiole; costa excentric or obsolete. *Veins* unilateral or radiating, forked;

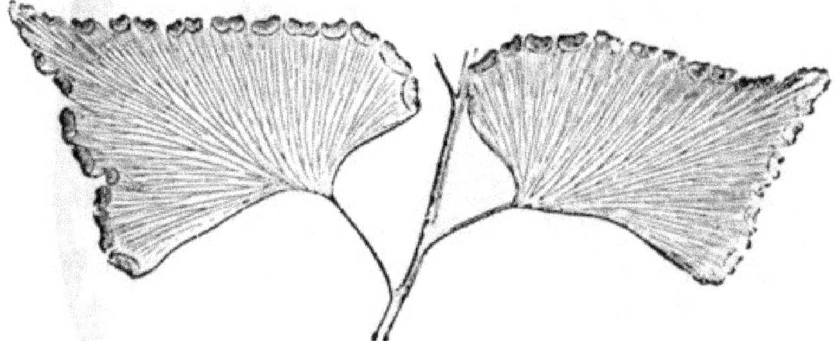

Genus 93.—Portion of mature frond, under side. No. 10.

venules free, terminating in the axis of a reflexed orbicular or elongated portion of the margin (indusium), which is altered in texture, venose and sporangiferous on its under side, ultimately becoming replicate.

1. *Reniforme group.*

Fronds simple, reniform.

1. A. **reniforme**, *Linn.; Schk. Fil. t.* 115; *Lodd. Bot. Cab. t.* 841; *Hook. Exot. Fl. t.* 104; *Sp. Fil.* 2, *t.* 71 *A; Hook. Fil. Exot. t.* 8; *Lowe's Ferns,* 3, *t.* 2 *B.*—Madeira, Teneriffe.

2. *Caudatum group.*

Fronds pinnate, pinnæ narrow.

2. A. **lunulatum**, *Burm.; Hook. et Grev. Ic. Fil. t.* 104; *Lowe's Ferns,* 3, *t.* 8 *B.* Adiantum arcuatum, *Sw.*—East Indies.

3. A. **dolabriforme**, *Hook. Ic. Pl. t.* 191.—East Indies, Malay and Pacific Islands, Brazil.

4. A. **caudatum**, *Linn.; Schk. Fil. t.* 117; *Hook. Exot. Fl. t.* 104; *Lowe's Ferns,* 3, *t.* 2 *A.*—East Indies.

3. *Macrophyllum group.*

Fronds pinnate, erect; pinnæ broad.

5. A. **lucidum**, *Sw.; Hook. Sp. Fil.* 2, 79 *C* (*excl. f.* 4); *Lowe's Ferns,* 3, *t.* 4 *A.*—Tropical America.

6. A. **obliquum**, *Willd.; Hook. Sp. Fil.* 2, *t.* 79 *A.*—Tropical America.

7. A. **Kaulfussii**, *Kunze.* Adiantum obliquum, *Kaulf.; Hook. et Grev. Ic. Fil. t.* 190 (*non Willd.*).—Tropical America.

8. A. **Wilsoni**, *Hook. Sp. Fil.* 2, *t.* 72 *A; Lowe's Ferns,* 3, *t.* 16; *Hook. Fil. Exot. t.* 14.—Jamaica.

9. A. **macrophyllum**, *Sw.; Hook. et Grev. Ic. Fil. t.* 132; *Hook. Fil. Exot, t.* 55.—Tropical America.

4. *Monosoratum group.*

Fronds uniformly bipinnate.

10. A. **villosum**, *Linn.; Schk. Fil. t.* 120. A. varium, *Presl; Lowe's Ferns,* 3, *t.* 18.—Tropical America.

11. A. **pulverulentum,** *Linn.; (Plum. Fil. t.* 55); *Schk. Fil. t.* 119; *Lowe's Ferns,* 3, *t.* 17. A. monosoratum, *Willd.*

Var.—Fronds small, pink when young.—A. rigidum, *Hort. Berol.*

Var.—Fronds small, green when young.—A. tetraphyllum, *Hort. Berol.*—Tropical America.

5. *Prionophyllum group.*

12. A. **intermedium,** *Sw.; Lowe's Ferns,* 3, *t.* 20. A. Brasiliense, *Link* (*non Radd.*). A. triangulatum, *Kaulf.; Klotz.*—Tropical America.

13. A. **prionophyllum,** *H. B K.* A. tetraphyllum, *Willd.*—Tropical America.

14. A. **fovearum,** *Radd. Fil. Bras. t.* 77.—Brazil.

6. *Trapeziforme group.*

Fronds large, tripinnate; pinnæ distant, alternate.

15. A. **subcordatum,** *Sw.;* A. betulinum, *Kaulf.* A. truncatum, *Radd. Fil. Bras. t.* 78, *f.* 1.—Brazil.

16. A. **trapeziforme,** *Linn.* A. rhomboideum, *Schk. Fil. t.* 122.—West Indies and Tropical America.

17. A. **pentadactylon,** *Lang. et Fisch. Ic. Fil. t.* 25; *Hook. et Grev. Ic. Fil. t.* 98.—Brazil.

18. A. **Mathewsianum,** *Hook. Sp. Fil.* 2, *t.* 84 *A.*—Peru.

19. A. **cultratum,** *J. Sm.; Lowe's Ferns,* 3, *t.* 21; *Moore, in Gard. Chron.* (1855), *p.* 660.—Tropical America.

20. A. **curvatum,** *Kaulf.; Hook. Sp. Fil.* 2, *t.* 84 *C. Lowe's Ferns,* 3, *t.* 6.—Brazil.

21. A. **polyphyllum,** *Willd.; Hook. Gard. Ferns, t.* 12. A. cardiochlæna, *Kunze; Hook. Sp. Fil.* 2, *t.* 83 *A.*—Tropical America.

22. A. **cristatum,** *Linn.; Jacq. Ic. Rar. t.* 646; *Lowe's Ferns,* 3, *t.* 22.—West Indies.

23. **A. Feei,** *Moore.* A. flexuosum, *Hook. 2nd Cent. Ferns, t.* 61.—Mexico.

7. *Formosum group.*

Fronds quadripinnate, decompound, with distant alternate pinnæ.

24. **A. Brasiliense,** *Radd. Fil. Bras. t.* 76.—Brazil.

25. **A. Wilesianum,** *Hook. Sp. Fil.* 2, *t.* 83 *C*; *Lowe's New Ferns, t.* 29.—Jamaica.

26. **A. affine,** *Willd.* A. trapeziforme, *Forst.* (*non Linn.*); *Schk. Fil. t.* 121 *B.*—New Zealand.

27. **A. Cunninghami,** *Hook. Sp. Fil.* 2, *t.* 86 *A*; *Lowe's Ferns,* 3, *t.* 12.—New Zealand.

28. **A. formosum,** *R. Br.*; *Hook. Sp. Fil.* 2, *t.* 88 *B*; *Lowe's Ferns,* 3, *t.* 11.—New Holland and New Zealand.

8. *Flabellatum group.*

Fronds dichotomously pedate-flabellate.

29. **A. flabellulatum,** *Linn.* (*Pluk. t.* 4, *f.* 3). A. fuscum, *Retz. Obs.* 2, *t.* 5. A. amœnum, *Wall.*; *Hook. et Grev. Ic. Fil. t.* 103.—East Indies.

30. **A. patens,** *Willd.*; *Hook. Sp. Fil.* 2, *t.* 87 *A.*—Tropical America.

31. **A. pedatum,** *Linn.*; *Schk. Fil. t.* 115; *Lowe's Ferns,* 3, *t.* 14.—North America, North Africa.

32. **A. hispidulum,** *Sw.* A. pubescens, *Schk. Fil. t.* 116; *Lowe's Ferns,* 3, *t.* 9.—*Var.* tenellum, *Moore, Ind.* A. hispidulum, *Hort.*; *Lowe's Ferns,* 3, *t.* 13 *A.*—East Indies, Malayan Archipelago, Australia, New Zealand, and Pacific Islands.

33. **A. setulosum,** *J. Sm. En. Fil. Hort. Kew, in Bot. Mag.* (1846). A. affine, *Hook. Sp. Fil.* (*excl. syn. Willd.*; *Forst.*; *Schk.*; *A. Cunn.*) — Norfolk Island, and Fijis.

34. **A. fulvum,** *Raoul*; *Hook. Sp. Fil. t.* 85 *A*; *Lowe's Ferns,* 3, *t.* 19.—New Zealand, and Fijis.

9. *Tenerum group.*

Fronds dichotomously decompound.

35. **A. tenerum**, *Sw.; (Plum. Fil. t.* 95); *Moore et Houlst. Gard. Mag. Bot.* 3, *f.* 22.—West Indies and Tropical America.

36. **A. sulphureum**, *Kaulf.; Kunze, Anal. t.* 22, *f.* 1; *Hook. Sp. Fil.* 2, *t.* 76 *A.*—Chili.

37. **A. Chilense**, *Kaulf.; Hook. et Grev. Ic. Fil. t.* 173; *Hook. Sp. Fil.* 2, *t.* 75 *B.*—Chili.

38. **A. emarginatum**, *Bory; Hook. Sp. Fil.* 2, *t.* 75 *A.*—South Africa, Mauritius.

10. *Capillus Veneris group.*

39. **A. concinnum**, *H. B. K. Nov. Gen. Fil. t.* 121 (*non Sw.*).—Tropical America.

40. **A. capillus-veneris**, *Linn.; Sm. Eng. Bot. t.* (1564); *Hook. Gen. Fil. t.* 66 *B*; *Moore's Nat. Print. Ferns, t.* 45; *Sowerby's Ferns, t.* 40; *Hook. Sp. Fil.* 2, *t.* 74 *B*; *Hook. Brit. Ferns, t.* 41. A. Moritzianum, *Klotz.*—Tropical and Temperate Zones of both Hemispheres.

41. **A. Æthiopicum**, *Linn.; Hook. Sp. Fil.* 2, *t.* 77 *A.* A. assimile, *Lowe's Ferns,* 3, *t.* 8 (*non Sw.*).—Tropics.

42. **A. cuneatum**, *Lang. et Fisch. Ic. Fil. t.* 26; *Radd. Fil Bras. t.* 78, *f.* 2; *Hook. et Grev. Ic. Fil. t.* 30.—Brazil.

43. **A. assimile**, *Sw. Syn. Fil. t.* 3, *f.* 4. A. trigonum, *Labil. Nov. Holl. t.* 248, *f.* 2.—Australia, Tasmania.

---

§ 2. *Pterideæ veræ. Sori marginal, transversely elongated. Indusium linear, plane.*

94. **OCHROPTERIS**, *J. Sm.; Hook. Gen. Fil. t.* 106 *A.*

Vernation decumbent. *Fronds* deltoid, long stipate, decompound, 2–3 feet high, smooth, glossy; stipes and racheæ pale, stramineous; ultimate pinnules and lobes oblong-elliptical, cuneiform, marginate, usually oblique. *Veins* pinnately forked, radiating; venules direct, apices of the sterile clavate, free, the

Genus 94.—Portion of frond, natural size; two ditto, enlarged. No. 1.

fertile 2–4 converging and transversely combined by a thick impressed, sporangiferous marginal receptacle. *Sori* oblong, rarely two on each lobe. *Indusium* formed of the reflexed margin, thick, coriaceous.

1. **O. pallens**, *J. Sm. Gen. Fil.* (1841); *Hook. Sp. Fil.* 2, *t.* 77; *Hook. et Bauer. Gen. Fil. t.* 106 *A.* Adiantum pallens, *Sw.*—Mauritius.

---

95. ONYCHIUM, *Kaulf.*

Vernation fasciculate and decumbent, or distant and sarmentose. *Fronds* decompound multifid, 1–3 feet high, smooth; sterile segments cuneiform. *Veins* forked, free, their apices clavate; fertile segments linear, apiculate, veins simple, short, their apices transversely combined by the receptacle. *Sorus* linear, in the axis of a linear slightly intramarginal indusium,

Genus 95.—Portion of frond, natural size; two ditto, enlarged. No. 1.

the inner free margin of which connives with the inner margin of the opposite indusium, ultimately becoming replicate, with the sporangia of both sori confluent.

1. **O. Japonicum,** *Kunze.* Trichomanes Japonicum, *Thunb.* O. Capense, *Kaulf. En. Fil. t.* 1, *f.* 8. O. lucidum, *Cat. Hort. Kew.* (1856) (*non Spreng.*) (*non Hook.*).—Japan.

2. **O. auratum,** *Kaulf.* Lomaria decomposita, *D. Don.* Pteris chrysocarpa, *Hook. et Grev. Ic. Fil. t.* 107.—East Indies and Malayan Archipelago.

---

96. PTERIS, *Linn.* (*in part*).

Vernation fasciculate and erect or decumbent, or uniserial and sarmentose. *Fronds* pinnate, bi-tri-quadripinnate, rarely simple, from a few inches to six or more feet high; the ultimate pinnæ entire, sinuose-lobed or pinnatifid. *Veins* forked; sterile venules free, the apices of the fertile transversely combined by the receptacle, constituting a linear, continuous or interrupted sorus. *Indusium* linear, marginal.

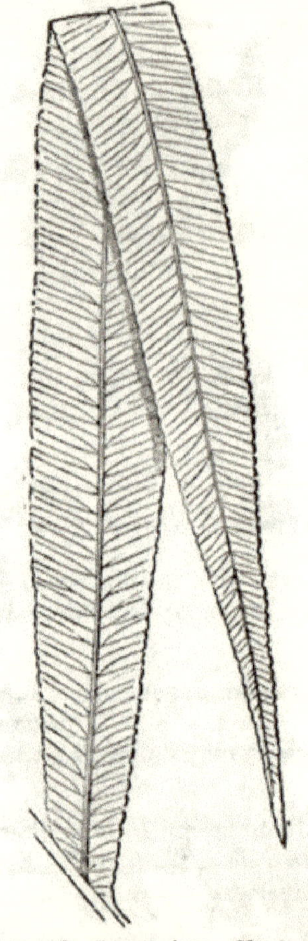

* *Vernation fasciculate, decumbent cæspitose.*

† *Fronds simply pinnate.*

1. P. longifolia, *Linn.*; (*Plum. Fil. t.* 69); *Schk. Fil. t.* 88; *Lowe's Ferns*, 3, *t.* 42. P. vittata, *Linn.* P. ensifolia, *Sw.* P. lanceolata, *Desf.* P. Alpini, *Desv.* P. obliqua, *Försk.* P. costata, *Bory.* P. æqualis, *Presl.* P. acuminatissimum, *Blume.* P. amplectans, *Wall.* P. Bahamensis, *Fée.*—Tropics and Northern Temperate Zone.

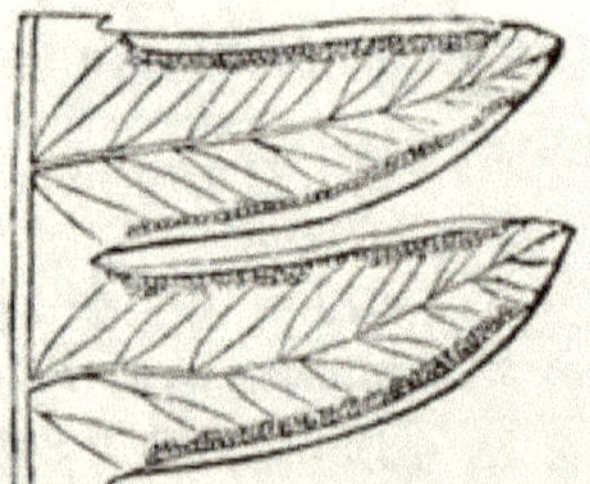

Genus 96.—Barren pinna. No. 4. Portion of fertile frond, natural size

†† *Fronds pinnate, the lower pair or more of pinnæ lobed or pinnate.*

2. P. Cretica, *Linn.*; *Schk. Fil. t.* 90; *Lowe's Ferns*, 3, *t.* 43. P. serraria, *Sw.* P. pentaphylla, *Willd.*

Var.—albo-lineata, *Hook. Bot. Mag. t.* 5194; *Lowe's New Ferns*, *t.* 25.

Tropics and North Temperate Zone.

3. **P. heterodactyla**, *Reinw.; J. Sm. En. Fil. Philipp.* Pteris Cretica, *Blume.*—Malayan Archipelago.

4. **P. umbrosa**, *R. Br.; Hook. Sp. Fil.* 2, *t.* 130 *B; Lowe's Ferns*, 3, *t.* 39.—Australia.

5. **P. serrulata**, *Linn.; Schk. Fil. t.* 91; *Lowe's Ferns*, 3, *t.* 40; β minor, *Moore et Houlst. Var.* cristata, *Moore, Gard. Chron.* (1863).—Tropics.

6. **P. crenata**, *Sw.; Burm. Fl. Zey. t.* 87; *Hook. Sp. Fil.* 2, *t.* 127 *A; Lowe's Ferns*, 3, *t.* 48. P. multidentata, *Wall.* P. Chinensis, *Hort. Ang.*—Tropics.

7. **P. heterophylla**, *Linn.; (Plum. Fil. t.* 37); *Hook. Bot. Mag. t.* 4925.—Jamaica.

8. **P. semipinnata**, *Linn.; Lowe's Ferns*, 4, *t.* 9; *Hook. Gard. Ferns, t.* 59. P. flabellata, *Schk. Fil. t.* 93. P. dimidiata, *Blume.*—East Indies, China, Malayan Archipelago.

9. **P. mutilata**, *Linn.; (Plum. Fil. t.* 51); *Hook. Sp. Fil. t.* 131 *A.* P. concinna, *Hew. in Mag. of Nat. Hist. N. Ser.* (1838).—Jamaica.

** *Vernation fasciculate; caudex erect, simple, rarely cæspitose.*

† *Fronds pinnate; pinnæ pinnatifid, the lower pair more or less bipartite or pinnate.*

10. **P. glauco-virens**, *Linden's Cat.* (1858) (name only).—Tropical America.

11. **P. pungens**, *Willd.; (Plum. Fil. t.* 13).—Tropical America.

12. **P. felosma**, *J. Sm. En. Fil. Hort. Kew.* (1846).—Jamaica.

13. **P. sulcata**, *Link; Lowe's Ferns*, 4, *t.* 5.—Brazil.

14. **P. pyrophylla**, *Blume.*—Java.

15. **P. quadriaurita**, *Retz.; Hook. Sp. Fil.* 2, *t.* 134 *B.*—East Indies.

16. **P. argyrea**, *Moore; Lowe's New Ferns, t.* 10. P. quadriaurita,—*var.* argyrea, *Hook. Bot. Mag. t.* 5183, *in part.*—East Indies.

17. **P. tricolor**, *Linden, Hort. Lind. t.* 12; *Lowe's New Ferns, t.* 9. Pteris quadriaurita,—*var.* tricolor, *Hook Bot. Mag. t.* 5183, *in part.*—East Indies.

18. **P. aspercaulis,** *Wall.; Lowe's Ferns,* 4, *t.* 8. P. pectinata, *Don,—var.* rubro-nervia. P. rubro-nervia, *Linden.*—East Indies.

19. **P. flabellata,** *Thunb.*—South Africa.

20. **P. arguta,** *Ait; Lowe's Ferns,* 3, *t.* 41. P. allosora. *Link.* P. palustris, *Poir.* Mongonia palustris, *Presl.*—Madeira.

21. **P. Kingiana,** *Endl.; Lowe's Ferns,* 3, *t.* 46.—Norfolk Island.

22. **P. tremula,** *R. Br.; Hook. Sp. Fil.* 2, *t.* 120 *B; Lowe's Ferns,* 3, *t.* 45. P. chrysocarpa, *Link.*—Australia and New Zealand.

23. **P. lata,** *Link; Lowe's Ferns,* 4, *t.* 6.—Brazil.

24. **P. paleacea,** *Roxb. in Beatson's Fl. of St. Helena, p.* 349; *Hook. Sp. Fil.* 3, *p.* 186.—St. Helena.

†† *Fronds tripartitely branched, deltoid, generally quadripinnate.*

25. **P. deflexa,** *Link, Enum. Hort. Berol.*—Tropical America.

26. **P. decussata,** *J. Sm. Enum. Fil. Philipp.* (1841). P. patens, *Hook. Sp. Fil.* 2, *p.* 177, *t.* 137.—Ceylon, Luzon.

27. **P. laciniata,** *Willd.; Presl, Pterid. t.* 5, *f.* 23; *Hook. Sp. Fil.* 2, *t.* 132 *B.*—West Indies.

28. **P. Gheisbeghtii,** *J. Sm.* Lonchitis Gheisbeghtii, *Linden, Cat.*—Tropical America.

*** *Vernation uniserial, distant, sarmentose.*

† *Sarmentum slender, epigæous.*

29. **P. scaberula,** *A. Rich; Hook. Sp. Fil. t.* 93 *A; Lowe's Ferns,* 4, *t.* 10.—New Zealand.

†† *Sarmentum thick, fleshy, hypogæous.*

30. **P. aquilina,** *Linn.; Schk. Fil. t.* 95 *et* 96; *Eng. Bot t.* 1679; *Lindl. and Moore's Brit. Ferns, t.* 44; *Sowerby's Ferns, t.* 38; *Hook. Brit. Ferns,* 38.—Tropical and Temperate Zones of both Hemispheres.

31. **P. esculenta,** *Forst.; Schk. Fil. t.* 97.—Australia and New Zealand.

---

97. LITOBROCHIA, *Presl; J. Sm.*

Vernation fasciculate and erect or decumbent, or uniserial and sarmentose. *Fronds* smooth, pinnate or bi-tripinnate, 1–8 feet high; ultimate pinnæ sinuose-lobed or pinnatifid. *Veins* elevated; only the lower venules anastomosing, or the

Genus 97.—Fertile pinna, under side. No. 7.

whole uniform reticulated. *Receptacles* marginal, transversely continued in the axis of a linear marginal indusium, constituting a linear continuous or interrupted sorus.

* *Vernation erect, caudex undefined, generally cæspitose.*

1. **L. denticulata,** *Presl.* Pteris denticulata, *Sw.; Hook. et Grev. Ic. Fil. t.* 28; *Lowe's Ferns,* 4, *t.* 1. Pteris Brasiliensis, *Radd. Fil. Bras. t.* 68 *bis.*—Brazil.
2. **L. leptophylla,** *Fée.* Pteris leptophylla, *Sw.; Lowe's Ferns,* 3, *t.* 47; *Hook. Gard. Ferns, t.* 23. Pteris spinulosa, *Radd. Fil. Bras. t.* 70. Cheilanthes spinulosa, *Link, in Hort. Berol.*—Brazil.
3. **L. macilenta,** *J. Sm.* Pteris macilenta, *A. Rich. Fl. Nov. Zeal. t.* 12.—New Zealand.

** *Vernation subsarmentose, epigæous.*

4. **L. grandifolia,** *J. Sm.* Pteris grandifolia, *Linn.* (*Plum. Fil. t.* 105); *Schk. Fil. t.* 89; *Hook. Sp. Fil. t.* 113 *B.*—Tropical America.

*** *Vernation contiguous, decumbent, hypogæous.*

5. **L. polita,** *J. Sm.* Pteris polita, *Link.*—Brazil.
6. **L. comans,** *Presl.* Pteris comans, *Forst.; Schk. Fil. t.* 92.—Polynesian Islands.
7. **L. macroptera,** *J. Sm.* Pteris macroptera, *Link.*—Brazil.
8. **L. Orizabæ,** *J. Sm.* Pteris Orizabæ, *Mart. et Gal. Fil. Mex. t.* 13. P. apicalis, *Sieb.*—Mexico.
9. **L. spinulifera,** *J. Sm.* Pteris spinulifera, *Schum.*—Tropical Western Africa.
10. **L. Kunzeana,** *J. Sm.* Pteris Kunzeana, *Agardh.; Hook. Sp. Fil.* 2, *t.* 139 (*excl. syn. Plum.*).—Jamaica.
11. **L. elata,** *Fée.* Pteris elata, *Agard.*—Tropical America.

**** *Vernation fasciculate, erect, caudiciform, subarborescent.*

12. **L. tripartita,** *J. Sm.* Pteris tripartita, *Sw.; Hook. Sp. Fil. t.* 138 *B.* ? Pteris linearis, *Poir.* ? Pteris intermedia, *Blume.*—East Indies, Java.

13. **L. podophylla,** *Presl; Hook. Gard. Ferns, t.* 55. Pteris podophylla, *Sw.* Lonchitis pedata, *Linn.; Brown, Jam. t.* 1.—West Indies.

14. **L. biaurita,** *J. Sm.* Pteris biaurita, *Linn.; (Plum. Fil. t.* 15); *Lowe's Ferns,* 3, *t.* 50. Campteria biaurita, *Hook. Gen. Fil. t.* 65 *A.* Pteris nemoralis, *Willd.; Wall. in part.*—West Indies.

***** *Vernation uniserial, distant; sarmentum elongating, generally epigæous and hirsute-squamose.*

15. **L. vespertilionis,** *Presl.* Pteris vespertilionis, *Labill. Nov. Holl. t.* 245; *Lowe's Ferns,* 3, *t.* 44.—Tropics and South Temperate Regions.

16. **L. aurita,** *J. Sm.* Pteris aurita, *Blume; Metten. Fil. Hort. Lips. t.* 14.—Malay Islands.

---

98. DORYOPTERIS, *J. Sm.; Fée.*

(Pteridis sp., *Auct.*)

Vernation fasciculate, erect, rarely uniserial sarmentose.

Genus 98.—Portion of mature frond, upper side. No. 3.

Fronds simple, cordate-hastate, palmate or bipinnate, smooth, opaque. *Veins* internal, reticulated. *Receptacles* transverse, marginal, continuous, in the axis of a linear, continuous indusium.

1. **D. sagittifolia,** *J. Sm.; Lowe's Ferns,* 3, *t.* 36. Pteris sagittifolia, *Radd. Fil. Bras. t.* 63, *f.* 1; *Hook. Fil. Exot. t.* 39. Litobrochia sagittifolia,—*var.* alcyonis, *Gard. Chron.* 1863.—Brazil.

2. **D. pedata,** *J. Sm.* Pteris pedata, *Linn.; (Plum. Fil. t.* 152); *Lang. et Fisch. Ic. Fil. t.* 20; *Schk. Fil. t.* 100; *Radd. Fil. Bras. t.* 65, *f.* 3 *et t.* 66 *B.; Hook. Bot. Mag. t.* 3247; *Hook. Fil. Exot. t.* 34.—Brazil.

3. **D. palmata,** *J. Sm. Gen. Fil.* (1841). Pteris palmata, *Willd., var.* lata, *Hook. Gard. Ferns, t.* 22.—Tropical America.

4. **D. collina,** *J. Sm.; Lowe's Ferns,* 3, *t.* 38. Pteris collina, *Radd. Fil. Bras. t.* 65,—*var.* nobilis, *Moore.*—Tropical America.

---

99. LONCHITIS, *Linn.*

Vernation fasciculate, erect, subarboreous. *Fronds* bi-tripinnate, villose, 2–6 feet long, the ultimate pinnæ sinuose-pinna-

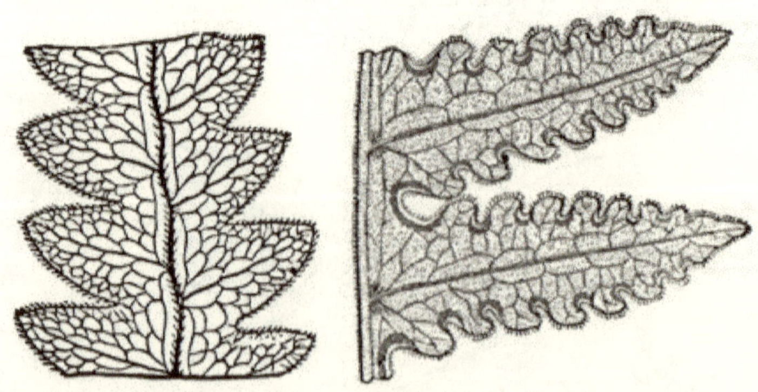

Genus 99.—Portion of barren and fertile fronds, natural size. No. 1.

tifid. *Veins* costæform, pinnate; venules anastomosing, forming irregular hexagonoid areoles. *Sporangia* produced on the apices of 4–5 venules, converging in the sinus of the laciniæ, forming an arcuate sorus in each sinus. *Indusium* linear, lunulate.

1. L. **Lindeniana,** *Hook. Sp. Fil.* 2, *t.* 89 *A.*—Tropical America.

2. L. **pubescens,** *Willd.; Hook. Gen. Fil. t.* 68 *A.*—Mauritius.

Tribe VIII.—BLECHNEÆ.

Sori intramarginal, medial or costal, transverse oblong linear, continuous or interrupted. *Indusium* lateral, linear, exteriorly attached, plane or vaulted, rarely obsolete. *Fronds* uniform, or the fertile contracted.

100. BLECHNUM, *Linn.; Presl.*

Vernation fasciculate, erect, caudiciform or cæspitose. *Fronds* simple, pinnatifid or pinnate, from a few inches to 4–6 feet high; pinnæ adherent or articulated with the rachis. *Veins* forked; the sterile venules free, or their apices thickened and forming a cartilaginous margin; the fertile veins combined near their base by a transverse, continuous, sporangiferous receptacle, constituting a linear, costal, or rarely extra-costal sorus. *Indusium* linear, plane.

* *Apices of the venules free.*

† *Sori costal.* (Blechnum, *Presl.*)

1. B. **Lanceola,** *Sw.; Lodd. Cab.* (1592); *Hook. Bot. Mag. t.* 3240; *Kunze, Fil. t.* 57, *f.* 1; *Hook. Ic. Pl. t.* 970. B. lanceolatum, *Radd. Fil. Bras. t.* 60, *f.* 3. B. trifoliatum, *Kaulf.*—Brazil.

2. B. **polypodioides,** *Radd. Fil. Bras. t.* 60, *f.* 2; *Kunze, Fil. t.* 58, *f.* 1; *Lowe's Ferns,* 4, *t.* 34.—Brazil.

3. **B. glandulosum**, *Link; Kaulf.* (*non Kunze*).—Brazil.

4. **B. cognatum**, *Presl.* B. glandulosum, *Kunze, Fil. t.* 58, *f.* 2.—Tropical America.

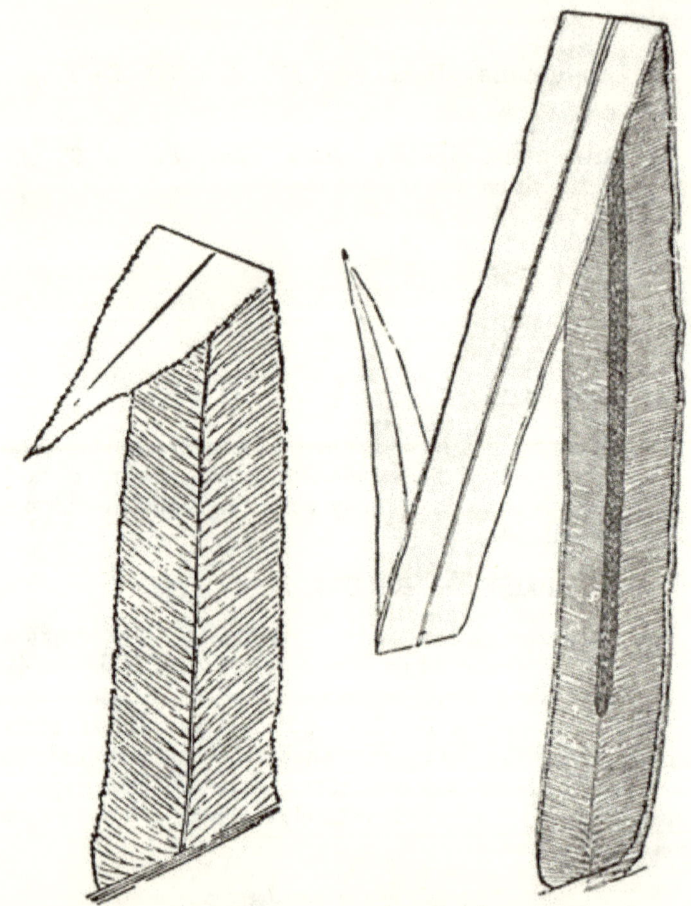

Genus 100.—Pinna of barren frond, under side. No. 13.

Genus 100.—Fertile pinna, under side. No. 17.

5. **B. triangulare**, *Link.* B. triangulatum, *J. Sm.; Lowe's Ferns*, 4, *t.* 35.—Tropical America.

6. **B. occidentale**, *Linn.;* (*Plum. Fil. t.* 62); *Jacq. Ic. Rar. t.* 644; *Hook. Gen. Fil. t.* 54; *Radd. Fil. Bras. t.* 53; *Lowe's Ferns*, 4, *t.* 39. B. conjugatum, *Klot.*—Tropical America.

7. **B. gracile**, *Kaulf.; Lodd. Cab. t.* (1905); *Lowe's Ferns*, 4, *t.* 36.—Tropical America.

8. **B. longifolium**, *H. B. K.; Hook. Bot. Mag.* 2818; *Hook. Sp. Fil. t.* 154; *Lowe's Ferns*, 4, *t.* 37.—Tropical America.

9. **B. campylotis**, *J. Sm.* Lomaria campylotis, *Kunze.*—Tropical America.

10. **B. intermedium**, *Link; Kunze, Fil. t.* 57, *f.* 2.—Tropical America.

11. **B. fraxineum**, *Willd.* B. latifolium, *Moritz.*—Tropical America.

†† *Sori extra-costal.* (Mesothema, *Presl.*)

12. **B. hastatum**, *Kaulf.* Lomaria hastata, *Kunze, Fil. t.* 55, *f.* 1.—Chili.

** *Apices of the venules thickened, forming a cartilaginous margin.* (Blechnopsis, *Presl.*)

† *Pinnæ adherent.*

13. **B. Brasiliense**, *Desv.; Hook. Sp. Fil.* 2, *t.* 157. B. Corcovadense, *Radd. Fil. Bras. t.* 61.—Brazil.

14. **B. striatum**, *R. Br.; Hook. Sp. Fil. t.* 159. B. stramineum, *Labill.*—Australia and Philippine Islands.

15. **B. lævigatum**, *Cav.; Hook. Sp. Fil.* 3, *t.* 160.—New South Wales.

16. **B. cartilagineum**, *Sw.; Metten. Fil. Hort. Leip. t.* 5; *Lowe's Ferns*, 4, *t.* 42.—Australia.

17. **B. orientale**, *Linn.; Schk. Fil. t.* 109; *Hook. Exot. Fil. t.* 77; *Lowe's Ferns*, 4, *t.* 40. B. latifolium, *Presl.*—East Indies, Malay Islands.

†† *Pinnæ articulated with the rachis.*

18. **B. serrulatum**, *Rich.; Schk. Fil. t.* 108; *Lowe's Ferns*, 4, *t.* 43. B. calophyllum, *Lang. et Fisch. Ic. Fil. t.* 23. B. angustifolium, *Willd.* B. stagninum, *Radd. Fil. Bras. t.* 62.—Tropical America.

101. DOODIA, *R. Br.*

Vernation fasciculate, erect. *Fronds* pinnatifid or subpinnate, the fertile sometimes subcontracted; segments serrate or spinulose. *Veins* forked, the lower venules transversely anastomosing and sporangiferous. *Receptacles* medial, elongated, constituting one, or sometimes two, transverse rows of oblong, straight, or arcuate sori. *Indusium* plane.

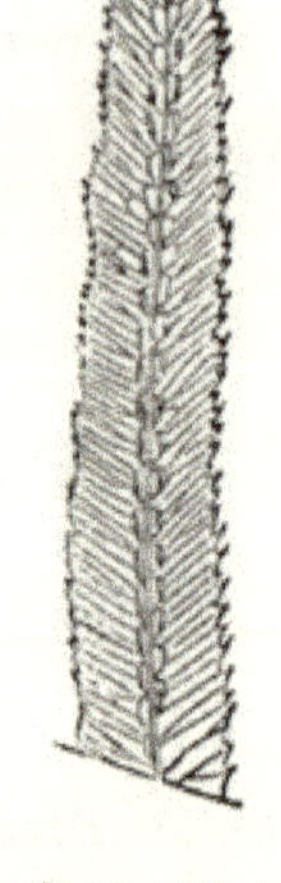

Genus 101.—Fertile pinna, underside. No. 1.

1. **D. aspera,** *R. Br.; Hook. Exot. Fil. t.* 8 *Hook. Gen. Fil. t.* 54; *Lowe's Ferns,* 4, *t.* 30.—Australia.
2. **D. blechnoides,** *A. Cunn.; Metten. Fil. Hort. Leip. t.* 6, *f.* 3. D. maxima, *Lowe's Ferns,* 4, *t.* 32; *J. Sm. in Loud. Hort. Brit.*—Australia.
3. **D. lunulata,** *R. Br. in Herb. Brit. Mus.; Lowe's Ferns,* 4, *t.* 31 *B.*—New Zealand.
4. **D. caudata,** *R. Br.; Hook. Exot. Fil. t.* 25; *Lowe's Ferns,* 4, *t.* 31 *A.* D. rupestris, *Kaulf.*—Australia.
5. **D. linearis.** *Vernation* fasciculate, erect, becoming cæspitose. *Fronds* linear, 6–10 inches long; the sterile sinuose-pinnatifid below, subentire above; the fertile linear, anfractose, rachiform, erect. D. caudata, *var.* confluens, *Hort.*—New Caledonia. (*C. Moore.*)
6. **D. dives,** *Kunze, Fil. t.* 105.—Ceylon.

---

102. LOMARIA, *Willd.*

Vernation uniserial and sarmentose, or fasciculate, erect, cæspitose, or sometimes subarboreous. *Fronds* simple pinnatifid, or pinnate, rarely bipinnatifid, 1–3 feet high, the fertile always contracted. *Veins* (of the sterile frond) forked; venules free, their apices usually clavate; fertile segments rachiform, veins obsolete, or more or less evident, and by their contiguity forming a broad, transverse, continuous, sporangiferous recep-

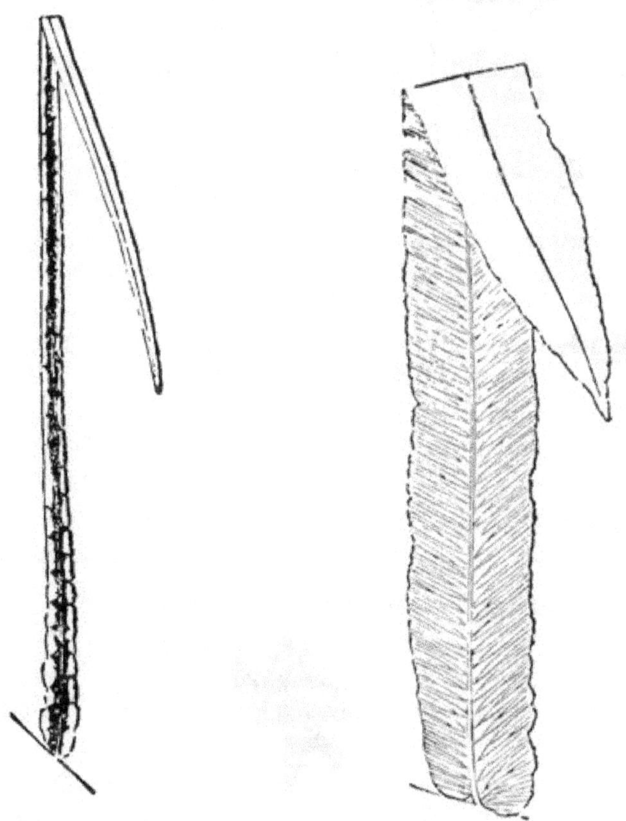

Genus 102.—Fertile pinna, under side. No. 22.

Genus 102.—Barren pinna, No. 22.

tacle, the sporangia becoming confluent over the whole disc of the segment. *Indusium* linear, sub-intramarginal, vaulted and revolute.

* *Fronds simple or pinnatifid.*

1. **L. Patersoni,** *Spreng.; Kunze, Fil. t.* 34; *Hook. Fil. Exot t.* 49; *Lowe's Ferns,* 4, *t.* 50. Stegania Patersoni, *R. Br.*—Australia.

2. **L. lanceolata,** *Spreng.; Hook. Ic. Pl. t.* 429; *Lowe's Ferns,* 4, *t.* 61. Stegania lanceolata, *R. Br.* — Australia, Tasmania, New Zealand, South America.

3. **L. blechnoides,** *Bory; Hook. Sp. Fil.* 3, *p.* 11.—Chili.

4. **L. L'Herminieri,** *Bory; Kunze, Fil. t.* 73; *Hook. Gard. Ferns, t.* 40; *Lowe's Ferns,* 4, *t.* 63.—Tropical America.

5. **L. nuda,** *Willd.; Lowe's Ferns,* 4, *t.* 51. Onoclea nuda, *Labill. Nov. Holl. t.* 246. Stegania nuda, *R. Br.*—Tasmania.

6. **L. discolor,** *Willd.; Lowe's Ferns,* 4, *t.* 65. Hemionitis discolor, *Schk. Fil. t.* 6.—New Zealand.

7. **L. vulcanica,** *Blume; Hook. Ic. Pl. t.* 969; *Hook. Sp. Fil.* 3, *p.* 12.—Java, Fiji and South Pacific Islands, Tasmania, New Zealand.

8. **L. attenuata,** *Willd.* Onoclea attenuata, *Sw.* Blechnum attenuatum, *Metten. Fil. Hort. Lips. t.* 3, *f.* 1–6.—Brazil.

9. **L. elongata,** *Blume.* Lomaria Colensoi, *Hook. fil. Ic. Pl. t.* 627–628; *Hook. Sp. Fil.* 3, *p.* 3.—New Zealand.

10. **L. onocleoides,** *Spreng.; Hook. Sp. Fil. t.* 146. Blechnum onocleoides, *Sw.*—West Indies and Tropical America.

11. **L. gibba,** *Labill. Sert. Aust. Caled. t.* 4–5.—New Caledonia.

12. **L. alpina,** *Spreng.; Hook. fil. Fl. Antarct. t.* 150; *Hook. Fil. Exot. t.* 32; *Lowe's Ferns,* 4, *t.* 52. Stegania alpina, *R. Br.* Lomaria antarctica, *Carm.*—Tasmania, New Zealand, Magellan. **H.**

13. **L. Spicant,** *Desv.* Osmunda Spicant, *Linn.* Blechnum Spicant, *Sw.; Lindl. and Moore's Nat. Print. Ferns, t.* 43. Blechnum boreale, *Sw.; Sm. Eng. Bot. t.* 1159; *Schk. Fil. t.* 110; *Hook. Brit. Ferns, t.* 40.—Europe, Madeira, North America.

14. **L. Banksii,** *Hook. fil. Fl. Nov. Zeal. t.* 76.—New Zealand.

** *Fronds pinnate.*

15. **L. nigra,** *Col. Hook. Ic. Plant. t.* 960; *Hook. Sp. Fil.* 3, *p.* 35.—New Zealand.

16. **L. fluviatilis,** *Spr.; Hook. fil. Fl. t.* 167. Stegania fluviatilis, *R. Br.* Lomaria rotundifolia, *Raoul, Pl. Nov. Zel. t.* 2 *B.*—Tasmania and New Zealand.

17. **L. australis,** *Link.* Blechnum australe, *Linn.; Schk. Fil. t.* 110 *B; Mett. Fil. Hort. Lips. t.* 3, *f.* 7. Lomaria pumila, *Kaulf.*—South Africa.

18. **L. punctulata,** *Kunze; Lowe's Ferns,* 4, *t.* 53. Blechnum punctulatum, *Sw.; Schlecht. Adumb. t.* 21, 22, *f.* 2, —β Krebsii, *J. Sm.* Scolopendrium Krebsii, *Kunze, Fil. t.* 74; *Hook. Bot. Mag. t.* 4768; *J. Sm. Cat. Cult. Ferns,* 1*st Ed. p.* 49 (abnormal form). Lomaria Australis, *Lowe's Ferns,* 4, *t.* 57, 58. Lomaria densa, *Kaulf.*—South Africa.

19. **L. Gilliesii,** *Hook. et Grev. Ic. Fil. t.* 207.—Chili.

20. **L. minor,** *Spreng.*—Tasmania.

21. **L. procera,** *Spreng.; Hook. Ic. Pl. t.* 127, 128; *Hook. fil. Fl. Nov. Zel. t.* 75. Osmunda procera, *Forst.* Blechnum procerum, *Labill. Nov. Holl. t.* 247.—Australia, Tasmania, New Zealand, Polynesia.

22. **L. Capensis,** *Willd.* Onoclea Capensis, *Linn.* Blechnum Capense, *Schlecht. Adumb. t.* 18.—South Africa.

23. **L. gigantea,** *Kaulf.; Schlecht. Adumb. t.* 20–22, *f.* 1.—South Africa.

24. **L. striata,** *Willd.* Onoclea striata, *Sw.* Lomaria Chilensis, *Kaulf; Hook. Gen. Fil. t.* 64 *B.* L. tuberculata, *J. Sm. Cat. Fil. Hort. Kew* (1856).—Tropical America.

25. **L. cycadifolia,** *Linden* (*Colla*).—Chili.

26. **L. Boryana,** *Willd.* Onoclea Boryana, *Sw.* Pteris osmundoides, *Bory, Itin.* 2, *t.* 32. L. Magellanica, *Desv.; Hook. Gard. Ferns, t.* 52. L. robusta, *Carm.* L. zamioides, *Gardn.* L. cinnamomea, *Kaulf.* L. setigera, *Gaud.* L. obtusifolia, *Presl.* Blechnum (Lomaria) Boryana, *Schlecht. Adumb. t.* 19.—Bourbon, South Africa, Brazil, Tierra del Fuego.

*** *Fronds bipinnatifid.*

27. **L. Fraseri,** *A. Cunn.; Hook. Ic. Pl. t.* 185.—New Zealand.

103. BRAINEA, *J. Sm.*

(Bowringia, *Hook. non Champ.*)

Vernation fasciculate, erect; caudex arboreous, 2–4 feet high. *Fronds* pinnate, rarely sub-bipinnate, 1–3½ feet long; pinnæ linear-lanceolate, 4–6 inches long, subsessile, base truncate

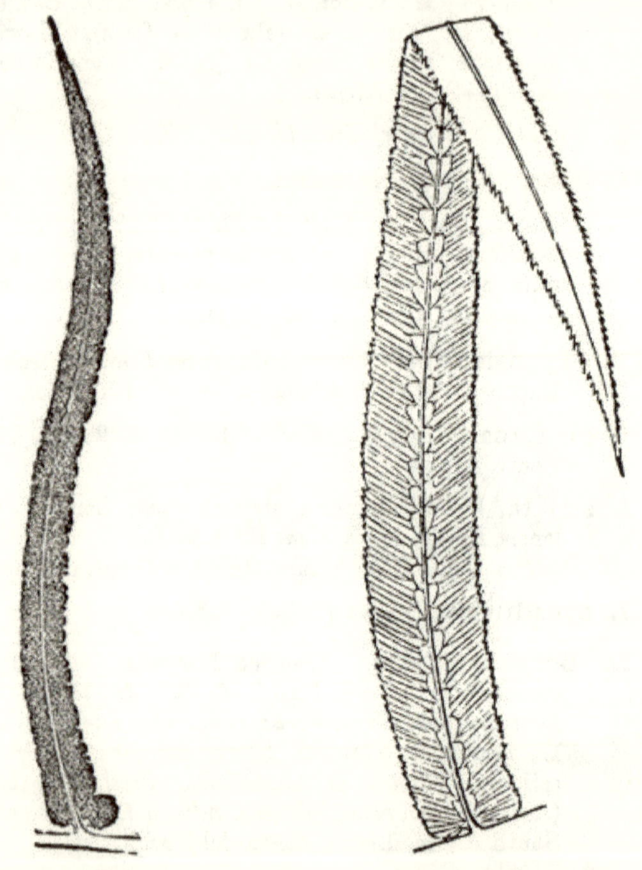

Genus 103.—Fertile pinna, under side. No. 1.

Genus 103.—Barren pinna, under side. No. 1.

above, auriculated below, margin crenate-serrulate. *Veins* flabellately forked, free exteriorly; the lower venules angularly anastomosing, forming a costal row of oblique, cuneiform areoles. *Sporangia* produced on the transverse anastomoses and bases of the free venules, confluent, constituting a broad, continuous or sub-interrupted, transverse, naked sorus.

1. **B. insignis**, *J. Sm. Cat. Fil. Hort. Kew* (1856); *Lowe's Ferns*, 4, *t.* 49. Bowringia insignis, *Hook. Journ. Bot. and Kew Miscell. v.* 5, *t.* 2.—Hong-kong and Khasia, East Indies.

---

104. WOODWARDIA, *Sm.*

Vernation fasciculate, decumbent. *Fronds* bipinnatifid, 4–6

Genus 104.—Fertile pinna. No. 1.

feet long. *Veins* reticulated, or the exterior venules free; the costal anastomoses transverse, elongated, and sporangiferous. *Receptacles* elongated, medial, constituting a costal row of oblong, linear, contiguous sori. *Sporangia* immersed. *Indusium* vaulted, revolute.

1. **W. radicans,** *Sm.; Schk. Fil. t.* 112; *Hook. Gen. Fil. t.* 17; *Lowe's Ferns,* 4, *t.* 44. Blechnum radicans, *Linn.* Woodwardia stans, *Sw.* β confluens. Woodwardia confluens, *Hort.* — South Europe, North India, Madeira, California.

2. **W. orientalis,** *Sw.* W. Fortunei, *Hort. Angl.* —Japan and China.

3. **W. Japonica,** *Sw.* Blechnum Japonicum, *Linn.; Thunb. Fl. Jap. t.* 35. — Japan and China.

---

105. ANCHISTEA, *Presl.*

Vernation uniserial; sarmentum hypogæous. *Fronds* bipinnatifid, 1–2 feet high. *Veins* flabellately forked, free exteriorly; the lower venules transversely anastomosing and sporangiferous. *Receptacles* elongated, medial. *Sori* oblong, contiguous, in a continuous costal row. *Indusium* linear, plane

1. **A. Virginica,** *Presl.* Blechnum Virginicum, *Linn.* Woodwardia Virginica, *Sm.; Metten. Fil. Hort. Lips. t.* 6, *f.* 1, 2; *Lowe's Ferns,* 4, *t.* 45. — North America. T.

Genus 105.—Fertile pinna, under side. No. 1

Genus 106.—Pinna of barren frond. No. 1.

106. ONOCLEA, *L.*

Vernation uniserial, distant; sarmentum hypogæous. *Sterile fronds* sub-bipinnatifid; veins reticulated. *Fertile fronds* bipinnate; veins free; pinnules contracted, sessile; margins conniving, forming unilateral spikes (pinnæ) of globose, bacciform segments, each compactly filled with sporangia, which rise from four to six punctiform, medial receptacles. *Special indusium* lateral, very membranous.

1. **O. sensibilis**, *Lin.; Schk. Fil. t.* 102; *Hook. Gen. Fil. t.*82; *Lowe's Ferns*, 6, *t.* 1. Onoclea obtusiloba, *Schk. Fil. t.* 103. — North America.

---

107. LORINSERIA, *Presl; Féc.*

Vernation uniserial, sarmentum hypogæous. *Fronds* distant, sinuose-pinnatifid or subpinnate, 1–1½ foot high, the fertile contracted. *Veins* of the sterile frond uniform reticulated; fertile segments rachiform, costal anastomoses transverse-elongated, sporangiferous. *Receptacles* elongated ,medial. *Sori* linear, contiguous, in a costal row. *Indusium* vaulted, involute, becoming reflexed.

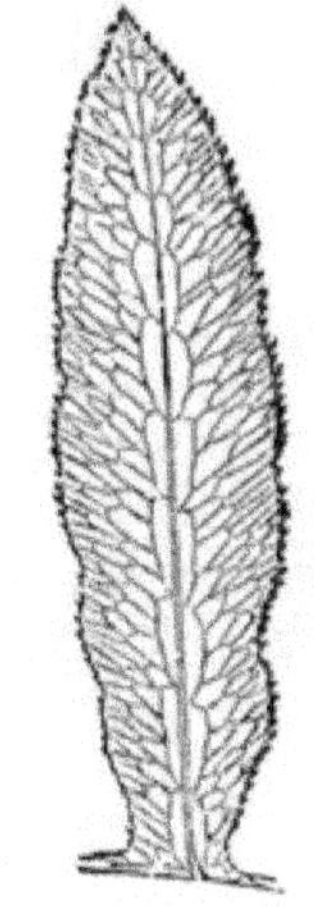

Genus 107.—Barren pinna, under side. No. 1.

1. **L. areolata,** *Presl.* Acrostichum areolatum, *Linn.* Woodwardia areolata, *Lowe's Ferns,* 4, *t.* 46. Woodwardia angustifolia, *Sm.; Metten. Fil. Hort. Lips. t.* 6, *f.* 6, 7. Woodwardia onocleoides, *Willd.* W. Floridana, *Schk. Fil. t.* 111.—North America. **T.**

---

108. STENOCHLÆNA, *J. Sm.*

Vernation uniserial, distant; sarmentum elongated, scandent,

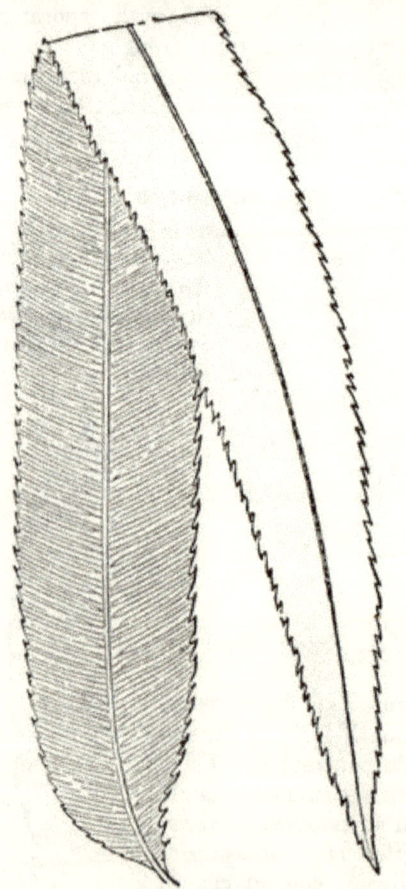

Geuns 108.—Barren pinna. No. 3.

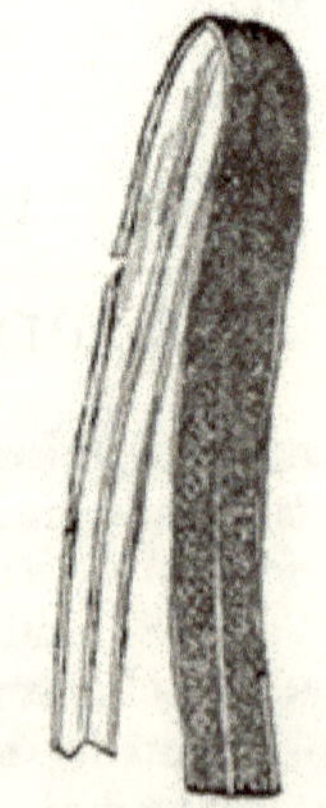

Genus 108.—Portion of fertile frond, natural size. No. 1.

smooth. *Fronds* of two forms, 2–3 feet long, the sterile pinnate, the fertile contracted, pinnate or bipinnate; pinnæ articulated with the rachis. *Veins* rising from an obscure, transverse vein continuous with, and close to, the costa; venules direct, their apices clavate, united, forming a pellucid, cartilaginous, spinulose margin; fertile segments linear, rachiform, margin membranaceous, revolute, indusiform. *Sporangiferous receptacle* linear, continuous; sporangia confluent.

* *Fertile fronds pinnate.*

1. S. **scandens,** *J. Sm.; Hook. Gen. Fil. t.* 105 B. Acrostichum scandens, *Linn.* Onoclea scandens, *Linn. Herb.*; *Schk. Fil. t.* 106. Lomaria scandens, *Willd.*—East Indies and Malayan Islands.

** *Fertile fronds bipinnate.*

2. S. **Meyeriana,** *J. Sm.; Lowe's Ferns,* 4, *t.* 47. 48. Lomaria Meyeriana, *Kunze.* Lomariobotrys Meyeriana, *Fée.* Stenochlæna tenuifolia, *T. Moore.* ? Lomaria tenuifolia, *Desv.* Stenochlæna scandens, *Hort.* Acrostichum Meyerianum, *Hook. Gard. Ferns, t.* 16.—South Africa.

---

109. SALPICHLÆNA, *J. Sm.*

Vernation subfasciculate, decumbent. *Fronds* bipinnate,

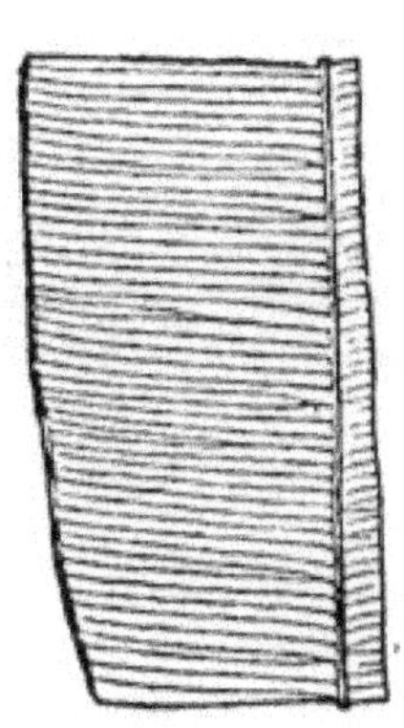

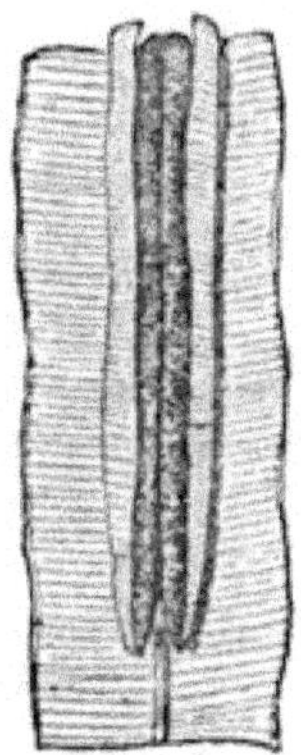

Genus 109.—Portion of barren and fertile pinna, natural size. No. 1.

flexuose, climbing to an indefinite height; pinnæ 1–2 feet long, adherent. *Veins* forked; venules combined by a transverse marginal vein, and in the fertile also near their base by a transverse, continuous, sporangiferous receptacle, forming a linear costal sorus. *Indusium* revolute, vaulted, cylindrical, sporangiferous along its base.

1. **S. volubile,** *J. Sm. in Hook. Gen. Fil. t.* 93. Blechnum volubile, *Kaulf.; Kunze, Anal. t.* 13; *Hook. Gard. Ferns, t.* 15. Blechnum scandens, *Bory, in Dup. Voy. t.* 36.—Tropical America.

Tribe IX.—ASPLENIEÆ.

Sori oblong or linear, oblique to the midrib or axis of venation. Furnished with a plane or vaulted lateral *indusium.*

110. ASPLENIUM, *Linn.*

Vernation fasciculate, erect or decumbent, rarely uniserial sarmentose. *Fronds* varying from simple-entire to decompound and from a few inches to 2–6 feet high, generally smooth *Veins* rayed, forked, or pinnate; venules free, sporangiferous on the superior side. *Sori* simple, oblong, or linear. *Indusium* plane or vaulted.

§ 1. Asplenium verum.—*Indusium plane.*

1. *Lanceum group.*

Vernation uniserial, sarmentum slender. Fronds distant, linear-lanceolate, 1–1½ *foot long. Sori simple, anti or opposite binate.* (Triblemma.)

1. **A. lanceum,** *Thunb. Ic. Plant. Jap. Dec.* 11, *t.* 18. A. subsinuatum, *Hook. et Grev. Ic. Fil. t.* 27. Diplazium lanceum, *Presl.* Scolopendrium dubium, *Don.*—India, China, Japan.

2. *Serratum group.*

Vernation fasciculate, erect. Fronds simple, broad, elliptical, or lanceolate, 1–2 *feet long.* (Phyllitis.)

2. **A. serratum,** *Linn.* (*Plum. Fil. t.* 124); *Schk. Fil. t.* 64; *Hook. Fil. Exot. t.* 70.—Tropical America.

3. **A. crenulatum,** *Presl.* A. Nidus, *Radd. Fil. Bras. t.* 53 (*non Linn.*). A. Brasiliense, *Hort.* (*non Sw.*); *Lowe's Ferns,* 5, *t.* 14 *B.*—Tropical America.

4. **A. sinuatum,** *Beauv. Fl. d'Oware,* 2, *t.* 79; *Hook. Fil. Exot. t.* 16.—West Tropical Africa.

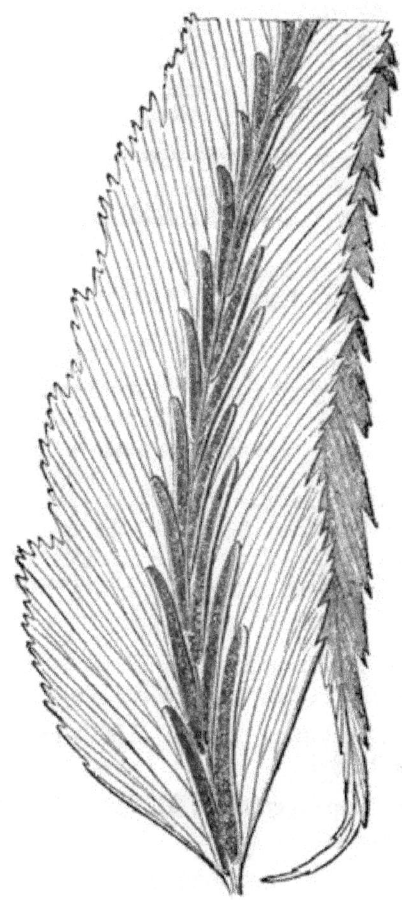

Genus 110.—Fertile pinna, under side. No. 69.

5. **A. stipitatum,** *J. Sm.* Neotopteris stipitata, *J. Sm. Cat. Cult. Ferns* (1857). Asplenium squamulatum, *var.* β Smithii, *Hook. Sp. Fil. p.* 83.—East Indies.

3. *Palmatum group.*

Vernation decumbent, subfasciculate. Fronds palmately lobed.

6. **A. Hemionitis,** *Linn.; Hook. Bot. Mag. t.* 4911. A. palmatum, *Lam.; Schk. Fil. t.* 66; *Lodd. Bot. Cab.* 868; *Lowe's Ferns,* 5, *t.* 6, β *var.* cristata, *Gard. Chron.* (*Jany.* 1865), Garden Sport.—South Europe, North Africa, Madeira.

4. *Trichomanes group.*

Vernation fasciculate, generally becoming cæspitose. Fronds pinnate, rarely pinnatifid only, linear, 3–12 *inches long; pinnæ short, often dimidiate, with the costa excentric.* (Asplenium verum.)

a. Fronds pinnatifid.

7. **A. alternans,** *Wall.; Hook. Gard. Ferns, t.* 38. A. Dalhousiæ, *Hook. Ic. Pl. t.* 105.—East Indies.

b. Fronds pinnate.

8. **A. Trichomanes,** *Linn.; Bolt. Fil. t.* 13; *Eng. Bot. t.* 576; *Sowerby's Ferns, t.* 30; *Lindl. and Moore's Brit. Ferns, t.* 39; *Lowe's Ferns,* 5, *t.* 22. A. anceps, *Soland.; Hook. et Grev. Ic. Fil. t.* 195. A. melanocaulon, *Willd.*

Var. incisum, *Moore; Lindl. and Moore's Brit. Ferns, t.* 39 *D, E; Schk. Fil. t.* 74, *f.*

Var. cristatum, *Moore; Lindl. and Moore's Brit. Ferns, t.* 39 *H.*

Var. depauperatum, *Wollast.; Lindl. and Moore's Brit. Ferns, t.* 39 *C.*

Var. multifidum, *Moore; Lindl. and Moore's Brit. Ferns, t.* 39 *G.*

Var. ramosum, *Moore; Lindl. and Moore's Brit. Ferns, t.* 39 *F.*

Europe, Madeira, South Africa, North India, Australia, North America.

9. **A. ebeneum,** *Ait.; Lodd. Bot. Cab. t.* 5; *Lowe's Ferns,* 5, *t.* 2. A. polypodioides, *Sw.; Schk. Fil. t.* 73.—North America.

10. **A. monanthemum,** *Linn.; Sm. Ic. ined. t.* 73; *Lodd. Bot. Cab. t.* 1700; *Metten. Fil. Hort. Lips. t.* 9, *f.* 7, 8; *Lowe's Ferns,* 5, *t.* 1 *A.*—Tropical and Sub-tropical America, South Africa, Madeira.

11. **A. Petrarchæ,** *De Cand.; Hook. et Grev. Ic. Fil. t.* 152; *Lowe's Ferns, t.* 5, 38 *A,* and *var.* lata, 38 *B.*—South of Europe.

12. **A. formosum,** *Willd.; Hook. Fil. Exot. t.* 16; *Lowe's Ferns,* 5, *t.* 43 *B.* A. subalatum, *Hook. et Arn. Beech Voy. t.* 71.—Tropical America.

13. **A. Brasiliense,** *Radd. Fil. Bras. t.* 51, *f.* 1. A. dimidiatum, *Lowe's Ferns,* 5, *t.* 13 *A.*—Tropical America.

14. **A. tenellum,** *Roxb. in Beat. St. Helena.* A. reclinatum, *Moore et Houlst.; Lowe's Ferns,* 5, *t.* 13 *B; J. Sm. Cat.* (1857). A. erectum,—*var.* proliferum, *Hook. Fil. Exot. t.* 72. A. radicans, *Prit. Cat. St. Helena, Pl.*—St. Helena. T.

15. **A. erectum,** *Bory, in Willd. Schlecht. Adum. t.* 15. A. dentax, *Lowe's Ferns,* 5, *t.* 43 *A.*—Islands of the Indian Ocean, South Africa.

5. *Auriculatum group.*

Vernation fasciculate, erect, cæspitose. Fronds pinnate; pinnæ 1–2 *inches in length, generally having a more or less evident lobe or auricle on the superior edge of their base.* (Asplenium verum.)

16. **A. hastatum,** *Klotzsch.; Hook. Sp. Fil.* 3, *t.* 172. A. fragrans, *Hook. Ic. Plant. t.* 88. A. odoratum, *Moore.*—Venezuela.

17. **A. salicifolium,** *Linn.* (*Plum. Fil. t.* 60); *Radd. Fil. Bras. t.* 50.—Tropical America, West Indies.

18. **A. compressum,** *Sw.; Hook. Fil. Exot. t.* 76; *Lowe's Ferns,* 5, *t.* 16. A. fœcundum, *Kunz.*—St. Helena.

19. **A. obtusifolium,** *Linn.* (*Plum. Fil. t.* 67); *Hook. et Grev. Ic. Fil. t.* 239.—West Indies.

20. **A. marinum,** *Linn.; Eng. Bot. t.* 392; *Schk. Fil. t.* 68; *Hook. Fl. Lond.* 4, *t.* 60; *Lindl. and Moore's Brit. Ferns, t.* 38; *Sowerby's Ferns, t.* 29; *Lowe's Ferns,* 5, *t.* 23; *Hook. Brit. Ferns, t.* 31. A. lætum, *Hort.; Lowe's Ferns,* 5, *t.* 21 *A* (*not of Sw.*).—Europe, Madeira.

Var. arcutum, *Moore.*

Var. crenatum, *Moore; Lindl. and Moore's Brit. Ferns, t.* 38 *G.*

Var. trapeziforme, *Huds.*

Var. ramosum, *Wollast.; Lindl. and Moore's Brit. Ferns, t.* 38 *H.*

Var. subpinnatum, *Moore.*—Europe, Madeira.

21. **A. elongatum,** *Sw.* Asplenium productum, *Presl, Reliq. Hænk. t.* 8, *f.* 1.—Ceylon.

22. **A. firmum,** *Kunze; Hook. Sp. Fil.* 3, *t.* 174. A. cultrifolium, *Hort.*—Tropical America.

23. **A. pumilum,** *Sw.* (*Plum. Fil. t.* 66 *A*); *Lowe's Ferns,* 5, *t.* 31 *B.*—Tropical America.

24. **A. dentatum,** *Linn.* (*Plum. Fil. t.* 101, *f. C*); *Hook. et Grev. Ic. Fil. t.* 72.—West Indies.

25. **A. pulchellum,** *Radd. Fil. Bras. t.* 52, *f.* 2. β Otites, *Metten.* Asplenium Otites, *Link.; Metten. Fil. Hort. Lips. t.* 9, *f.* 1–4. A. pulchellum, *Hort.; Lowe's Ferns,* 5, *t.* 31 *A.*—Brazil.

26. **A. alatum,** *Humb.; Hook. et Grev. Ic. Fil. t.* 137; *Lowe's New Ferns, t.* 12 *B.*—Tropical America.

6. *Lucidum group.*

Fronds pinnate; pinnæ generally oblique, cuneiform at the base. (Asplenium verum.)

27. **A. lucidum,** *Forst.; Schk. Fil. t.* 72; *Metten. Fil. Hort. Lips. t.* 13, *f.* 12.—New Zealand.

28. **A. heterodon,** *Blume; Metten. Fil. Hort. Lips. t.* 8, *f.* 1–2; *Lowe's New Ferns, t.* 3.—Java.

29. **A. gemmiferum,** *Schrad.* A. lucidum, *Schlecht. Fil. t.* 14 *A.*—South Africa.

30. **A. emarginatum**, *Beauv. Fl. d'Oware*, 2, *t.* 61; *Hook. 2nd Cent. Ferns*, *t.* 78 (letterpress 80).—West Africa.

31. **A. obtusatum**, *Forst.; Schk. Fil. t.* 68; *Labill. Nov. Holl. t.* 242, *f.* 2; *β.* difforme, *J. Sm.; Hook. Fil. Exot. t.* 46; *Lowe's Ferns*, 5, *t.* 5 *B.* A. difforme, *R. Br.* Asplenium consimile, *Remy, in Gay. Chil.*—Chili, Tasmania, New Zealand.

32. **A. obliquum**, *Forst.; Schk. Fil. t.* 71; *Labill. Nov. Holl. t.* 242, *f.* 1.—Polynesia.

33. **A. oligophyllum**, *Kaulf.*—Brazil.

7. *Flaccidum group.*

Vernation fasciculate, decumbent, or erect. Fronds pinnate, bipinnate, or decompound; segments bifidly laciniated; laciniæ unisorous or linear. (Darea *of Willd.* Cænopteris, *Berg.*)

34. **A. brachypteron**, *Kunze*; *Hook. Fil. Exot. t.* 44; *Lowe's Ferns*, 5, *t.* 15 *B.*—Sierra Leone.

35. **A. prolongatum**, *Hook. Sp. Fil.* 3, *p.* 209; *2nd Cent Ferns*, *t.* 42.—Ceylon.

36. **A. rutæfolium**, *Presl.* Darea rutæfolia, *Sm.* — South Africa.

37. **A. lineatum**, *Sw.* *α. Fronds simply pinnate.* A. plumosum, *Bory*; *β.* bipinnatum; *fronds* bipinnatifid or bipinnate. Darea inæqualis, *Willd.* Asplenium inæquale, *Kunze.* Darea bifida, *Kaulf.* A. bifidum, *Presl.; J. Sm. Cat. Cult. Ferns* (1857).—Mauritius.

38. **A. Belangeri**, *Kunze; Hook. Fil. Exot. t.* 41; *Metten. Fil. Hort. Lips. t.* 13, *f.* 1–2; *Lowe's Ferns*, 5, *t.* 5 *A.* Darea Belangeri, *Bory.* Asplenium scandens, *Hort.* Asplenium Veitchianum, *Moore.*—Java.

39. **A. flaccidum**, *Forst.; Lowe's Ferns*, 5, *t.* 19. Cænopteris flaccida, *Thunb.; Schk. Fil. t.* 82. Cænopteris Odontites, *Thunb.; Sw.* Asplenium Odontites, *R. Br.* Cænopteris Novæ-Zelandiæ, *Spreng.; Schk. Fil. t.* 82.—New Zealand and Tasmania.

40. **A. bulbiferum,** *Forst.; Schk. Fil. t.* 79; *Hook. Ic. Pl. t.* 423; *Metten. Fil. Hort. Lips. t.* 13, *f.* 10–11; *Lowe's Ferns,* 5, *t.* 11.—New Zealand.

41. **A. Fabianum,** *Hombr. et Jacq. Voy. t.* 3, *bis.* Cænopteris Fabiana, *Bory.* Asplenium fœniculaceum, *Hort.* (*non H. et B.*).—Mauritius, Australia, Pacific Islands.

42. **A. appendiculatum,** *Labill.; Lowe's Ferns,* 5, *t.* 18. Cænopteris appendiculata, *Labill. Nov. Holl.* 2, *t.* 243. Asplenium laxum, *R. Br.; Hombr. et Jacq. Voy. t.* 3, *f.* 1.—Australia, Tasmania.

43. **A. Richardi,** *Hook. fil. Nov. Zeal.* **A. adiantoides,**—*var.* Richardi, *Hook. fil. in Hook. Ic. Plant. t.* 977. A. adiantoides,—*var.* Colensoi, *Hook. fil. in Hook. Ic. Plant.* 984. A. Colensoi, *Hook. fil.*—New Zealand.

44. **A. Hookerianum,** *Colenso.* A. adiantoides, *Raoul.* (*non Radd.*). A. adiantoides,—*var.* minus, *Hook. fil. in Hook. Ic. Pl. t.* 983.—New Zealand.

45. **A. dimorphum,** *Kunze; Hook. 2nd Cent. Ferns, t.* 36. A. diversifolium, *A. Cunn.* (*non Blume*); *Lowe's Ferns* 5, *t.* 17.—Norfolk Island.

46. **A. viviparum,** *Presl; Hook. Fil. Exot. t.* 64; *Lowe's Ferns, t.* 9. Cænopteris vivipara, *Sw.*—Mauritius.

8. *Rhizophorum group.*

Vernation fasciculate, cæspitose, or erect and solitary. Fronds pinnate or bi-tripinnate, the apex often long, caudate, flagelliform and viviparous; segments small, unisorous. (Darea, *Willd.*)

47. **A. viride,** *Huds.; Schk. Fil. t.* 73; *Eng. Bot. t.* 2257; *Lindl. and Moore's Brit. Ferns, t.* 40; *Sowerby's Ferns, t.* 31; *Hook. Brit. Ferns, t.* 30;—*var.* multifidum *Moore; Lindl. and Moore's Brit. Ferns. t.* 400.—Europe, North India. T.

48. **A. fontanum,** *Bernh.; Lindl. and Moore's Brit. Ferns, t.* 35 *A; Hook. Brit. Ferns, t.* 34; *Sowerby's Ferns t.* 26. Polypodium fontanum, *Linn.* Aspidium fontanum, *Sw.; Schk. Fil. t.* 53; *Eng. Bot. t.* 2024. Aspidium Halleri, *Willd.* Asplenium Halleri, *Spreng.* β refractum. Asplenium refractum, *Moore, Nat. Print. Ferns, sub t.* 35 *A; Lowe's Ferns, t.* 35 *A.*—Europe, North India. T.

49. **A. flabellifolium,** *Cav.; Sw. Syn. Fil. t.* 3, *f.* 2; *Lodd. Bot. Cab. t.* 1567; *Hook. Ex. Fl. t.* 208; *Lowe's Ferns,* 5, *t.* 1 *B.*—Australia, Tasmania.

50. **A. obtusilobum,** *Hook. Ic. Plant.* 1000.—Fiji Islands.

51. **A. cicutarium,** *Sw.* (*Plum. Fil. t.* 48 *A*); *Hook. Gen. Fil. t.* 6; *Metten. Fil. Hort. Lips. t.* 13, *f.* 3–9; *Lowe's Ferns,* 5, *t.* 20. Darea cicutaria, *Sm.* Asplenium dissectum, *Link.*—Tropical America.

52. **A. myriophyllum,** *Presl, Reliq. Hænk.* Cænopteris myriophylla, *Sw.* A. cicutarium, *J. Sm. Cat.* (1857); *Hook. Sp. Fil.* 3, *p.* 201 (*non Linn.*).—West Indies, Tropical America.

53. **A. divaricatum,** *Kunze; Schk. Fil. Supp. t.* 139. A. flabellulatum, *Hort.*—Peru.

54. **A. rhizophorum,** *Linn.; Hook. Sp. Fil.* 3, *t.* 187 *A.* A. radicans, *Sw.; Lowe's Ferns,* 5, *t.* 12 *B*—*a. var.* bipinnatum, *Hook. Sp. Fil. t.* 187 *C, b.* A. cyrtopteron, *Kunze; Hook. Sp. Fil. t.* 187 *B; Metten. Fil. Hort. Lips. t.* 10, *f.* 3–4.—Venezuela and Jamaica.

55. **A. cirrhatum,** *Rich.; Willd.* A. Karstenianum, *Klot.* A. comptum, *Moore et Houlst.*—Tropical America.

56. **A. rachirhizon,** *Radd. Fil. Bras. t.* 56; *Lowe's Ferns,* 5, *t.* 34.—Tropical America.

57. **A. pinnatifidum,** *Nutt.; Hook. Ic. Plant. t.* 972; *Metten. Fil. Hort. Lips. t.* 10, *f.* 1, 2; *Lowe's New Ferns, t.* 4 *B.* —United States.

9. *Adiantum nigrum group.*

Vernation fasciculate, erect or decumbent. Fronds bi-tripinnate; ultimate segments or laciniæ with two or more sori. (Tarachea, *Presl.*)

58. **A. Adiantum-nigrum,** *Linn.; Bolt. Fil. t.* 17; *Schk. Fil. t.* 80; *Eng. Bot. t.* 1950; *Lindl. and Moore's Brit. Ferns, t.* 36; *Sowerby's Ferns, t.* 28; *Hook. Brit. Ferns, t.* 28-33; *Lowe's Ferns, t.* 25.—Europe, South Africa, Madeira, North India.

59. A. **lanceolatum**, *Huds.; Eng. Bot. t.* 240; *Lindl. and Moore's Brit. Ferns, t.* 35 *B; Sowerby's Ferns, t.* 27; *Lowe's Ferns, t.* 26; *Hook. Brit. Ferns, t.* 32;—*var.* microdon, *Moore; Lowe's New Ferns, t.* 11 *B.*—Europe.

60. A. **acutum**, *Bory.* A. adiantum-nigrum,—*var.* acutum, *Lindl. and Moore's Brit. Ferns, t.* 37. A. productum, *R. T. Lowe.*—South Europe, Madeira.

61. A. **auritum**, *Sw.; Schk. Fil. t.* 130 *B; Lowe's Ferns,* 5, *t.* 32.—Tropical America.

62. A. **dispersum**, *Kunze; Metten. Fil. Hort. Lips. t.* 9, *f.* 5, 6. A. bipartitum, *Link.* A. bissectum, *Hort.*—Tropical America.

63. A. **macilentum**, *Kunze.* A. auritum,—*var.* obtusum, *Kunze; Metten. Fil. Hort. Lips. t.* 8, *f.* 3–6.—Tropical America.

64. A. **fragrans**, *Sw.* (*non Hook.*). A. planicaule, *Lowe's Ferns,* 5, *t.* 10 (*non Wall.*).—Jamaica.

65. A. **Mexicanum**, *Mart. et Gal. Fil. Mex. t.* 15, *f.* 4. A fœniculaceum, *J. Sm. Cat.* (1857) (*non H. et B.*).—Mexico.

10. *Falcatum group.*

Vernation fasciculate, erect, or decumbent. Fronds pinnate; pinnæ 1–6 *inches long, lanceolate or elliptical, acuminate or subdeltoid, entire, serrated, or erosely laciniated; angle of venation generally acute with the costæ.* (Tarachia, *Presl.*)

66. A. **attenuatum**, *R. Br.; Hook. et Grev. Ic. Fil. f.* 220; *Hook. Ic. Plant. t.* 914; *Lowe's Ferns,* 5, *t.* 35 *B.*—New South Wales, Queensland.

67. A. **longissimum**, *Blume; Hook. Sp. Fil.* 3, *t.* 190.—Java, Malacca, Mauritius.

68. A. **nitens**, *Sw.; Hook. Sp. Fil.* 3, 195. A. macriophyllum, *J. Sm. Cat.* (1857) (*non Sw.*); *Lowe's Ferns,* 5, *t.* 42.—Mauritius.

69. A. **serra**, *Lang. et Fisch. Ic. Fil. t.* 19; *Lowe's Ferns,* 5, *t.* 8.—Brazil.

70. **A. polyodon,** *Forst.; Lowe's Ferns,* 5, *t.* 33 *B.*—New Zealand.

71. **A. falcatum,** *Lam.* Trichomanes adiantoides, *Linn.; Burm. Fl. Zey. t.* 43.—Tropics.

72. **A. caudatum,** *Forst.; Schk. Fil. t.* 77; *Lowe's Ferns,* 5, *t.* 44.—Polynesia.

73. **A. paleaceum,** *R. Br.; Hook. Sp. Fil. t.* 199.—Tropical North-east Australia.

11. *Erosum group.*

Vernation fasciculate, generally cæspitose, rarely subsarmentose. Fronds bi-tripinnate, decompound, rarely linear or simply forked; segments rarely otherwise than cuneiform, with erose apices. Venation often flabellate, the costa being obsolete or evanescent. (Tarachia, *Presl.*)

74. **A. septentrionale,** *Schk. Fil. t.* 65; *Eng. Bot. t.* 1017; *Lindl. and Moore's Brit. Ferns, t.* 41 *C; Sowerby's Ferns, t.* 34; *Lowe's Ferns,* 5, *t.* 3 *A; Hook. Brit. Ferns, t.* 26. Acrostichum septentrionale, *Linn.; Bolt. Fil. t.* 8. Acropteris septentrionalis, *Link* (1833). Amesium septentrionale, *Newm.*—Europe, North India. **T.**

75. **A. Germanicum,** *Weis.; Lindl. and Moore's Brit. Ferns, t.* 41 *B; Hook. Brit. Ferns, t.* 27. A. alternifolium, *Wulf. Jacq. Misc. t.* 5, *f.* 2; *J. Sm. Cat. Ferns* (1857); *Eng. Bot. t.* 2259; *Sowerby's Ferns, t.* 33. A. Breynii, *Retz.; Schk. Fil. t.* 81.—Europe. **T.**

76. **A. Seelosii,** *Leibold. Flora* (1855), *t.* 15; *Hook. 2nd Cent. Ferns, t.* 26; *Hook. Sp. Fil.* 3, *p.* 175. — South Tyrol. **T.**

77. **A. Ruta-muraria,** *Linn.; Schk. Fil. t.* 80 *B; Eng. Bot. t.* 150; *Bolt. Fil. t.* 16; *Hook. Gen. Fil. t.* 30; *Lowe's Ferns,* 5, *t.* 27; *Lindl. and Moore's Brit. Ferns, t.* 41 *A; Sowerby's Ferns, t.* 32; *Hook. Brit. Ferns, t.* 28;—*var.* elatum, *Moore, Nat. Print. Ferns, oct. edit. t.* 79, *f. D.*—Europe, North India.

78. **A. cuneatum,** *Sloan. Jam.* 1, *t.* 46, *f.* 2; *Schk. Fil. t.* 78.—Jamaica.

79. **A. præmorsum,** *Sw.* A. laceratum, *Desv.* A. cuneatum, *Hook. et Grev. Ic. Fil. t.* 189; β Canariense. A. Canariense, *Willd.; Webb. Phyt. Canar.* 3, *t.* 251; *Lowe's Ferns*, 5, *t.* 25, *f.* 1–2–3. A. Maderense, *Penny.*—Mauritius, Tropical America, Madeira.

80. **A. furcatum,** *Thunb.; Schk. Fil. t.* 79. A. præmorsum, *Lowe's Ferns*, 5, *t.* 7.—South Africa, India, Madeira.

81. **A. laserpitiifolium,** *Lam.; Hook. Sp. Fil.* 3, *t.* 203; *Lowe's New Ferns*, *t.* 13.—Malay Islands.

82. **A. dimidiatum,** *Sw.* A. zamiæfolium, *Lodd. Bot. Cab. t.* 852; *Lowe's Ferns*, 5, *t.* 33 *A; J. Sm. Cat. Cult. Ferns* (1857) (*non Willd.*).—West Indies, Venezuela.

83. **A. contiguum,** *Kaulf.; Hook. Sp. Fil.* 3, *t.* 194.—Ceylon.

84. **A. erosum,** *Linn.; Hook. Sp. Fil.* 3, *t.* 198.—West Indies.

85. **A. nitidum,** *Sw.; Schk. Fil. t.* 81; *Lowe's New Ferns*, *t.* 18.—South Africa, East Indies.

12. *Actiniopteris group.*

Vernation fasciculate, erect. Fronds stipitate, flabellate, 6–8 *inches high, rigid; segments linear-rachiform. . Veins radiating and dichotomous.*

86. **A. radiatum,** *Sw.; Hook. Ic. Pl. t.* 9756. Acropteris radiata, *Fée.* Actiniopteris radiata, *Link; Hook. Sp. Fil.* 3, *p.* 275.—India, Ceylon.

§ 2. *Athyriæ* (Athyrium, *Roth*).—*Indusium vaulted.*

87. **A. Filix-fœmina,** *Bernh.; Hook. Brit. Ferns*, *t.* 35. Polypodium Filix-fœmina, *Linn.* Aspidium Filix-fœmina, *Sw.; Schk. Fil. t.* 58, 59; *Eng. Bot. t.* 282. Athyrium Filix-fœmina, *Roth; Sowerby's Ferns*, *t.* 25; *Lindl. and Moore's Brit. Ferns*, *t.* 30;—*var.* rhæticum, *Lindl. and Moore, Nat. Print. Ferns*, *t.* 31 *A.* Polypodium rhæticum, *Linnæan Herb.*

Var. latifolium, *Lindl. and Moore, Nat. Print. Ferns*, *t.* 31 *B.*

Var. marinum, *Lindl. and Moore's Nat. Print. Ferns, t.* 31 *C.*

Var. polydactylon, *Lindl. and Moore's Nat. Print. Ferns, under t.* 30.

Var. multifidum, *Lindl. and Moore's Nat. Print Ferns, t.* 33.

Var. depauperatum, *Lindl. and Moore's Nat. Print. Ferns, t.* 34 *B.*

Var. crispum, *Lindl. and Moore's Nat. Print. Ferns, t.* 34 *A.* A. (Filix-fœmina), *Hort.*

Var. corymbiferum, *Moore, Hand-bk. Brit. Ferns, p.* 145.

Var. Victoriæ, *Moore, Gard. Chron.* (1864).

Var. plumosum (*Moore*); *Lowe's New Ferns, t.* 14.

Var. dissectum (*Wollast.*); *Lindl. and Moore's Brit. Ferns, t.* 34 *C.*

Var. ovatum (*Roth.*); *Lindl. and Moore's Brit. Ferns, t.* 32.

Var. rhæticum (*Linn.*); *Lindl. and Moore's Brit. Ferns, t.* 31 *A.*

Var. Fieldiæ (*Moore*); *Gard. Chron.* (1861), *p.* 1046, *f. c.*

Var. Frizelliæ (*Moore*); *Gard. Chron.* (1861), *p.* 1046, *f. c.*

Var. acrocladon (*Clapham*); *Lowe's New Ferns t.* 40.

—Temperate Regions of Northern Hemisphere.

88. **A. Michauxii**, *Spreng.; Lowe's Ferns*, 5, *t.* 37. Nephrodium Filix-fœmina, *Michx.* Aspidium angustum, *Willd.* Asplenium Athyrium, *Spreng.; Schk. Fil. t.* 78. Nephrodium asplenoides, *Michx.*— North America.

89. **A. eburneum**, *J. Sm.* Aspidium eburneum, *Wall. Cat.* 389. Lastrea eburnea, *Cat. Hort. Kew.* (1846). Polypodium oxyphyllum, *Wall. Cat.* 324. Athyrium oxyphyllum, *Moore.*—Nepal.

90. **A. denticulatum,** *J. Sm.* Allantodia denticulata, *Wall.* Asplenium setulosum, *Hort.* Asplenium strigillosum, *Lowe's Ferns*, 5, *t.* 36. Athyrium tenuifrons, *Moore.*—Nepal.

91. **A. macrocarpum,** *Blume, in Herb.* Athyrium foliolosum, *Moore.*—Java and Ceylon.

92. **A. Ceylonense,** *Klot.* Athyrium Ceylonense, *Moore.*—Ceylon.

93. **A. umbrosum,** *J. Sm.; Lowe's Ferns*, 5, *t.* 1. Polypodium umbrosum, *Ait.* Aspidium umbrosum, *Sw.; Schk. Fil. t.* 61. Allantodia umbrosa, *R. Br.*—Madeira.

94. **A. axillare,** *Webb.* Polypodium axillare, *Ait.* Aspidium axillare, *Sw.* Allantodia axillaris, *Kaulf.*—Madeira.

95. **A. Brownii,** *J. Sm.; Hook. Ic. Pl. t.* 978. Allantodia Australis, *R. Br.* Athyrium Australe, *Presl; Hook. Gen. Fil. t.* 16.—Australia.

96. **A. decurtatum,** *Link; Metten. Fil. Hort. Lips. t.* 13, *f.* 17, 18. A. pubescens, *Houlst. and Moore.*—Brazil.

---

111. DIPLAZIUM, *Sm.*

Vernation fasciculate, erect or decumbent. *Fronds* simple, pinnate, or bi-tripinnate, 1–5 feet high. *Veins* forked or pin-

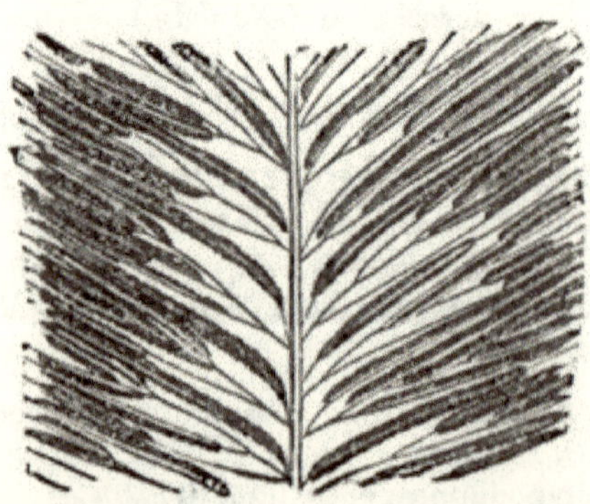

Genus 111.—Portion of fertile frond, under side. No. 1.

nate; venules free, sporangiferous on both sides, constituting binate linear sori. *Indusium* plane, binate.

* *Vernation erect.*

† *Fronds simple.*

1. **D. plantagineum**, *Sw.; Schk. Fil. t.* 15; *Lodd. Bot. Cab.* 1588; *Lowe's Ferns,* 5, *t.* 46. Asplenium plantagineum, *Linn.* Diplazium acuminatum, *Radd. Fil. Brass. t.* 57, *f.* 2.—Tropical America.

†† *Fronds pinnatifid.*

2. **D. Zeylanicum**, *J. Sm.* Asplenium (Eudiplazium) Zeylanicum, *Hook. Sp. Fil.* 3, *p.* 237; *Hook. 2nd Cent. Ferns, t.* 16.—Ceylon.

††† *Fronds pinnate.*

3. **D. grandifolium**, *Sw.*—Tropical America.

4. **D. juglandifolium**, *Sw.; Schk. Fil. t.* 85; *Hook. Fil. Exot. t.* 100. Asplenium juglandifolium, *Lam.*—Jamaica and Venezuela.

5. **D. alternifolium**, *Blume; Hook. Fil. Exot. t.* 17. Asplenium alternifolium, *Metten. Fil. Hort. Lips. t.* 12, *f.* 1–2. Diplazium integrifolium, *J. Sm. Cat.* (1857); *in Herb. J. Sm.*—Java.

†††† *Fronds bipinnatifid or bipinnate.*

6. **D. Shepherdi**, *Link.* Asplenium Shepherdi, *Hort.* Diplazium ambiguum, *J. Sm. Cat. Cult. Ferns* (1857); *Lowe's Ferns,* 5, *t.* 47.—West Indies.

7. **D. coarctatum**, *Link.* D. Shepherdi, *Presl.*—Brazil.

8. **D. striatum**, *Presl; Lowe's Ferns,* 5, *t.* 48. Asplenium striatum, *Linn.* (*Plum. Fil. t.* 18, 19).—Tropical America.

9. **D. expansum**, *Willd.* D. subalatum, *Hew.* — Tropical America.

10. **D. diversifolium**, *Wall. Herb.* (*fide spec. in Herb. J. Sm.*).—East Indies.

11. D. **conchatum**, *J. Sm.* Athyrium conchatum, *Fée, Gen. Fil. t.* 17 *C, f.* 1. Hypochlamys pectinata, *Fée, Gen. Fil. t.* 17 *C, f.* 3. Diplazium brevisorum, *J. Sm Cat. Cult. Ferns* (1857) (*non J. Sm. Enum. Fil. Philipp.*).—Jamaica, St. Domingo.

12. D. **polypodioides**, *Blume.* D. marginatum, *Hort.* Asplenium polypodioides, *Metten.; Hook. Sp. Fil.*—East Indies, Malayan and Pacific Islands.

13. D. **Klotzschii**, *Moore.* Asplenium Klotzschii, *Metten.* Lotzea diplazioides, *Klot. et Karst.* — Tropical America.

14 D. **costale** *Presl.* Asplenium costale, *Sw.* Diplazium fabæfolium, *J. Sm. Ms. in Herb.*—West Indies.

** *Vernation decumbent.*

15. D. **sylvaticum**, *Sw.; Schk. Fil. t.* 85 *B.; Lowe's Ferns,* 5, *t.* 49. Callipteris sylvatica, *Bory.* Anisogonium sylvaticum, *Hook. Gen. Fil. t.* 56 *B.* Asplenium acuminatum, *Wall.* Diplazium acuminatum, *Presl; J. Sm. Cat. Cult. Ferns* (1857).—East Indies.

16. D. **dilatatum.** *Blume.*—East Indies, Malayan Islands.

17. D. **arborescens**, *Sw.; J. Sm. Cat. Fil. Hort. Kew.* (1856). Asplenium arborescens, *Metten. Fil. Hort. Lips. t.* 13, *f.* 19, 20.—St. Helena.

18. D. **decussatum**, *J. Sm.; Lowe's Ferns,* 5, *t.* 50. Asplenium decussatum, *Well.* D. lasiopteris, *Kunze.*—East Indies.

19. D. **Thwaitesii**, *J. Sm.* Asplenium Thwaitesii, *A. Br.; Hook. 2nd Cent. Ferns, t.* 45.—Ceylon.

20. D. **thelypteroides**, *Presl; Lowe's Ferns,* 5, *t.* 51. Asplenium thelypteroides, *Michx.; Schk. Fil. t.* 76 *B.*—North America. T.

†† *Fronds deltoid, decompound.*

21. D. **Franconis**, *Lieb.* Asplenium Franconis, *Metten. Asplen. p.* 66, *t.* 5, *f.* 30.—Mexico, Jamaica.

112. SCOLOPENDRIUM, *Sm.*

Vernation fasciculate, erect. *Fronds* simple entire, lobed, or pinnate; frequently abnormally forked, plain, undulate, or with a comose, crested apex; from 6 inches to 2–3 feet long. *Veins* forked; venules free, the superior and inferior branch of each fascicle contiguous, parallel, and sporangiferous on their proximate sides, constituting two linear, confluent sori, each furnished with a linear indusium, the free margins of which connive.

1. **S. vulgare,** *Sm.; Eng. Bot. t.* 1150*; Lindl. and Moore's Brit. Ferns, t.* 40; *Sowerby's Ferns, t.* 35; *Lowe's Ferns,* 5, *t.* 55; *Hook. Brit. Ferns, t.* 37. S. officinarum, *Sw.; Schk. Fil. t.* 83; *Hook. Gen. Fil. t.* 57 *B.* Asplenium Scolopendrium, *Linn.*—Europe, Madeira.

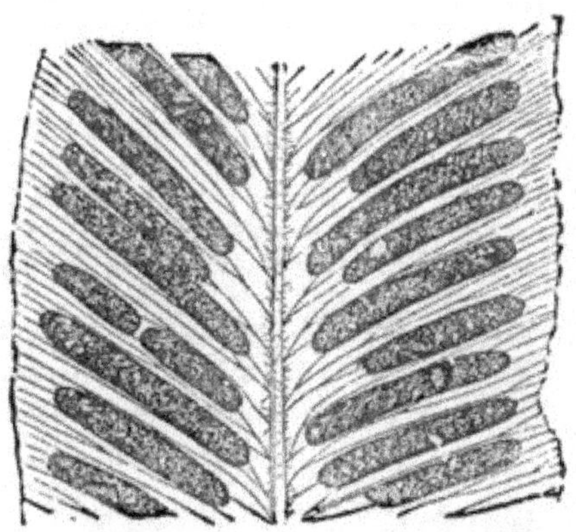

Genus 112.—Portion of mature frond, under side. No. 1.

Var. polyschides, *Lindl. and Moore, l. c. t.* 42, *f.* 2.

Var. cornutum, *Lindl. and Moore, l. c.*

Var. marginatum, *Lindl. and Moore, l. c. t.* 42, *f.* 3.

Var. crispum, *Lindl. and Moore, l. c. t.* 42, *f.* 4.

Var. multifidum, *Lindl. and Moore, l. c.*

Var. laceratum, *Lind. and Moore, l. c. t.* 42, *f.* 10.

Var. incisum (*Roth.*) *Lindl.; and Moore's Brit. Ferns, t.* 30.

Other Varieties: — Macrosorum; fissum; obtusidentatum; crenato-lobatum; resectum; sinuatum; inæquale; rimosum; inops; irregulare; spirale; compositum; nudicaule; abruptum; variabile; striatum; subvariegatum; apicilobum; lanceolum; sagittifolium; sagittato-cristatum; retinervium; pachyphyllum; coriaceum; pocilliferum; peraferum; muricatum; jugosum; papillosum; scalpturatum; imperfectum; siciforme; submarginatum; proliferum; fimbriatum; bimarginatum; supralineatum; supralineato-resectum; multiforme; chelæfrons; crista-galli; digitatum; glomeratum; flabellatum; cristatum; lacerato-marginatum; ramo-marginatum; ramosum-majus.—Moore's "Handbook of British Ferns."*

---

113. NEOTTOPTERIS, *J. Sm.*

Vernation fasciculate, erect. *Fronds* simple, linear or broad elliptical-lanceolate, smooth, 1–4 feet long by 2–8 inches wide.

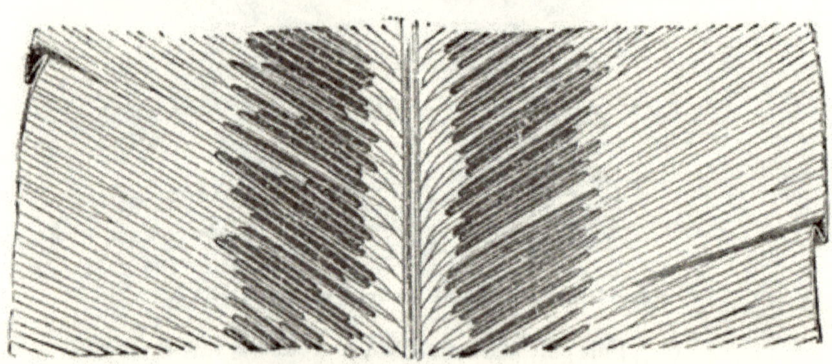

Genus 113.—Portion of mature frond, under side. No. 1.

Veins forked; venules parallel, sporangiferous on their superior side, their apices combined by a transverse, continuous, marginal vein. *Sori* unilateral. *Indusium* plane.

---

* See page xii of Preface to Second Edition.

1. **N. Nidus,** *J. Sm.; Hook. Gen. Fil. t.* 113. Asplenium Nidus, *Linn.; Bot. Mag. t.* 3101; *Lowe's Ferns, t.* 36.—East Indies.

2. **N. Australasica,** *J. Sm. Cat. Cult. Ferns* (1857). Asplenium Australasicum, *Hook. Fil. Exot. t.* 88. Asplenium Nidus, *R. Br.* (*non Linn.*); *Lowe's Ferns*, 5, *t.* 15.—New South Wales.

3. **N. phyllitidis,** *J. Sm. En. Fil. Philipp.* Asplenium Phyllitidis, *Don. Prod. Fl. Nep.*—India.

---

114. ANTIGRAMMA, *Presl; J. Sm.*

Vernation fasciculate, erect. *Fronds* simple, cordate-lanceolate, 4–18 inches high. *Veins* forked; venules angularly anastomosing, reticulated, the marginal veinlets free. *Sporangia* produced on the proximate sides of the primary venules of each fascicle, constituting two linear, confluent sori, each furnished with a linear indusium, the free margins of which connive.

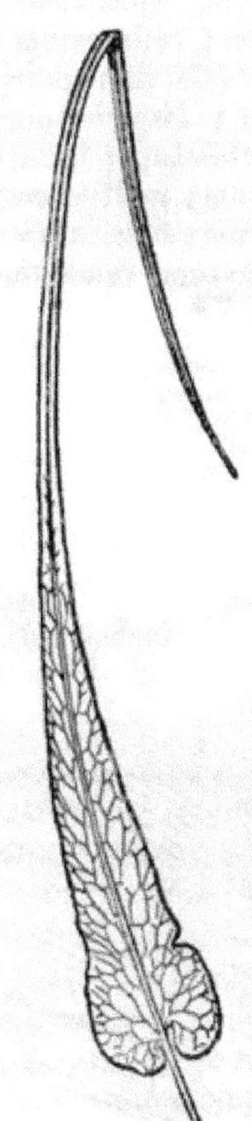

Genus 114. — Mature frond, upper side. No. 1.

1. **A. rhizophylla,** *J. Sm.* Asplenium rhizophyllum, *Linn.* Camptosorus rhizophyllus, *Link; Hook. Gen. Fil. t.* 57 *C*; *Hook. Fil. Exot. t.* 85; *Metten. Fil. Hort. Lips. t.* 5, *f.* 6. Scolopendrium rhizophyllum, *Hook.* — North America. **T.**

2. **A. Brasiliensis,** *Moore.* Asplenium Brasiliense, *Sw.* Scolopendrium Brasiliense, *Kunze.* S. ambiguum, *Radd. Fil. Bras. t.* 57, *f.* 1. Antigramme repanda, *Presl; Hook. Gen. Fil. t.* 57 *A*; *Hook. Ic. Pl. t.* 183.—Brazil.

---

115. CALLIPTERIS, *Bory; J. Sm.*

Vernation fasciculate, erect. *Fronds* simple and pinnate, bipinnatifid or bipinnate, 2–5 feet high. *Veins* uniform and

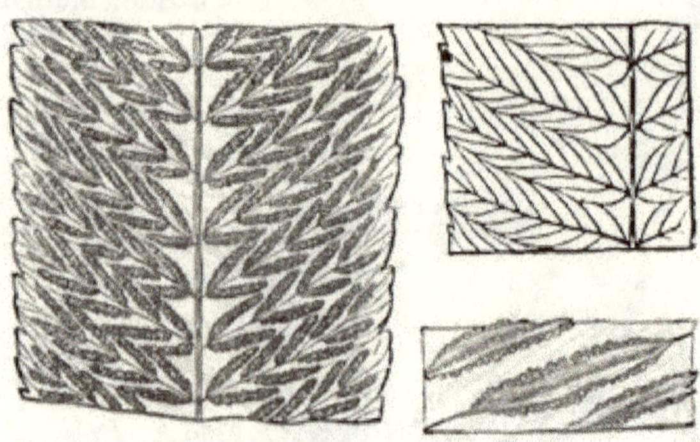

Genus 115.—Portion of barren and fertile fronds, natural size; fertile enlarged. No. 1.

forked, or costæform pinnate; the whole, or the lower venules only, anastomosing, sporangiferous on both sides, constituting binate, linear, decussate sori. *Indusium* plane.

1. **C. prolifera,** *Bory.* Asplenium proliferum, *Lam.; Metten. Fil. Hort. Lips. t.* 11, *f.* 7. Diplazium proliferum, *Kaulf.* Asplenium decussatum, *Sw.* Anisogonium decussatum, *Presl; Hook. Gen. Fil. t.* 56 *A.*—East Indies, Malayan Archipelago.

2. **C. esculenta,** *J. Sm.* Hemionites esculenta, *Retz.* Diplazium esculentum, *Sw.* Anisogonium esculentum, *Presl.* Microstegia esculenta, *Presl, Epim. Bot.* Digrammaria esculenta, *Fée.* Asplenium ambiguum, *Sw.; Schk. Fil. t.* 75 *B* (*Rheede, Mal.* 12, *t.* 15). Digrammaria ambigua, *Presl*; *Hook. Gen. Fil. t.* 56 C. Microstegia ambigua, *Presl, Epim. Bot.* Diplazium Malabaricum, *Spreng.* Callipteris Malabarica, *J. Sm. Cat. Cult. Ferns* (1857). Diplazium Serampurense, *Spreng.* Anisogonium Serampurense, *Presl.* Callipteris Serampurense, *Fée.* Diplazium pubescens, *Link.*—East Indies, Malayan Archipelago.

---

116. HEMIDICTYUM, *Presl.*

Vernation fasciculate, erect. *Fronds* pinnate, 10–14 feet long; pinnæ 1–2 feet long, 3–5 inches wide. *Veins* forked; venules parallel till near the margin, then anastomosing and reticulated,

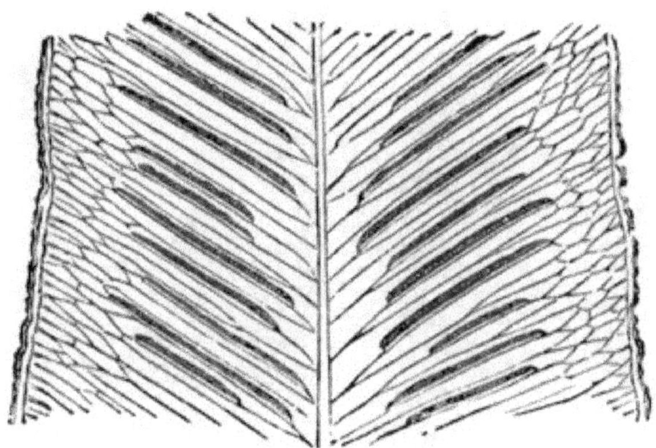

Genus 116.—Portion of fertile pinna, under side. No. 1.

combined by a transverse, continuous, marginal vein. *Sporangia* produced on the superior side of the parallel veins, constituting unilateral, linear sori. *Indusium* plane.

1. **H. marginatum,** *Presl; Hook. Gen. Fil. t.* 55 *A.* Asplenium marginatum, *Linn.* (*Plum. Fil. t.* 106); *Hook. Fil. Exot. t.* 73; *Lowe's Ferns,* 5, *t.* 53. Diplazium giganteum, *Hort. Linden.*—Tropical America.

---

117. CETERACH, *Willd.; J. Sm.*

Vernation fasciculate, erect, cæspitose. *Fronds* 2–12 inches long, sinuose-pinnatifid or pinnate, the under side densely squamose. *Veins* forked, anastomosing. *Sporangia* unilateral, protruding through the dense squamæ, forming oblong sori. *Indusium* obsolete.

1. **C. officinarum**, *Willd.; Hook. Gen. Fil. t.* 113; *Lindl. and Moore's Brit. Ferns, t.* 43 *A; Lowe's Ferns,* 5, *t.* 54. Asplenium Ceterach, *Linn.; Hook. Brit. Ferns, t.* 36.

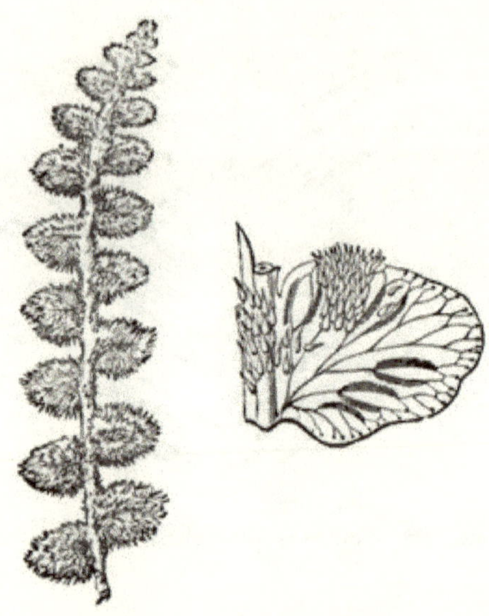

Genus 117.—Frond natural size, and portion enlarged. No. 1.

Grammitis Ceterach, *Sw.; Schk. Fil. t.* 7 *B; Lodd. Bot. Cab. t.* 15. Scolopendrium Ceterach, *Eng. Bot.* 1244.—Europe, North India.

TRIBE X.—DICKSONIEÆ.

Sori marginal, round, or linear and transverse. *Indusium* lateral, interiorly attached, its extrorse margin free and conniving more or less with the changed margin of the frond, which becomes an accessory indusium, the two forming a cucullate or bivalved round cyst, or elongated grove, containing the sporangia.

§ 1. *Lindsæeæ.*

Receptacles combined, forming a linear, continuous or interrupted marginal sorus, or rarely punctiform or binate. Indusium linear or sub-rotund.

* *Receptacles elongated.*

118. LINDSÆA, *Dry.*

Vernation fasciculate, erect or decumbent. *Fronds* simple, pinnate, or bi-tripinnate; pinnæ oblong, dimidiate, upper margin fertile only; costa excentric or obsolete. *Veins* radiating, forked; venules free, their apices combined by an elongated transverse receptacle. *Sori* linear, continuous or interrupted. *Indusium* linear, usually shorter than the indusiform margin.

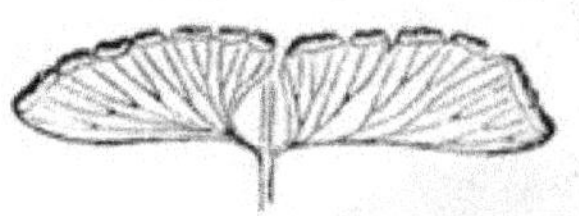

Genus 118.—Portion of fertile frond, under side. No. 1.

* *Occidental species.*

1. **L. reniformis,** *Dry. in Linn. Trans.* 3, *t.* 7, *f.* 1; *Kunze, in Schk. Fil. Suppl. t.* 16, *f.* 2.—French Guiana.*
2. **L. sagittata,** *Dry.; Hook. et Grev. Ic. Fil. t.* 87.—French Guiana.
3. **L. Leprieusii,** *Hook. Sp. Fil.* 1, *t.* 62 *D.*—French Guiana.
4. **L. falcata,** *Dry. in Linn. Trans.* 3, *t.* 7, *f.* 2.—Guiana.
5. **L. trapeziformis,** *Dry. in Linn. Trans.* 3, *t.* 9; *Hook. et Bauer. Gen. Fil. t.* 63 *A.*—West Indies and Tropical America.
6. **L. Guianensis,** *Dry.; Hook. Sp. Fil.* 1, *t.* 62.—Guiana.
7. **L. stricta,** *Dry.; Schk. Fil. t.* 114. L. Javitensis, *H. B. K.; Radd. Fil. Bras. t.* 75, *f.* 1. L. elegans, *Hook. Ic. Pl. t.* 98.—West Indies and Tropical America.
8. **L. crenata,** *Klot.; Hook. Sp. Fil.* 1, *p.* 208.—British Guiana.
9. **L. dubia,** *Spr.; Hook. Sp. Fil.* 1, *t.* 64 *C.*—French Guiana.

** *Indian and Malayan species.*

10. **L. cultrata,** *Sw.; Schk. Fil. t.* 114; *Hook. et Grev. Ic. Fil. t.* 114; *Hook. Fil. Exot. t.* 67; *Lowe's New Ferns, t.* 16 *B.*—East Indies.

* See Appendix to Second Edition.

11. **L. obtusa,** *J. Sm. En. Fil. Philipp.; Hook. Sp. Fil.* 1, *p.* 224.—Malacca.

*** *Polynesian and Australian species.*

12. **L. linearis,** *Sw. Syn. Fil. t.* 3, *f.* 3; *Kunze, Fil. t.* 16; *Lowe's New Ferns, t.* 16 *C.*—Australia and Tasmania.

13. **L. trichomonoides,** *Dry. in Linn. Trans.* 3, *t.* 11; *Schk. Fil. t.* 14, *f.* 3.—New Zealand.

14. **L. microphylla,** *Sw.; Hook. et Grev. Ic. Fil. t.* 194.—New South Wales and Queensland.

---

119. SCHIZOLOMA, *Gaud.*

Vernation fasciculate. *Fronds* pinnate; pinnæ oblong or linear-lanceolate; costa central. *Veins* forked; venules anastomosing, forming oblique, elongated areoles, transversely combined by an elongated *Receptacle* on both margins. *Sori* linear, continuous. *Indusium* linear, usually equal with the indusiform margin.

1. **S. ensifolia,** *J. Sm.* Lindsæa ensifolia, *Sw.; Hook. et Grev. Ic. Fil. t.* 3. Lindsæa lanceolata, *Labill. Nov. Holl. t.* 248, *f.* 1.—Malayan and Polynesian Islands.

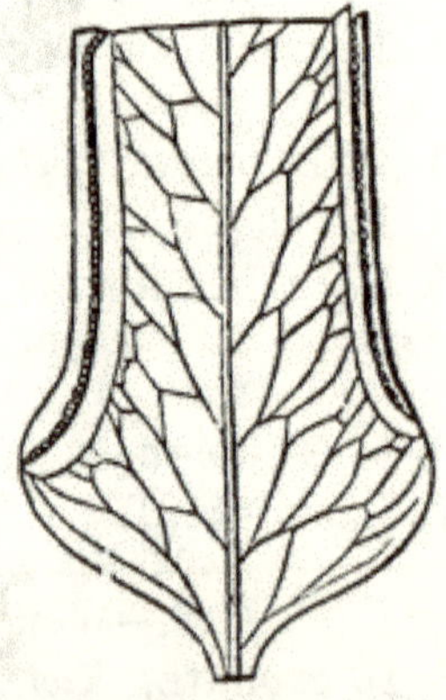

Genus 119.—Portion of fertile frond; natural size. No. 1.

** *Receptacles punctiform.*

120. ODONTOSORIA, *J. Sm.*

Vernation uniserial, distant and sarmentose, contiguous and sub-fasciculate. *Fronds* bi-tripinnatifid, lanceolate or deltoid, 1–5 feet long, erect or flexuose and scandent; ultimate segments cuneiform, entire, lobed or laciniated. *Veins* dichotomously forked; venules free. *Receptacles* terminal, punctiform.

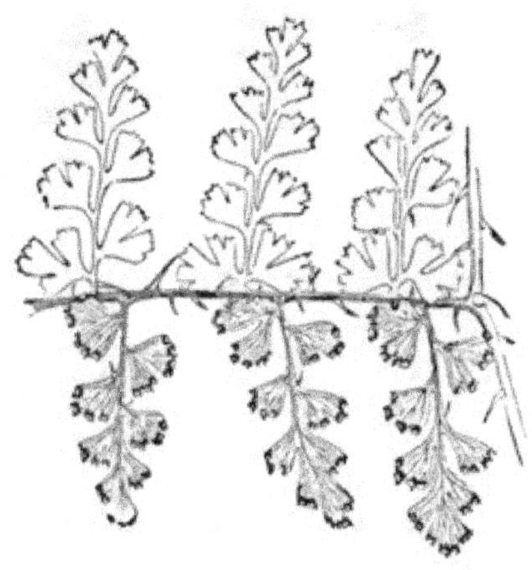

Genus 120.—Portion of fertile pinna, under side. No. 2.

Sori simple or binate. *Special* and *accessory indusia* forming a vertical, urceolate or, by confluence, oblong, sporangiferous, marginal cyst.

1. **O. tenuifolium**, *J. Sm.* Davallia tenuifolia, *Sw.; Lowe's Ferns*, 8, *t.* 14. Stenoloma tenuifolium, *Fée.*—East Indies and Malayan Archipelago.

2. **O. aculeatum**, *J. Sm.* Davallia aculeata, *Sm.; Hook. Sp. Fil. t.* 54 *B; Lowe's Ferns*, 8, *t.* 26. Adiantum aculeatum, *Linn.* (*Plum. Fil. t.* 94). Stenoloma aculeatum, *Fée, Gen. Fil. t.* 27, *f.* 4.—West Indies.

---

121. MICROLEPIA, *Presl.*

Vernation uniserial, sarmentose. *Fronds* pinnate or bi-tripinnatifid, 1–6 feet high, deltoid. *Veins* simply or pinnately forked; venules free, the exterior one or more soriferous. *Sori* simple, often anti-marginal. *Receptacles* terminal, punctiform. *Indusium* attached by its broad base only, or by its base and sides, constituting a simple, cucullate or semi-urceolate, vertical cyst.

Genus 121.—Fertile pinna, under side. No. 4.

1. **M. tricosticha,** *J. Sm.* Davallia tricosticha, *Hook.; Lowe's Ferns,* 8, *t.* 29.—Philippine Islands.

2. **M. scabra,** *J. Sm.* Davallia scabra, *Don.* Davallia villosa, *Wall.; Hook. Sp. Fil. t.* 48 *A.*—India, Japan.

3. **M. cristata,** *J. Sm. En. Fil. Philipp.* Davallia Khasyana, *Hook. Sp. Fil. t.* 47 *A,* 5–7 *A.*—East Indies.

4. **M. platyphylla,** *J. Sm.* Davallia platyphylla, *D. Don.* Davallia Lonchitidea, *Wall.; Hook. Sp. Fil. t.* 46 *B; Lowe's Ferns,* 8, *t.* 30; *Hook. Fil. Exot. t.*19. Davallia majuscula, *Lowe's Ferns,* 8, *t.* 33.—East Indies.

5. **M. polypodioides,** *Presl; Hook. Gen. Fil. t.* 58. Davallia polypodioides, *D. Don.* Polypodium nudum, *Forst.* Davallia rhomboidea, *Wall.* . Davallia flaccida, *R. Br.* —East Indies, Polynesia.

6. **M. strigosa,** *Moore.* Davallia strigosa, *Sw.* Trichomanes strigosa, *Thunb.*—Japan.

7. **M. Novæ-Zelandiæ,** *J. Sm.* Davallia Novæ-Zelandiæ, *Colenso; Hook. Sp. Fil. t.* 51 *B; Hook. Gard. Ferns, t.* 51. Davallia hispida, *Hew.* Acrophorus hispidus, *Moore.*—New Zealand.

122. LOXSOMA, *R. Br.*

Vernation uniserial, sarmentose. *Fronds* long stipitate, deltoid, decompound, 1–1½ foot high, glaucous beneath; laciniæ lanceolate, dentate. *Veins* simple or forked; venules free, their apices prolonged, forming a free columnar receptacle. *Special*

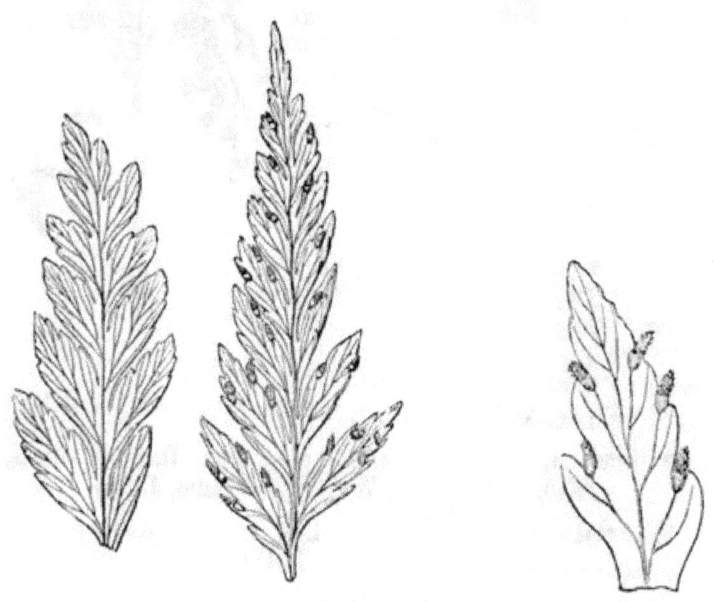

Genus 122.—Portions of barren and fertile frond, natural size; ditto, enlarged. No. 1.

and *Accessory Indusia* united, forming a vertical, urceolate, extrorse cyst. *Sporangia* obovate, pedicellate, seated round the receptacle, which is elongated beyond the mouth of the indusium. Ring of sporangium oblique.

1. **L. Cunninghamii,** *R. Br.; Hook. et Bauer. Gen. Fil. t.* 15; *Comp. to Bot. Mag. t.* 31, 32; *Hook. Gard. Ferns, t.* 31.—New Zealand.

§ 2. *Eudicksonieæ.*

Receptacles punctiform. Special and accessory indusia conniving, forming an urceolate or bivalved, reflexed cyst.

* *Vernation uniserial, sarmentose, or rarely sub-fasciculate and erect or decumbent, naked or thinly furnished with scales.*

123. SACCOLOMA, *Kaulf.*

Vernation fasciculate, erect. *Fronds* 4–6 feet high, pinnate, 1–2 feet broad, smooth; pinnæ linear-lanceolate, acuminate, 8–12 inches long, serrated at the apex. *Veins* simple, rarely forked, direct, parallel, free. *Receptacles* punctiform, terminal. *Sori* punctiform, contiguous, laterally coalescing and forming a compound, marginal, continuous sorus. *Special indusium* small, transverse, elongated, sub-scyphiform; accessory one universal, formed of the continuous, reflexed margin.

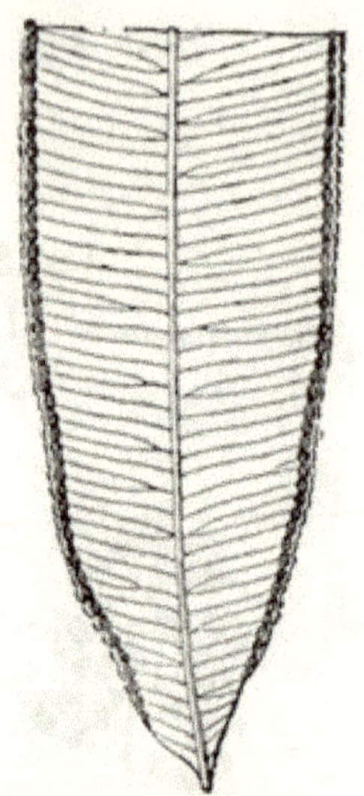

Genus 123.—Portion of fertile pinna. No. 1.

1. **S. elegans,** *Kaulf. En. Fil. t.* 1, *f.* 12; *Hook. Gen. Fil. t.* 58, *f.* 1, 2; *Kunze, Fil. t.* 41. Davallia saccoloma, *Spreng.* —West Indies.

---

124. DEPARIA, *Hook. et Grev.*

Vernation subfasciculate, decumbent. *Fronds* bipinnatifid,

Genus 124.—Portion of barren and fertile frond. No. 1.

1–2½ feet long. *Veins* pinnate; venules free. *Receptacles* punctiform, terminal. *Sori* exserted. *Special* and *accessory indusia* conniving, and forming a calyciform, pedicellate, vertical extrorse cyst.

1. **D. prolifera,** *Hook. et Grev. Ic. Fil.* (*corrig.*); *Hook. Gen. Fil. t.* 44 *B*; *Hook. Fil. Exot. t.* 82; *Lowe's Ferns*, 8, *t.* 38. Dicksonia prolifera, *Kaulf.* Deparia Macræi, *Hook. et Grev. Ic. Fil. t.* 154.—Sandwich Islands.

---

125. SITOLOBIUM, *Desv.*

Vernation uniserial and sarmentose. *Fronds* bi-tripinnatifid, 2–6 feet high. *Veins* simple or pinnately forked, the exterior venule, or more, soriferous. *Receptacles* punctiform, terminal. *Sori* globose, reflexed. *Special* and *accessory indusia* united and forming a reflexed, entire, or sub-bilabiate cup.

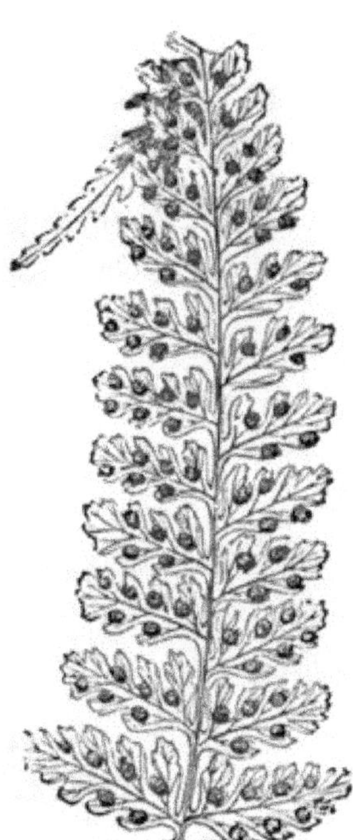

Genus 125.—Portion of fertile frond, under side. No. 2.

1. **S. punctilobum,** *J. Sm.* Nephrodium punctilobulum, *Michx.* Aspidium punctilobulum, *Sw.* Dicksonia punctiloba, *Hook.*; *Lowe's Ferns*, 8, *t.* 42. Dicksonia pubescens, *Schk. Fil. t.* 131. Dicksonia pilosiuscula, *Willd.* Sitolobium pilosiusculum, *Desv.*; *J. Sm. Gen. Fil.* Dennstædtia punctilobula, *Moore.*—North America.

2. **S. adiantoides,** *J. Sm.* Dicksonia adiantoides, *Humb.*; *Hook. Sp. Fil. t.* 26 *B.* Polypodium globuliferum, *Poir* (*Plum. Fil. t.* 30). Dennstædtia adiantoides, *Moore.*—Tropical America.

3. **S. Pavoni,** *J. Sm.* Dicksonia Pavoni, *Hook. Sp. Fil.* 1, *t.* 26 *A.* Dennstædtia Pavoni, *Moore.* — Tropical America.

4. **S. dissectum,** *J. Sm.* Dicksonia dissecta, *Sw.; Schk. Fil. t.* 130 *B.* Dennstædtia dissecta, *Moore.* Dennstædtia tenera, *Moore.*—West Indies.

5. **S. cicutarium,** *J. Sm.* Dicksonia cicutaria, *Sw.; Lowe's Ferns,* 8, *t.* 40 (*Plum. Fil. t.* 31). Dennstædtia cicutaria, *Moore.*—Tropical America.

6. **S. anthriscifolium,** *J. Sm.* Dicksonia anthriscifolia, *Kaulf.; Hook. Sp. Fil.* 1, *t.* 27 *B.* Dennstædtia anthriscifolia, *Moore.*—Tropical America.

7. **S. davallioides,** *J. Sm.* Dicksonia davallioides, *R. Br.; Lowe's Ferns,* 8, *t.* 41. Dennstædtia davallioides, *Moore.*—Australia.

8. **S. rubiginosum,** *J. Sm.* Dicksonia rubiginosa, *Kaulf.; Hook. Sp. Fil. t.* 27 *A; Lowe's Ferns,* 8, *t.* 45. Dennstædtia rubiginosa, *Moore.* Dicksonia nitidula, *Kunze.* Dennstædtia nitidula, *Moore.* — Tropical America.

9. **S. Moluccanum,** *J. Sm.* Dicksonia Moluccana, *Blume; Lowe's Ferns,* 8, *t.* 46. Dennstædtia Moluccana, *Moore.*—Malayan Archipelago.

** *Vernation fasciculate, erect and arboreous, or rarely decumbent, densely criniferous.*

126. BALANTIUM, *Kaulf.*

Vernation fasciculate, decumbent, densely criniferous. *Fronds* deltoid, tripinnate, smooth; ultimate segments dentate. *Veins* pinnate; venules free, simple or forked. *Receptacles* punctiform, terminal. *Sori* transversely oblong, large, exserted in thrysiform clusters. *Indusium* bivalved, coriaceous, the two valves nearly equal, concave, reniform.

1. B. Culcita, *Kaulf.* Dicksonia Culcita, *L'Hérit.; Lowe's*

Genus 126.—Portions of barren and fertile frond, natural size. No. 1.

Ferns, 8, *t.* 39. Culcita macrocarpa, *Presl; Hook. Gen. Fil. t.* 60 *A.*—Madeira, Azores, and Tropical America.

---

127. DICKSONIA, *L'Hérit.*

Vernation fasciculate, erect, arborescent, criniferous. *Fronds*

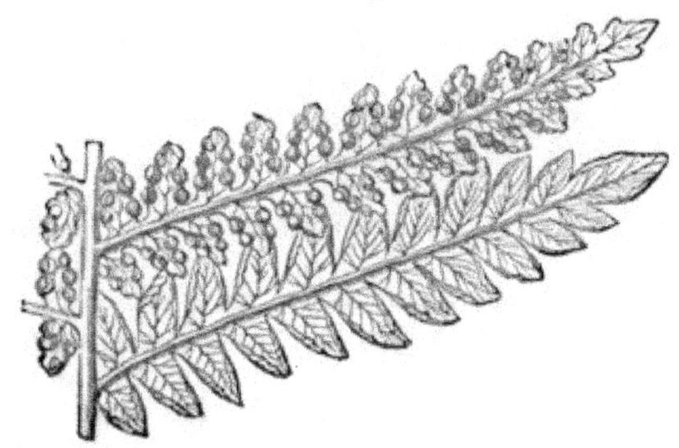

Genus 127.—Portions of barren and fertile fronds, natural size. No. 1.

bi-tripinnate, 5–15 feet long. *Veins* pinnate; venules free, simple. *Receptacles* punctiform, terminal. *Sori* globose, large, reflexed. *Indusium* bivalved, coriaceous; the outer valve (*accessory indusium*) concave, cucullate, conniving with the smaller, usually less cucullate, inner valve or special indusium, forming an unequal valved cyst.

1. D. **arborescens,** *L'Hérit.; Hook. Sp. Fil. t.* 22 *A.* D. auricoma, *Spreng.* Balantium auricomum, *Kaulf.; Presl.* Dicksonia integra, *Sw.* Balantium arborescens, *Hook. Gen. Fil. t.* 30.—St. Helena.
2. D. **antarctica,** *Labill. Nov. Holl. t.* 249. Balantium antarcticum, *Presl.* Cibotium Billardieri, *Kaulf.*—Australia.
3. D. **Sellowiana,** *Hook. Sp. Fil.* 1, *t.* 22 *B.* Balantium Sellowiana, *Presl.*—Tropical America.
4. D. **squarrosa,** *Sw.; Schk. Fil. t.* 130.—New Zealand.
5. D. **lanata,** *Colenso; Hook. Sp. Fil. t.* 23 *C.*—New Zealand.

---

128. CIBOTIUM, *Kaulf.*

Vernation fasciculate, decumbent, or erect and arborescent, densely criniferous. *Fronds* tripinnatifid, 5–15 feet long, generally glaucous beneath. *Veins* forked or pinnate; venules free.

Genus 128.—Portions of barren and fertile fronds, natural size.

Receptacles punctiform, terminal. *Sori* dentiform, reflexed. *Indusium* bivalved, horny, the outer valve (*accessory indusium*) concave, cucullate, adnate to the margin; the inner (*special indusium*) smaller, and conniving with the outer, forming an unequal bivalved cyst.

1. **C. Schiedei,** *Schlecht.; Hook. Sp. Fil. t.* 30 *A*; *Hook. Gen. Fil. t.* 25; *Lowe's Ferns*, 8, *t.* 35.—Mexico.

2. **C. Barometz,** *J. Sm. Gen. of Ferns.* Polypodium Barometz, *Lour.* Cibotium glaucescens, *Kunze, Fil. t.* 31. Cibotium Cumingi, *Kunze.*—China.

3. **C. Menziesii,** *Hook. Sp. Fil. t.* 29 *C.*—Sandwich Islands.

---

129. THYRSOPTERIS, *Kunze.*

Vernation fasciculate, arborescent. *Fronds* decompound-multifid, the fertile portion contracted. *Sori* paniculate. *Veins* pinnate; venules free. *Accessory* and *special indusia* equal,

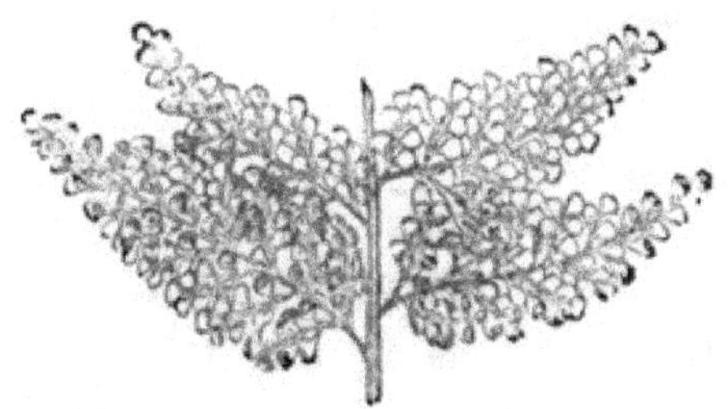

Genus 129.—Portion of fertile pinna. No. 1.

uniform, constituting a calyciform cyst, including sessile, compressed sporangia seated on an elevated, globose receptacle.

1. **T. elegans,** *Kunze, Fil. t.* 1; *Hook. Gen. Fil. t.* 44 *A; Lowe's Ferns*, 8, *t.* 34.—Juan Fernandez.

TRIBE XI.—CYATHEÆ.

Sori round, intra-marginal. *Receptacles* elevated, globose or columnar. *Indusium* calyciform, semi-calyciform or squamiform, or altogether absent.

130. SCHIZOCÆNA, *J. Sm.*

Vernation fasciculate, erect, slender, arborescent. *Fronds* simple, pinnate or bipinnatifid, 2–8 feet long, smooth, stipes adherent pinnæ articulated with the rachis. *Veins* pinnately forked; venules free. *Sori* medial. *Receptacles* globose. *Indusium* calyciform, ultimately deeply laciniated.

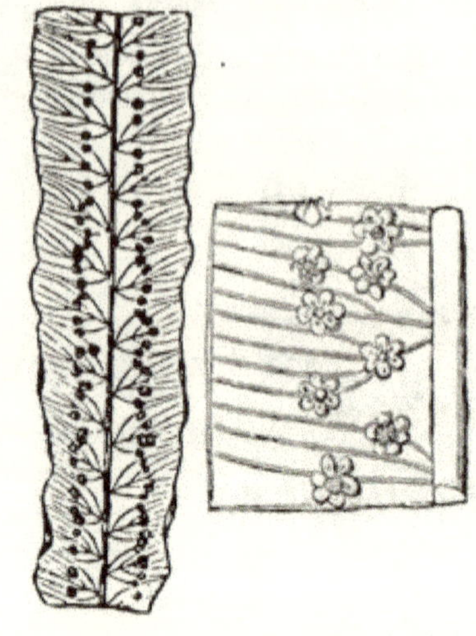

Genus 130.—Portion of fertile pinna, natural size; ditto enlarged. No. 1.

1. **S. sinuata**, *J. Sm. Gen. of Ferns* (1841). Cyathea sinuata, *Hook. et Grev. Ic. Fil. t.*106.—Ceylon.

---

131. CYATHEA, *Sm.*

Vernation fasciculate, erect, arborescent. *Fronds* bi-tripinnatifid, 5–15 feet long; pinnæ and pinnules in some species articulated with the rachis. *Veins* forked; venules free. *Sori* axillary. *Receptacles* columnar. *Indusium* complete calyciform, its margin entire or unequally laciniated.

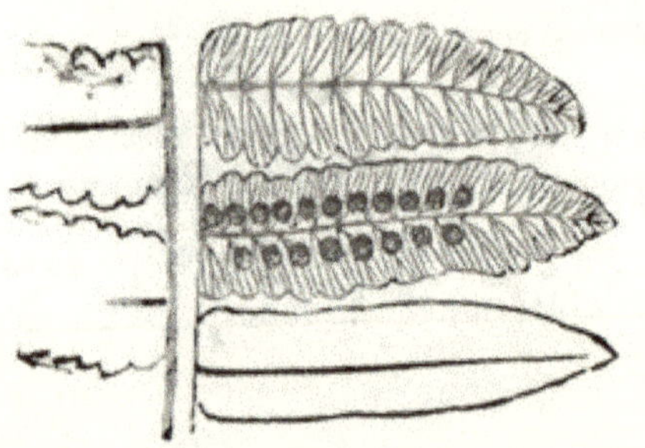

Genus 131.—Portion of fertile pinna, under side. No. 5.

* *West Indian and American Species.*

1. **C. arborea**, *Sm.* Polypodium arboreum, *Linn.* (*Plum. Fil. t.* 1 *et* 2). Disphenia arborea, *Presl.* Cyathea elegans, *Hew; Hook. Gen. Fil. t.* 23. Cyathea Grevilleana, *Mart.* Disphenia Grevilleana, *Kunze.*—West Indies.

2. **C. serra,** *Willd.; Hook. Sp. Fil.* 1, *t.* 9 *A.*—West Indies.

3. **C. aculeata,** *Willd.* Disphenia aculeata, *Presl.*—West Indies.

4. **C. nigrescens,** *J. Sm.* C. arborea, *var.* nigrescens, *Hook.*—Jamaica.

5. **C. muricata,** *Willd.* (*Plum. Fil. t.* 4).—West Indies.

** *African species.*

6. **C. canaliculata,** *Willd.; Hook. Sp. Fil.* 1, *t.* 11 *B; Lowe's Ferns,* 8, *t.* 55.—Mauritius.

7. **C. excelsa,** *Sw.; Hook. Sp. Fil.* 1, *t.* 12 *B; Lowe's Ferns,* 8, *t.* 56.—Mauritius.

8. **C. Dregei,** *Kunze; Hook. Sp. Fil. t.* 10 *B.*—South Africa.

9. **C. Manniana,** *Hook. Syn. Fil. p.* 21.—Fernando Po.

*** *Indian and Malayan species.*

10. **C. Hookeri,** *Thw. Enum. Plant. Zeyl.*—Ceylon.

11. **C. integra,** *J. Sm. En. Fil. Philipp.; Hook. Sp. Fil.* 1, *p.* 26.—Philippine Islands.

**** *Australian and Polynesian species.*

12. **C. medullaris,** *Sw.; Schk. Fil. t.* 133; *Hook. Gard. Ferns, t.* 25. Polypodium medullare, *Forst.*—Pacific Isles and New Zealand.

13. **C. Smithii,** *Hook. fil. Fl. New Zeal. t.* 72.—New Zealand.

14. **C. dealbata,** *Sw.; A. Rich. Fl. Nou. Zel. t.* 10; *Lowe's Ferns,* 8, *t.* 58. Polypodium dealbatum, *Forst.*—New Zealand.

15. **C. Cunninghamii,** *Hook. fil. Fl. New Zeal.* 2, *p.* 7; *Hook. fil. Hook. Ic. Pl. t.* 985.—New Zealand.

---

132. HEMITELIA, *R. Br.*

Vernation fasciculate, erect, arborescent. *Fronds* bipinnatifid, 4–8 feet long; stipes smooth or aculeated. *Veins* simply or pinnately forked; venules all free, or the lower pair of the lowest fascicles angularly anastomosing, forming a costal arch or more or less acute angle. *Sori* medial. *Receptacles* globose. *Indusium* semicalyciform.

1. **H. speciosa,** *Kaulf.; Hook. Sp. Fil. t.* 13 *B; Hook. Fil. Exot. t.* 66. Cyathea speciosa, *Humb.* Hemitelia integrifolia, *Klot.*—Tropical America.
2. **H. grandifolia,** *Spreng.; Hook. Sp. Fil. t.* 14 *B; Lowe's Ferns,* 8, *t.* 59. Cyathea grandifolia, *Willd.* (*Plum. Fil. t.* 26).—West Indies.
3. **H. horrida,** *R. Br.; Hook. Sp. Fil. t.* 15; *Hook. Fil. Exot. t.* 69; *Lowe's Ferns,* 8, *t.* 60. Polypodium horridum, *Linn.* (*Plum. Fil. t.* 8). Cyathea horrida, *Sm.* Cnemidaria horrida, *Presl.; Hook. Gen. Fil. t.* 4.—West Indies.
4. **H. obtusa,** *Kaulf.; Hook. Sp. Fil.* 1, *t.* 14. Hemitelia speciosa, *Mart. Ic. Crypt. Bras. t.* 48, *f.* 2. Cnemidaria speciosa, *Presl.*—West Indies.
5. **H. Karsteniana,** *Klot.; Kunze, Ind. Fil.*—Venezuela.
6. **H. Imrayana,** *Hook. Sp. Fil.* 1, *p.* 33; *Hook. Ic. Pl. t.* 669.—Dominica.

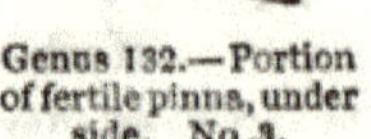

Genus 132.—Portion of fertile pinna, under side. No. 3.

---

133. ALSOPHILA, *R. Br.*

Vernation fasciculate, erect, arborescent. *Fronds* bi-tripinnatifid, 5–15 feet long. *Veins* simple or forked, free. *Sori* axillary or medial. *Receptacles* globose or columnar. *Indusium* semicalyciform, or small and squamiform, or trichiform, often obsolete.

§ 1. Hymenostegia, *J. Sm Gen. Fil.* (1841).

Indusium nearly complete calyciform or very small and scale-like.

* *African species.*

Genus 133.—Portion of fertile pinna, under side. No. 1.

1. **A. Capensis,** *J. Sm. Gen. Fil.* (1841). Polypodium capense, *Linn.* Cyathea Capensis, *Sm.* Hemitelia Capensis, *R. Br.* Amphicosmia Capensis, *Moore.*—South Africa.

** *Tropical America and West Indian species.*

2. **A. lævis,** *J. Sm. Gen. Fil.* Amphicosmia lævis, *Moore.* Hemitelia Guianensia, *Hook. Ic. Pl. t.* 648.—British Guiana.

3. **A. Hostmanni,** *J. Sm.* Hemitelia Hostmanni, *Hook. Ic. Pl. t.* 646; *Lowe's Ferns,* 8, *t.* 61. Amphicosmia Hostmanni, *Moore.*—Guiana.

4. **A. Surinamensis,** *J. Sm.* Hemitelia Surinamensis, *Miquel.*—Guiana and Martinique.

5. **A. radens,** *Kaulf.; Metten. Fil. Hort. Lips.*—Brazil.

6. **A. Beyrichiana,** *J. Sm.* Cyathea Beyrichiana, *Presl; Hook. Ic. Pl. t.* 623. Amphicosmia Beyrichiana, *Moore.*—Brazil.

§ 2. Tricostegia, *J. Sm. Gen. Fil.* (1841).

Indusium absent. Sori furnished with articulated hairs, or naked.

* *Tropical American and West Indian species.*

7. **A. aspera,** *R. Br.; Hook. et Grev. Ic. Fil. t.* 213, 214, 215; *Hook. Gen. Fil. t.* 21; *Hook. Sp. Fil. t.* 19 *B.* Polypodium asperum, *Linn.* (*Plum. Fil. t.* 3).—Tropical America and West Indies.

8. **A. ferox,** *Presl; Hook.* A. armata, *Mart. Ic. Crypt. Bras. t.* 48 (*non Presl*). Polypodium aculeatum, *Radd. Fil. Bras. t.* 42.—Tropical America and West Indies.

9. **A. aculeata**, *J. Sm.* Polypodium aculeatum, *Radd. Fil. Bras. t.* 42. Alsophila ferox, *Presl.*—West Indies and Tropical America.

10. **A. armata**, *Presl.* Polypodium armatum, *Sw.*—Tropical America.

11. **A. procera**, *Kaulf.* Polypodium procerum, *Willd.*—Brazil.

12. **A. villosa**, *Presl.* Cyathea villosa, *H. B. K. Nov. Gen. t.* 670.—Tropical America.

13. **A. paleolata**, *Mart. Ic. Crypt. Bras. t.* 43. A. munita, *Hort. Berol.*—Brazil.

** *Indian and Malayan species.*

14. **A. glauca**, *J. Sm. Gen. Fil.* Chnoophora glauca, *Blume.* Alsophila contaminans, *Wall.; Hook. Sp. Fil. t.* 18 *B.*—Malayan, Molucca, and Philippine Islands.

15. **A. gigantea**, *Wall.; Hook. Sp. Fil.* 1, *p.* 53.—India, Ceylon.

*** *Australian and Polynesian species.*

16. **A. Australis**, *R. Br.; Hook. Sp. Fil. t.* 19 *A.*—East and South Australia and Tasmania.

17. **A. excelsa**, *R. Br.; Hook. Gen. Fil. t.* 9; *Hook. Sp. Fil. t.* 18 *A; Backhouse's Narrative, p.* 265, *with table.* A. Cooperi, *Hook. Mss.*—Norfolk Island and Queensland.

18. **A. Moorei**, *J. Sm. Mss. Hort. Kew.* (1854). Stem slender, black, 3–5 feet high; fronds bipinnate, 3–4 feet long; pinnules deeply pinnatifid; laciniæ elliptical, obtuse, entire; stipes and main rachis muricate; sori small, naked. A. Macarthuri, *Hook. Mss.*—New South Wales.

---

134. TRICHOPTERIS, *Presl.*

Vernation fasciculate, erect, arborescent. *Fronds* bipinnate, 4–6 feet long, smooth, stipes articulated with the axis; pinnæ distant; pinnules lanceolate, coriaceous, articulated with the rachis. *Veins* pinnately forked; venules free, their apices clavate. *Sori* medial, criniferous, oblong, laterally contiguous

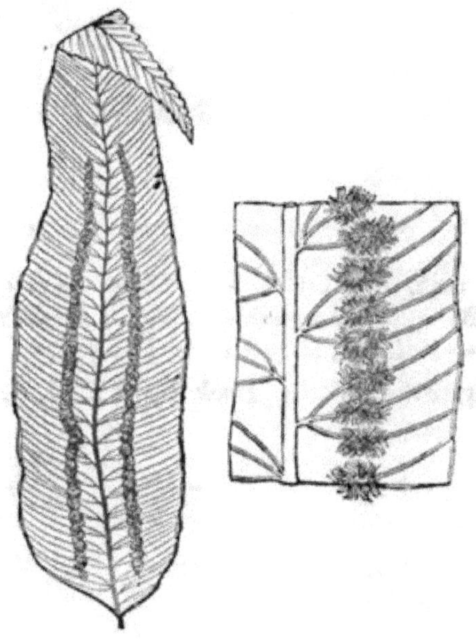

Genus 134.—Fertile pinna, natural size; portion of ditto enlarged. No. 1.

and confluent, forming a transverse row. *Receptacles* scarcely elevated. *Indusium* absent.

1. **T. excelsa**, *Presl.* Alsophila excelsa, *Mart. Ic. Crypt Bras. t.* 37.—Brazil.

---

135. LOPHOSORIA, *Presl.*

Vernation fasciculate, erect, arborescent, densely criniferous. *Fronds* tripinnatifid, 3–5 feet long, glaucous beneath. *Veins* forked; venules free, their apices thickened. *Sori* medial, criniferous. *Receptacle* scarcely elevated. *Indusium* absent.*

1. **L. pruinata,** *Presl.* Polypodium pruinatum, *Sw.* Alsophila pruinata, *Kaulf.* Polypodium griseum, *Schk. Fil. t.* 25 *B.*—Tropical America.

Genus 135.—Portions of fertile frond, natural size. No. 1.

2. **L. affinis,** *Presl.* Alsophila affinis, *Fée.* A. Deckeriana, *Klot.*—Venezuela.

SUB-ORDER II.—GLEICHENIACEÆ.

Sporangia globose or pyriform, furnished with a transverse or sub-oblique ring. *Fronds* rigid, opaque. *Sori* punctiform, intramarginal, naked or rarely with a peltate indusium.

136. GLEICHENIA, *R. Br.*

Vernation uniserial and sarmentose. *Fronds* 1–6 feet high once or more times dichotomously branched; pinnæ linear,

---

* Setting aside the arborescent character of the stem, this genus is more naturally related to the section *Desmopodium* of *Phegopteris* than to *Alsophila*.

pinnatifid, ultimate divisions small, ovate, orbicular, and often revolute or larger, linear and plane. *Veins* simply or pinnately

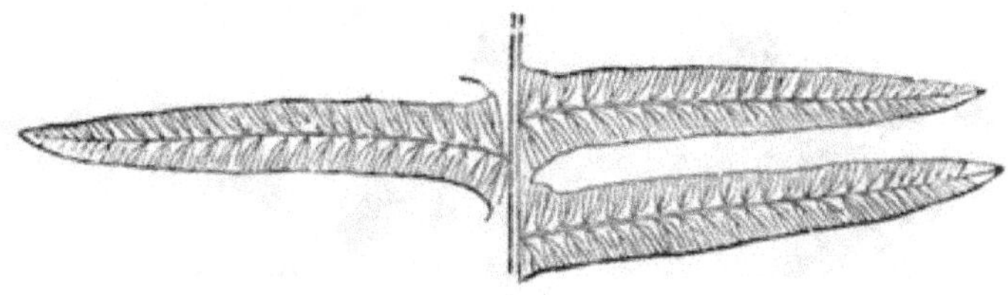

Genus 136.—Portion of barren frond, under side. No. 9.

forked; venules free, the exterior one fertile. *Sori* punctiform, terminal or medial. *Sporangia* few, 2–8, superficial or immersed.

* *Ultimate divisions small, concave or cucullate.* (Euglcichenia.)

1. **G. microphylla,** *R. Br.; Lowe's Ferns,* 8, *t.* 47.—New South Wales and Tasmania.
2. **G. dicarpa,** *R. Br.; Hook. Sp. Fil.* 1, *t.* 1 *C; Kunze, Fil. t.* 70; *Lowe's Ferns,* 8, *t.* 48.—Tasmania.
3. **G. semivestita,** *Labill. Sert. Nov. Caled. t.* 11; *Lowe's Ferns,* 8, *t.* 54; *Hook. Sp. Fil.* 1, *t.* 2 *A.*—New Caledonia and Malacca.
4. **G. hecistophylla,** *A. Cunn.; Hook. Sp. Fil.* 1, *t.* 2 *B; Lowe's Ferns,* 8, *t.* 52.—New Zealand.
5. **G. rupestris,** *R. Br.; Hook. Sp. Fil.* 1, *t.* 1 *B; Lowe's Ferns,* 8, *t.* 35.—New South Wales.
6. **G. alpina,** *R. Br.; Hook. et Grev. Ic. Fil. t.* 58.—Tasmania.
7. **G. speluncæ,** *R. Br.; Hook. Sp. Fil.* 1, *t.* 1 *A; Lowe's Ferns,* 8, *t.* 94.—New South Wales and Tasmania.

** *Ultimate divisions plane.* (Mertensia, *Willd.*)

8. **G. flabellata,** *R. Br.; Labill. Sert. Nov. Caled. t.* 12; *Lowe's Ferns,* 8, *t.* 50; *Hook. Fil. Exot. t.* 71. Mertensia flabellata, *J. Sm.*—Australia and Tasmania.

9. **G. dichotoma,** *Hook.; Lowe's Ferns*, 8, *t.* 21. Mertensia dichotoma, *Willd.; Schk. Fil. t.* 148; *Lang. et Fisch. Ic. Fil. t.* 29. Polypodium dichotomum, *Thunb. Fl. Jap. t.* 37. Gleichenia Hermanni, *R. Br.*— General throughout the Tropical and Subtropical regions of the Southern Hemisphere.

10. **G. furcata,** *Spreng.; Lowe's New Ferns, t.* 60. Acrostichum furcatum, *Linn.* (*Plum. Fil. t.* 28).—West Indies.

11. **G. pectinata,** *Presl.* Mertensia glaucescens, *Willd.* Gleichenia Hermanni, *Hook. et Grev. Ic. Fil. t.* 14 (*non R. Br.*).—West Indies.

12. **G. pubescens,** *Kunth.* Mertensia pubescens, *H. B. K.* Gleichenia immersa, *Spreng.; Hook. et Grev. Ic. Fil. t.* 15.—Tropical America.

13. **G. cryptocarpa,** *Hook. Sp. Fil.* 1, *t.* 6 *A.*—Chili.

14. **G. Cunninghami,** *Hew.; Hook. Sp. Fil.* 1, *t.* 6 *B; Hook. fil. Fl. New Zeal.* 6, *t.* 71.—New Zealand.

SUB-ORDER III.—HYMENOPHYLLACEÆ.

Sporangia globose or oblate, furnished with a horizontal or sub-oblique ring. *Fronds* thin, membranaceous, pellucid. *Sori* marginal. *Indusium* an urceolate, sub-bivalved, extrorse, open cyst.

137. HYMENOPHYLLUM, *Sm.*

Vernation uniserial and sarmentose. *Fronds* varying from simple to decompound-multifid, membranaceous and pellucid, smooth, or bearing simple, forked, or stellate hairs. *Veins* simple or forked, free. *Sori* terminal. *Indusium* short, urceolate, bilabiate or bivalved. *Receptacle* short, included within the indusium.

* *Fronds glabrous.*

† *Segments entire, plane or undulated.*
Stipes and rachis rarely pilose.

1. **H. asplenioides,** *Sw.; Hook.* 1st *Cent. Ferns, t.* 56.—Jamaica, Brazil.

2. **H. abruptum,** *Hook. Sp. Fil.* 1, *t.* 31 *B.*—West Indies.

3. **H. polyanthos,** *Sw.; Hedw. Fil. cum Ic.; Lowe's Ferns,* 8, *t.* 8 *A.* H. sanguinolentum, *Sw.; Schk. Fil. t.* 135 *C.*—West Indies, Tropical America, India, Philippines, New Zealand, &c.

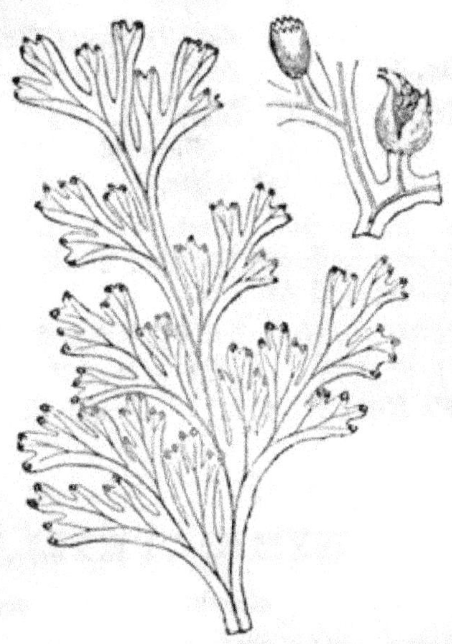

Genus 137.—Portion of fertile frond, natural size; ditto slightly enlarged. No. 13.

4. **H. rarum,** *R. Br.* H. semibivalve, *Hook. et Grev. Ic. Fil. t.* 83.—Tasmania, New Zealand, Chili, South Africa, Ceylon.

5. **H. flabellatum,** *Labill. Nov. Holl. t.* 250. H. nitens, *Hook. et Grev. Ic. Fil. t.* 197.—Tasmania, New Zealand.

6. **H. demissum,** *Sw.; Schk. Fil. t.* 135 *C.*—Tasmania, New Zealand.

7. **H. scabrum,** *A. Rich. Fl. Nov. Zel. t.* 14, *f.* 1; *Lowe's New Ferns, t.* 179.—New Zealand.

8. **H. crispatum,** *Wall.; Hook. et Grev. t.* 77; *Lowe's New Ferns, t.* 69 *B.*—East Indies, Philippines, Tasmania, New Zealand.

9. **H. flexuosum,** *A. Cunn.; Hook. Ic. Pl. t.* 962; *Lowe's New Ferns, t.* 178.—New Zealand.

10. **H. caudiculatum,** *Mart. Ic. Crypt. Bras. t.* 67; *Lowe's New Ferns, t.* 68.—Brazil.

11. **H. fuciforme,** *Sw.; Hook. Sp. Fil. t.* 36 *D; Lowe's New Ferns, t.* 72.—Chili, Juan Fernandez.

12. **H. pulcherrimum,** *Colenso; Hook. Sp. Fil.* 1, *t.* 37 *A; Hook. fil. Fl. Nov. Zealand,* 2, *t.* 74; *Lowe's New Ferns, t.* 71.—New Zealand.

13. **H. dilatatum,** *Sw.; Schk. Fil. t.* 135; *Hook. et Grev. Ic. Fil. t.* 60; *Lowe's New Ferns, t.* 70.—New Zealand.

†† *Segments dentate or spinulose-serrate, often undulate.*

14. **H. Tunbridgense,** *Sm.; Eng. Bot. t.* 162; *Hook. Gen. Fil. t.* 32; *Hook. Fl. Lond. t.* 71; *Sowerby's Ferns, t.* 42; *Hook. Brit. Ferns, t.* 43; *Lindl. and Moore's Nature-printed Ferns, t.* 49 *A.* H. cupressiforme, *Labill. Nov. Holl. t.* 250, *f.* 2.—Temperate regions of both hemispheres.

15. **H. unilaterale,** *Willd.; Lindl. and Moore's Nature-printed Ferns, t.* 49 *B; Sowerby's Ferns, t.* 43. H. Wilsoni, *Hook. Eng. Bot. t.* 2686; *Hook. Brit. Ferns, t.* 44.—Temperate regions of both hemispheres.

16. **H. multifidum,** *Sw.; Schk. Fil. t.* 135 *B; Hook. et Grev. Ic. Fil. t.* 167.—New Zealand.

17. **H. dichotomum,** *Cav.; Hook. Sp. Fil.* 1, *t.* 36 *A.*—Chili.

18. **H. fucoides,** *Sw.; Hook. Ic. Pl. t.* 963.—West Indies and Tropical America.

** *Fronds pilose.*

19. **H. hirsutum,** *Sw.; Radd. Fil. Bras. t.* 79, *f.* 1; *Hook. et Grev. Ic. Fil.* 84.—West Indies, Brazil.

20. **H. ciliatum,** *Sw.; Hook. et Grev. Ic. Fil. t.* 35; *Lowe's New Ferns, t.* 69 *C.*—West Indies and Tropical America.

21. **H. hirtellum,** *Sw.; Hook. Sp. Fil.* 1, *t.* 31.—Jamaica.

22. **H. Chiloense,** *Hook. Sp. Fil.* 1, *t.* 32 *A; Lowe's New Ferns, t.* 69 *A.*—Chili.

23. **H. valvatum,** *Hook. et Grev. Ic. Fil. t.* 219.—Columbia.

24. **H. lineare,** *Sw.* H. trifidum, *Hook. et Grev. Ic. Fil. t.* 196. H. elegans, *Spreng.* — West Indies and Tropical America.

25. **H. sericeum,** *Sw.* (*Plum. Fil. t.* 73). — West Indies and Tropical America.

26. **H. æruginosum,** *Carm.; Hook. Sp. Fil. t.* 34 *A.*—Tristan d'Acunha, New Zealand.

H. venolum.

138. TRICHOMANES, *Linn.*

Vernation fasciculate and erect, or uniserial and sarmentose. *Fronds* varying from simple to decompound-multifid, membranaceous and pellucid, smooth, or bearing simple, forked, or stellate hairs. *Veins* simple or forked, free. *Sori* terminal, often sub-pedicellate. *Indusium* urceolate or tubular. *Receptacle* continued beyond the sporangia and mouth of the indusium, often elongated and filiform.

Genus 138.—Fertile frond, natural size. No. 11.

* *Vernation uniserial, sarmentose.*

1. **T. reniforme,** *Forst.; Hook. et Grev. Ic. Fil. t.* 31; *Hook. Fil. Exot. t.* 76. — New Zealand.

2. **T. membranaceum,** *Linn.* (*Plum. Fil. t.* 101, *f. A*); *Hook. Exot. Fl. t.* 76. — West Indies.

3. **T. punctatum,** *Poir.; Hook. et Grev. Ic. Fil. t.* 236. — West Indies.

4. **T. reptans,** *Sw.; Hook. et Grev. Ic. Fil. t.* 32.—West Indies.

5. **T. Bojeri,** *Hook. et Grev. Ic. Fil. t.* 155.—Mauriti

6. **T. muscoides,** *Sw.; Hook. et Grev. Ic. Fil. t.* 179.—West Indies.

7. **T. pusillum,** *Sw.; Hedw. Fil. cum Ic.; Lowe's New Ferns, t.* 163.—West Indies.

8. **T. Kraussii,** *Hook. et Grev. Ic. Fil. t.* 149; *Lowe's New Ferns, t.* 164.—West Indies and Tropical America.

9. **T. venosum,** *R. Br.; Hook. et Grev. Ic. Fil. t.* 78.—New South Wales and Tasmania.

10. **T. sinuosum,** *Rich.; Hook. et Grev. Ic. Fil. t.* 13; *Lowe's Ferns,* 8, *t.* 10 *C.*—West Indies.

11. **T. pyxidiferum,** *Linn.* (*Plum. Fil. t.* 20 *C*); *Hook. et Grev. Ic. Fil. t.* 206; *Lowe's New Ferns, t.* 161.—West Indies.

12. **T. Filicula,** *Bory.* T. bilabiatum, *Nees, in Nov. Act. Cur.* (1823), *t.* 13, *f.* 2. T. bilingue, *J. Sm.* Hymenophyllum alatum, *Schk. Fil. t.* 135 *B.*—East Indies, Mauritius, Philippine and Polynesian Islands.

13. **T. angustatum,** *Carm.; Hook. et Grev. Ic. Fil. t.* 166; *Lowe's New Ferns, t.* 67 *A.*—Tristan d'Acunha.

14. **T. exsectum,** *Kunze, Anal. Pterid. t.* 29. *f.* 2; *Lowe's New Ferns, t,* 64 *A.*—Chili and Juan Fernandez.

15. **T. trichoideum,** *Sw.; Hook. et Grev. Ic. Fil. t.* 199; *Lowe's New Ferns, t.* 67 *B.* T. pyxidiferum, *Schk. Fil. t.* 134.—West Indies.

16. **T. radicans,** *Sw.; Lindl. and Moore's Nature-printed Ferns, t.* 48; *Hook. Brit. Ferns, t.* 42; *Sowerby's Ferns, t.* 41. T. brevisetum, *R. Br.* T. speciosum, *Willd.* T. pyxideferum, *Huds.* (*non Linn.*). T. alatum, *Hook. in Fl. Lond. t.* 53 (*non Sw.*). T. Europæum, *Sm. in Rees' Cyclop.* T. Hibernicum, *Spreng.* Hymenophyllum alatum, *Sm. Eng. Bot. t.* 1417;—β Andrewsii, *Lindl. and Moore's Nat. Print. Ferns, t.* 48 *C.* Trichomanes Andrewsii, *Newm.*—Tropical and Temperate regions of the Northern Hemisphere.

17. **T. scandens,** *Linn.; Sloane's Jam.* 1, *t.* 58; *Lowe's New Ferns, t.* 62 *A.*—West Indies.

18. T. **incisum,** *Kaulf.; Bory, in Dup. Voy. t.* 38, *f.* 1.—Brazil.

19. T. **pluma,** *Hook. Ic. Pl. t.* 997; *Lowe's New Ferns, t.* 63 *A.*—Borneo.

** *Vernation fasciculate.*

20. T. **crispum,** *Linn.* (*Plum. Fil. t.* 86); *Hook. et Grev. Ic. Fil. t.* 12; *Hook. Gard. Ferns, t.* 27. T. pilosum, *Radd. Fil. Bras. t.* 79.—West Indies and Tropical America.

21. T. **pennatum,** *Hedw. Fil. t.* 4, *f.* 1; *Hook. Gard. Ferns, t.* 8. T. floribundum, *H. B. K.; Hook. et Grev. Ic. Fil. t.* 9. T. Vittaria, *Dec.; Hook. Lond. Journ. Bot.* 1, *t.* 5.—West Indies and Tropical America.

22. T. **Kaulfussii,** *Hook. et Grev. Ic. Fil. App.; Lowe's New Ferns, t.* 63 *B.* T. lucens, *Hook. et Grev. Ic. Fil. t.* 10.—West Indies.

23. T. **fimbriatum,** *Backhouse, Cat.* (1861); *Gard. Chron.* (1862), *p.* 44.—West Indies.

24. T. **attenuatum,** *Hook. Sp. Fil.* 1, *t.* 39 *C; Lowe's New Ferns, t.* 66.—West Indies.

25. T. **alatum,** *Sw.* (*Plum. Fil. t.* 50, *f.* 1); *Hook. et Grev. Ic. Fil. t.* 21.—West Indies.

26. T. **Bancrofti,** *Hook. et Grev. Ic. Fil. t.* 204; *Hook. Gard. Ferns, t.* 56. T. coriaceum, *Kunze.*—West Indies.

27. T. **crinitum,** *Sw.; Hedw. Fil. cum Ic.*—West Indies.

28. T. **superbum,** *Backhouse, Cat.* (1861); *Gard. Chron.* (1862), *p.* 44.—Borneo.

*** *Vernation fasciculate, erect.*

29. T. **Javanicum,** *Blume; Hook. et Grev. Ic. Fil. t.* 210; *Hook. Gard. Ferns, t.* 37. T. alatum, *Bory, in Dup. Voy. t.* 38, *f.* 2 (*non Sw.*). T. rhomboideum, *J. Sm.* T. curvatum, *J. Sm.*—Malayan and Pacific Islands.

30. T. **Leprieurii,** *Kunze, Hook. Gard. Ferns, t.* 11. T. anceps, *Hook. Sp. Fil.* 1, *t.* 40 *C; Lowe's New Ferns, t.* 68. T. Mandioccana, *Radd. Fil. Bras. t.* 79.—Brazil and West Indies.

31. **T. rigidum,** *Sw.; Hedw. Fil. cum Ic.* T. obscurum, *Blume.*—Tropics.

32. **T. fœniculaceum,** *Bory.* T. meifolium, *Kaulf. En. Fil. t.* 2 (*non Bory*).—Mauritius and Bourbon.

33. **T. meifolium,** *Bory.* T. Bauerianum, *Endl.*—East Indies, Bourbon, Malayan, Philippine and Pacific Islands, Norfolk Island.

34. **T. elongatum,** *A. Cunn.; Hook. Ic. Pl. t.* 701.—New Zealand.

35. **T. setigerum,** *Backhouse, Cat.* (1861); *Gard. Chron.* (1862).—Borneo.

36. **T. saxatile,** *Moore, in Gard. Chron.* (1862). T. rupestre, *Backhouse, Cat.* (1861).—Borneo.

37. **T. tricophyllum,** *Moore, in Gard. Chron.* (1862).—Borneo.

---

139. FEEA, *Bory.*

Vernation fasciculate, erect. *Fronds* of two forms, 2–6 inches high; the sterile pinnatifid or sub-pinnate; the fertile contracted, rachiform, stipate, longer than the sterile. *Veins* simple or forked; venules free. *Sori* terminal, free, pedicellate, in a row along both sides of the rachis.* *Indusium* tubular, calyciform. *Receptacle* filiform, continued beyond the sporangia and mouth of the *indusium.*

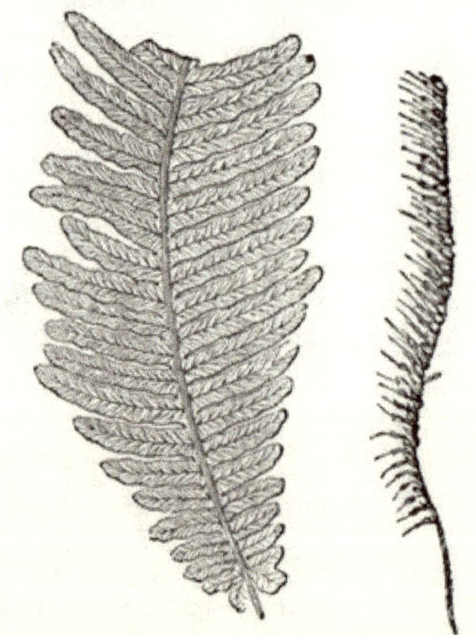

Genus 139.—Barren and fertile frond, natural size. No. 1.

* Not one-sided as shown in figure

1. **F. spicata,** *Presl.* Trichomanes spicatum, *Hedw.; Hook. Gard. Ferns, t.* 60; *Lowe's New Ferns, t.* 67 *C.* T. elegans, *Rudge* (*in part*); *Hook. Exot. Fil. t.* 52. Feea polypodina, *Bory, in Dict. Sc. Nat. cum Ic.*—West Indies.

2. **F. nana,** *Bory.* Trichomanes nanum, *Bory; Hook. Sp. Fil.* 1, *p.* 115.—Guiana.

---

140. HYMENOSTACHYS, *Bory.*

Vernation fasciculate, erect. *Fronds* of two forms, 6–10 inches high; the sterile pinnatifid; the fertile contracted, linear, longer than the sterile. *Veins* in the sterile forked; venules anastomosing, forming oblique elongated areoles; in

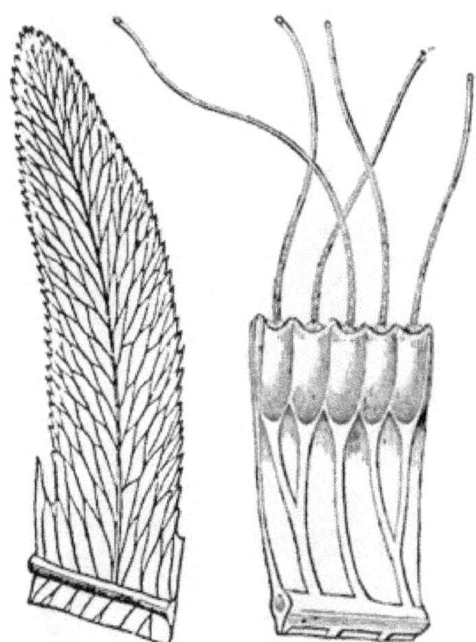

Genus 140.—Portion of barren frond, natural size; ditto fertile enlarged. No. 1.

the fertile, simple or forked, free. *Sori* terminal, immersed contiguous in a row along both margins of the fertile frond. *Indusium* urceolate-calyciform. *Receptacle* filiform, continued beyond the sporangia and mouth of the indusium.

1. H. elegans, *Presl.* Trichomanes elegans, *Rudge, Guian. t.* 35 (*in part*); *Hook. Gen. Fil. t.* 108; *Hook. Gard. Ferns, t.* 2.—Guiana, Trinidad, Pacific side of Central America.

SUB-ORDER IV.—OSMUNDACEÆ.

Sporangia globose, oval or oblong, opening vertically; apex striated, the striæ forming a more or less complete ring, which is sometimes rudimentary only.

TRIBE I.—SCHIZÆÆ.

Sporangia oval or oblong, opening on the exterior side, produced on contracted racemes, or on terminal or marginal spike-like appendices, or the fertile frond is wholly contracted, or sub-contracted. Apical ring complete.

141. LYGODIUM, *Sw.*

Vernation uniserial, distant and sarmentose, or contiguous and cæspitose. *Fronds* scandent, twining, extending to an indefinite length; pinnæ conjugate, palmate-lobed, pinnatifid or pinnate. *Veins* forked, free. *Sporangiferous* spikelets marginal, composed of two rows of imbricate indusiate cysts, each cyst containing an oval resupinate sporangium attached by its inner side, and opening longitudinally on its outer side.

1. L. palmatum, *Sw.; Schk. Fil. t.*140; *Lowe's Ferns,* 8, *t.* 74; *Hook. Fil. Exot. t.* 24.—North America. T.

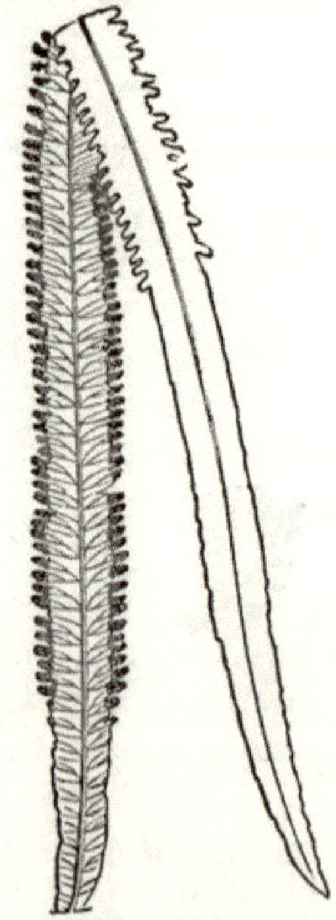

Genus 141.—Portion of fertile frond, under side. No. 2.

2. **L. flexuosum,** *Sw.* Ophioglossum flexuosum, *Linn.* Lygodium dichotomum, *Sw.; Hook. et Grev. Ic. Fil. t.* 55.—East Indies and Malayan Archipelago.

3. **L. circinnatum,** *Sw.*—Malayan and Philippine Islands.

4. **L. scandens,** *Sw.* Ophioglossum scandens, *Linn.*—East Indies.

5. **L. Japonicum,** *Sw.* Ophioglossum Japonicum, *Thunb.*—China and Japan.

6. **L. articulatum,** *A. Rich. in Voy. d'Astrolabe, t.* 15.—New Zealand.

7. **L. polystachyum,** *Wall.*—East Indies.

8. **L. microphyllum,** *R. Br.*—Tropical Australia, Polynesian Islands.

---

142. LYGODICTYON, *J. Sm.*

Vernation and general habit the same as in *Lygodium.* *Veins* reticulated.

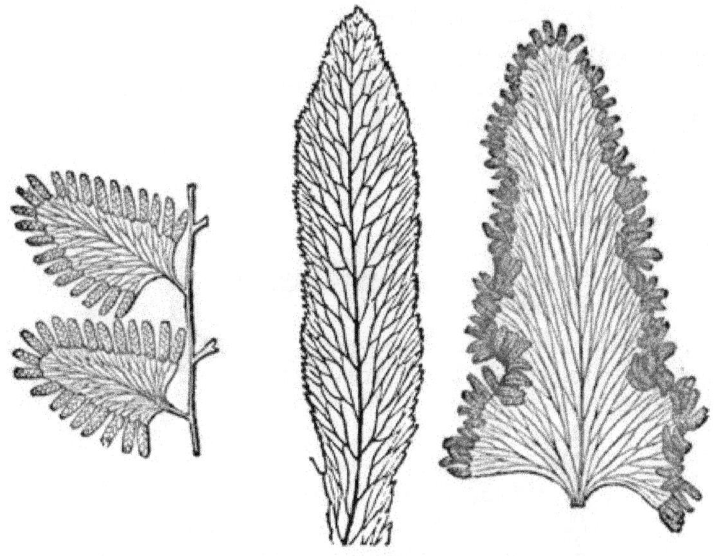

Genus 142.—Barren and fertile pinna, natural size; fertile enlarged. No. 1.

1. **L. Forsteri,** *J. Sm. in Hook. Gen. Fil. t.* 111 *B.* Lygodium reticulatum, *Schk. Fil. t.* 139. Hydroglossum polycarpum, *Willd.* Ophioglossum scandens, *Forst.* (*non Linn.*).—Polynesian Islands.

2. **L. heterodoxum,** *J. Sm.* Lygodium heterodoxum, *Kunze, Fil. t.* 113. Hydroglossum heterodoxum, *Moore.* Lygodium Lindeni, *Hort.*—Guatemala.

---

143. ANEMIA, *Sw.*

Vernation fasciculate, erect or decumbent. *Fronds* pinnate or bi-tripinnatifid; the fertile always tripartite, the two opposite segments contracted, erect, constituting two sporangiferous racemes, the third segment sterile, with forked free *veins. Sporangia* oval, attached by the base, opening vertically on the exterior side.

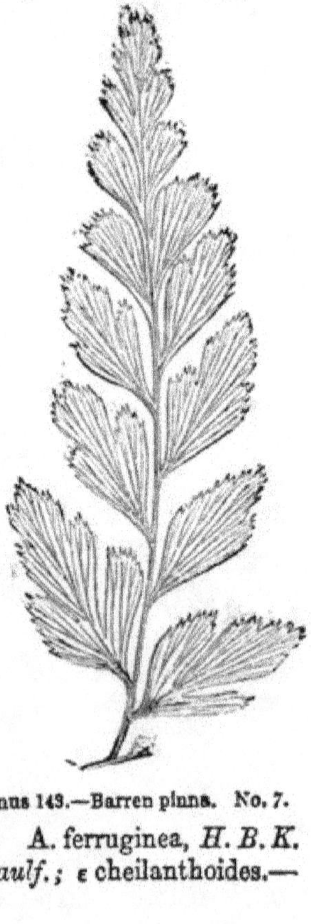

1. **A. Dregeana,** *Kunze, Fil. t.* 20; *Hook. Icon. Pl. t.* 236.—South Africa.

2. **A. collina,** *Radd. Fil. Bras. t.* 12; *Hook. Fil. Exot. t.* 1. Anemia hirta, *Hort.* (*non Sw.*). — Tropical America.

3. **A. Mandioccana,** *Radd. Fil. Bras. t.* 9, *f.* 1; *Hook. Gard. Ferns, t.* 36.—Brazil.

4. **A. tomentosa,** *Sw.* Osmunda tomentosa, *Lam.* Anemia flexuosa, *Sw.; Radd. Fil. Bras. t.* 13; *Hook. Fil. Exot. t.* 30. A. villosa, *H. et B.; Presl.* A. raddiana, *Link.* A. ferruginea, *H. B. K.* Anemia cheilanthoides, *Kaulf.;* ε cheilanthoides.—Tropical America, Brazil.

Genus 143.—Barren pinna. No. 7.

5. **A. fulva,** *Sw.; Schk. Fil. t.* 142; *Hook. Fil. Exot. t.* 126.—Tropical America.

6. **A. hirsuta,** *Sw.* Osmunda hirsuta, *Linn.* (*Plum. Fil. t.* 162). Anemia repens, *Radd. Fil. Bras. t.* 2 *B.*—Tropical America.

7. **A. adiantifolia,** *Sw.; Lowe's New Ferns, t.* 32. Osmunda adiantifolia, *Linn.* (*Plum. Fil. t.* 158). Anemia cicutaria, *Moore et Houlst.* Anemia adiantifolia, β asplenifolia, *Hook. et Grev. Ic. Fil. t.* 16. A. asplenifolia, *Sw.*—Tropical America.

---

144. ANEMIDICTYON, *J. Sm.*

Vernation and habit the same as in *Anemia;* but with reticulated *veins.*

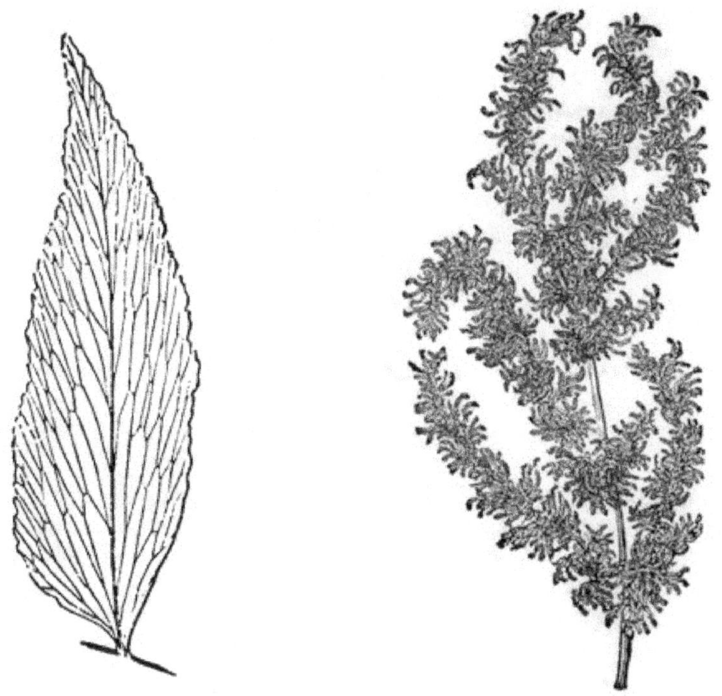

Genus 144.—Barren pinna, natural size; fertile spike, ditto. No. 1.

1. A. **Phyllitidis**, *J. Sm. in Hook. Gen. Fil. t.* 103. Osmunda Phyllitidis, *Linn.* (*Plum. Fil. t.* 156). Anemia Phyllitidis, *Sw.*; β longifolium. Anemia longifolia, *Radd. Fil. Bras. t.* 8. Anemidictyon Phyllitidis, *Lowe's Ferns*, 8, *t.* 71; γ fraxinifolium. Anemia fraxinifolia, *Radd. Fil. Bras. t.* 8 *bis*; δ densum. Anemia densa, *Link.*—Tropical America.

---

145. MOHRIA, *Sw.*

Vernation fasciculate, decumbent. *Fronds* bipinnate, 6–12 inches high; pinnæ entire, laciniated, or multifid; the fertile generally contracted, constituting a sporangiferous raceme, or sub-contracted, with the margin of the segments inflexed, indusiform, and sporangiferous. *Veins* free. *Sporangia* sessile, oval or nearly globose, opening vertically on their exterior side.

Genus 145.—Portion of fertile frond, under side. No. 1.

1. M. **thurifraga**, *Sw. Syn. Fil. t.* 5; *Schk. Fil. t.* 143; *Hook. Gen. Fil. t.* 104 *B*; *Lowe's Ferns*, 8, *t.* 70. Osmunda thurifraga, *Linn.*; β achillæfolia, *Lowe's New Ferns, t.* 42 *B*. Mohria achillæfolia, *Hort.*—South Africa.

---

146. SCHIZÆA, *Sm.*

Vernation fasciculate, erect or decumbent, rarely distant. *Fronds* linear, simple, stipitiform, simply forked, or dichotomously flabellate, 2–20 inches high. *Veins* forked, free. *Fertile appendices* terminal, pinnate, cristæform; segments induplicate, each bearing on its inner side two compact rows of sessile oval sporangia, attached by their base and opening vertically.

1. S. **pusilla**, *Pursh*; *Hook. et Grev. Ic. Fil. t.* 47.—United States and New Zealand.

2. **S. elegans,** *Sw.* Lophidium elegans, *Presl.* Schizæa elegans; *a* latifolia, *Hook. Gard. Ferns, t.* 34.—Tropical America.

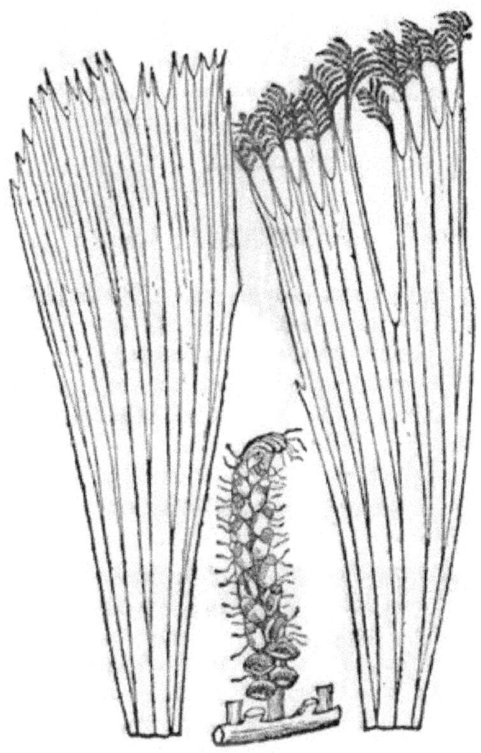

Genus 146.—Portion of barren and fertile frond, natural size, fertile spikelet enlarged. No. 2.

3. **S. rupestris,** *R. Br.; Hook. et Grev. Ic. Fil. t.* 48; *Hook. Gard. Ferns, t.* 42.—Australia.

---

147. ACTINOSTACHYS, *Wall.*

Vernation uniserial, contiguous; sarmentum short. *Fronds* linear, simple, stipitiform, compressed, triquetrous at the base, 9-18 inches high. *Costa* prominent, continuous. *Fertile appendices* terminal, digitato-flabellate, cristæform; segments induplicate, each bearing on its inner side four rows of sessile, oval sporangia, attached by their base and opening vertically.

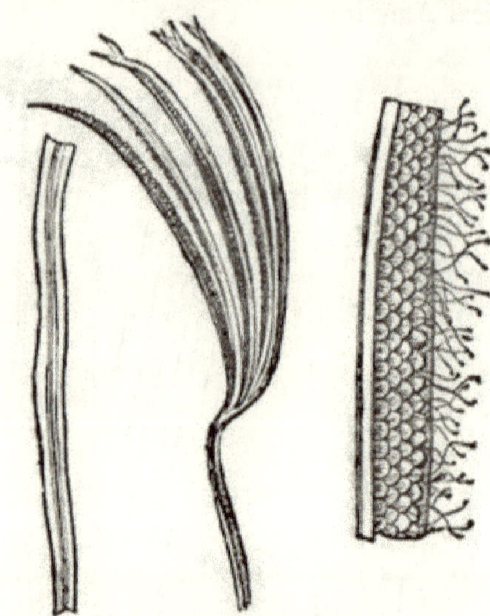

Genus 147.—Portion of fertile frond, natural size; spikeret enlarged. No. 1.

1. **A. digitata**, *Wall.* Schizæa digitata, *Sw.*; *Hook. Gard. Ferns, t.* 49.—East Indies; Malayan, Philippine, and Fiji Islands.

Tribe II.—OSMUNDEÆ.

Sporangia globose, reticulated, short-pedicellate, oblique and gibbous at the apex, opening by a vertical slit, sub-bivalved. *Ring* incomplete or obsolete.

148. OSMUNDA, *Linn.*

Vernation fasciculate, erect, subarboreous. *Fronds* pinnate or bipinnate, 3–10 feet high; pinnæ articulated with the rachis. *Veins* forked; venules free. *Fertile fronds* wholly, or the upper or middle portion, contracted, forming simple or compound sporangiferous panicles.

* *Fertile frond wholly contracted.* (Osmundastrum.)

1. **O. cinnamomea,** *Linn.; Schk. Fil. t.* 146; *Hook. Gard. Ferns, t.* 45; *Lowe's Ferns,* 8, *t.* 1.—North and South America, East Indies.

** *Lateral pinnæ of fertile frond contracted.* (Plenasium.)

2. **O. Claytoniana,** *Linn.; Lowe's Ferns,* 8, *t.* 2. O. interrupta, *Michx.; Schk. Fil. t.* 144.—North America.

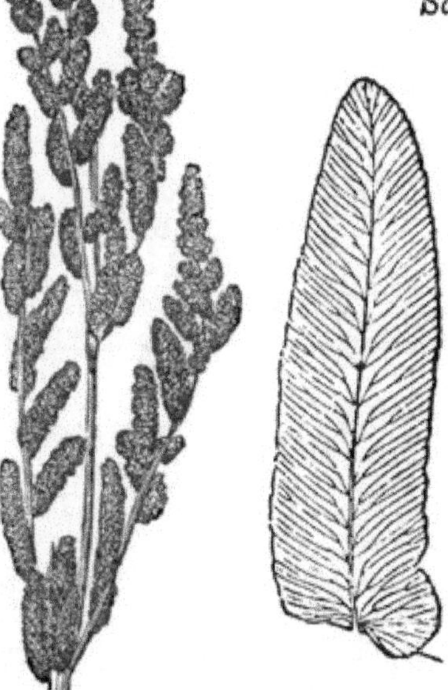

Genus 148.—Pinnule of barren frond, and fertile spike, natural size. No. 3.

*** *Terminal pinnæ of fertile frond contracted.* (Euosmunda.)

3. **O. regalis,** *Linn.* (*Plum. Fil. t. B, f.* 4); *Schk. Fil. t.* 145; *Hook. Gen. Fil. t.* 46 *A*; *Eng. Bot. t.* 209; *Lindl. and Moore's Brit. Ferns, t.* 50; *Hook. Brit. Ferns, t.* 45; *Sowerby's Ferns, t.* 44;—*var.* cristatus, *Moore, in Gard. Chron.* (1863).—North Temperate Zone, Brazil.

4. **O. spectabilis,** *Willd.* O. regalis, *var.* β, *Linn.*—North America.

5. **O. gracilis,** *Link.*—North America.

---

149. TODEA, *Willd.*

Vernation asciculate, erect, sub-arboreous. *Fronds* bipinnatifid, 2–6 feet high; pinnæ coriaceous or membranaceous; fertile frond sub-contracted. *Veins* forked; venules free.

Receptacles medial. *Sori* oblong, linear, simple or forked, naked, often confluent.

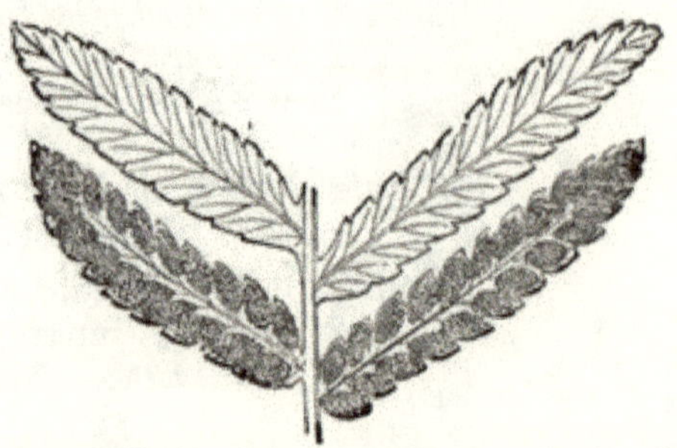

Genus 149.—Portion of fertile frond, under side. No. 1.

* *Fronds coriaceous, opaque.* (Eutodea.)

1. **T. Africana,** *Willd.; Schk. Fil. t.* 147; *Hook. Gen. Fil. t.* 46 *B, f.* 1; *Hook. fil. Fl. Tasm. t.* 168; *Lowe's Ferns*, 8, *t.* 67. Acrostichum barbarum, *Linn.* Todea barbara, *Moore's Synop.* T. rivularis, *Sieb.; Kunze, Anal. t.* 4. T. Australasica, *A. Cunn.*—South Africa, Australia, Tasmania.

** *Fronds membraneous.* (Leptopteris.)

2. **T. hymenophylloides,** *Rich. Voy. d'Astrolabe, t.* 16; *Hook. Gen. Fil, t.* 46 *B, f.* 7; *Hook. Gard. Ferns, t.* 54. Leptopteris hymenophylloides, *Presl.* Todea pellucida, *Carm.; Hook. Ic. Pl. t.* 8.—New Zealand.
3. **T. superba,** *Colenso.* Leptopteris superba, *Hook. Ic. Pl. t.* 910.—New Zealand.
4. **T. Fraseri,** *Hook. et Grev. Ic. Fil. t.* 101. Leptopteris Fraseri, *Presl.*—New Holland.

---

§ 2. *Exannulatæ.*

Sporangia coriaceous, opaque, destitute of a ring.

ORDER II.—MARATTIACEÆ, *Kaulf.*

Sporangia dorsal, exannulate, opaque, horny, generally sessile, distinct, or connate and forming synangia, opening by a pore or longitudinal slit.

150. MARATTIA, *Sm.*

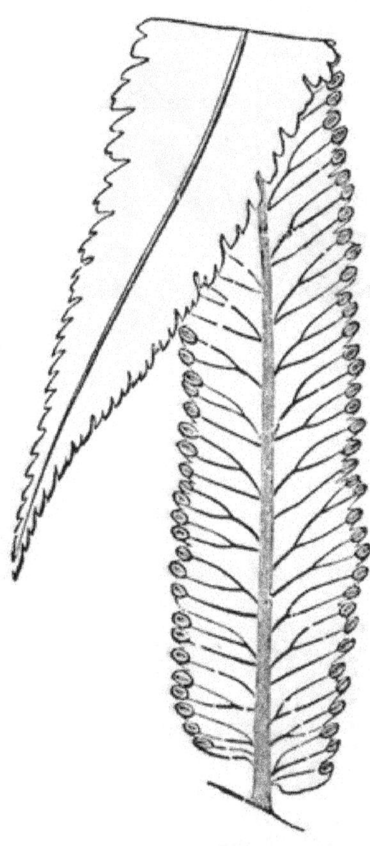

Genus 150.—Fertile pinnule, under side. No. 5.

Vernation fasciculate, erect, subarboreous; each frond rising from between two fleshy stipulæform appendages (which sometimes assume the character of abnormal fronds); base of the stipes clavate, pseudo-articulated with the axis. *Fronds* bi-tripinnate, 6–18 feet long; pinnules articulated with the rachis. *Veins* simple or forked, free. *Synangia* sessile, superficial, oblong, distant, sub-terminal, longitudinally bivalved, each valve consisting of 3–12 laterally-connate sporangia, which open by a slit on their interior side. *Receptacles* sometimes furnished with an indusioid fimbriate membrane.

1. **M. alata,** *Sm. Ic. ined. t.* 46; *Schk. Fil. t.* 152; *Hook. Gen. Fil. t.* 26. Discostegia alata, *Presl.* Marattia Lauchiana, *Hort.* — West Indies.

2. **M. cicutæfolia,** *Kaulf.; Mart. Ic. Crypt. Bras. t.* 69, 71, 72. Gymnotheca cicutæfolia, *Presl.*—Brazil.

3. **M. elegans,** *Endl.*—Norfolk Island, New Zealand.

4. **M. fraxinea,** *Sm. Ic. ined. t.* 48; *Schk. Fil. t.* 152.—Mauritius, West Africa.

5. **M. laxa,** *Kunze; Schk. Supp.* 1, *t.* 95; *Lowe's Ferns,* 8 *t.* 77. Gymnotheca laxa, *Presl.* Marattia macrophylla, *Hort.*—Mexico.

6. **M. purpurascens**, *De Vriese; Hook. Fil. Exot. t.* 65. Marattia Ascensionis, *J. Sm. Cat. Cult. Ferns* (1857); *var.* cristata, *J. Sm. Cat. Cult. Ferns* (1857).—Island of Ascension.

7 **M. Verschaffeltiana**, *J. Sm.* Gymnotheca Verschaffeltiana, *De Vriese.*

---

151. EUPODIUM, *J. Sm.*

Vernation and general character as in *Marattia*, differing in the *synangia* being pedicellate.

Genus 151.—Portion of fertile frond, natural size; ditto enlarged. No. 1.

1. **E. Kaulfussii**, *J. Sm.; Hook. Gen. Fil. t.* 118. Marattia Kaulfussii, *J. Sm.; Hook. 2nd Cent. Ferns, t.* 95; *Lowe's New Ferns, t.* 17. Marattia lævis, *Kaulf.* (*non Sm.*).—Brazil.

---

152. ANGIOPTERIS, *Hoffm.*

Vernation fasciculate, erect, subarboreous; each frond rising from between two fleshy stipulæform appendages; base of the stipes clavate, pseudo-articulated with the axis. *Fronds* 6–18 feet high, bipinnate; pinnules articulated with the rachis. *Veins* simple or forked, free. *Receptacles* oblong, linear, subterminal. *Sporangia* 7–24, biserial, sessile, free, opening by a slit on the inner side. *Sori* oblong, laterally contiguous, forming a broad, submarginal, transverse band.

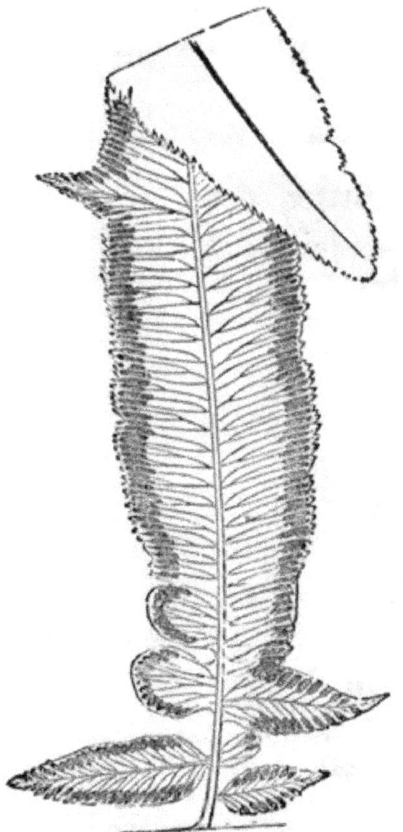

Genus 152.—Fertile pinnule, under side. No. 2.

1. **A. evecta,** *Hoffm.; Schk. Fil. t.* 150; *Hook. Fil. Exot. t.* 15. Polypodium evectum, *Forst.*—Islands of the Pacific Ocean and Ceylon.
2. **A. Teysmanniana,** *De Vriese, Mon. Maratt. t.* 1–2; *Lowe's Ferns,* 8, *t.* 76.—Java.
3. **A. Brongniartiana,** *De Vriese, Mon. Maratt. t.* 3, *f.* 5.—Tahiti.
4. **A. pruinosa,** *Kunze, Fil. t.* 91.—Java.
5. **A. hypoleuca,** *De Vriese.*—Java.
6. **A. Miqueliana,** *De Vriese.* A. longifolia, *Miq. et Hort.*—Malayan Islands.

153. DANÆA, *Sm.*

Vernation uniserial, contiguous, decumbent; sarmentum thick, fleshy; each frond rising from between two short stipulæform appendages, and having the stipes once or more times articulated. *Fronds* pinnate, rarely simple, 1–3 feet high, the fertile usually somewhat contracted; pinnæ opposite, linear-lanceolate, articulated with the rachis. *Veins* forked; venules

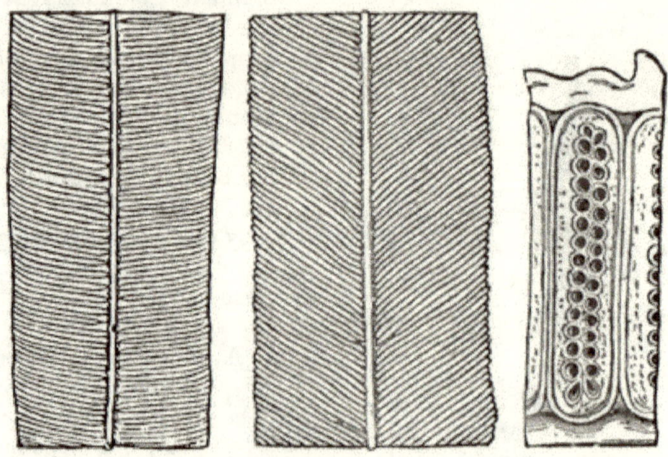

Genus 153.—Barren and fertile frond, natural size; synangium enlarged. No. 1.

parallel, their apices arcuate and anastomosing at the margin. *Synangia* sessile, immersed, linear, contiguous, occupying nearly the whole length of the venules, and covering the whole under surface of the fertile fronds; each consisting of two rows of numerous laterally and oppositely connate sporangia united into a concrete mass, forming linear synangia, each cell opening by a circular pore at their apices.

1. **D. alata,** *Sm.; Hook. et Grev. Ic. Fil. t.* 18; *Hook. Gen. Fil. t.* 7.—West Indies and Tropical America.

2. **D. nodosa,** *Sm.; Schk. Fil. t.* 152; *Hook. et Grev. Ic. Fil. t.* 51. Asplenium nodosum, *Linn.* (*Plum. Fil. t.* 108). —West Indies and Tropical America.

---

154. KAULFUSSIA, *Blume.*

Vernation uniserial, contiguous, decumbent; sarmentum thick, fleshy; each frond rising from between two short, fleshy, stipulæform appendages. *Fronds* broad, pinnately-trifoliate, with the two lower pinnæ sometimes bipartite, 1–2 feet high, long, stipate, pale underneath, and furnished with numerous concave dots. *Veins* costæform, parallel; venules compound anastomosing, with free veinlets terminating within the areoles. *Synangia* sessile, compital, superficial, distant, orbicular, concave-hemispherical, each consisting of 10–20 sporangia, laterally connate in a circular series, and opening by a slit on their interior side.

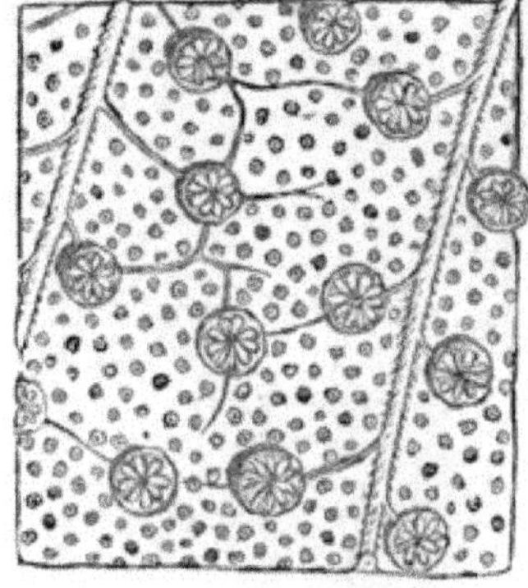

Genus 154.—Portion of fertile frond, slightly enlarged. No. 1.

1. **K. æsculifolia,** *Blume; Hook. et Grev. Ic. Fil. t.* 229; *Hook. Gen. Fil. t.* 59 *A.*—Java and Philippine Islands.

ORDER III.—OPHIOGLOSSACEÆ.

Flowerless plants consisting of straight evolved fronds, which produce from their disk or base a rachiform, simple or compound paniculate spike, of unilocular, sessile, connate, homogeneous cases (sporangia), opening by a vertical slit in two valves, containing reproducing spores.

155. OPHIOGLOSSUM, *Linn.*

Fronds erect or pendulous, with dissimilar fertile and sterile segments, or rarely the whole frond rachiform. *Sterile frond or segments* foliaceous, simple, entire, palmately lobed or dichotomously branched. *Veins* reticulated. *Fertile segments* spike-like, simple or rarely forked. *Sporangia* connate, in two rows, forming a distichous, synangeous spike, each cell opening horizontally in two equal valves.

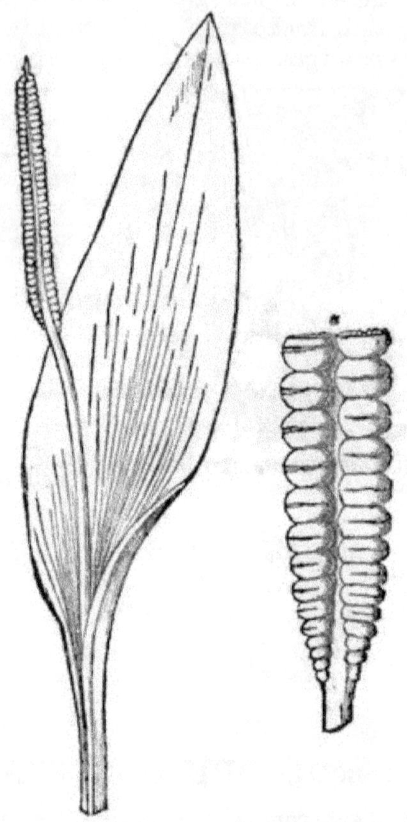

Genus 155.—Plant natural size; fertile spike enlarged. No. 3.

Fronds erect, the sterile segments ovate or linear (Terrestrial).
(Euophioglossum.)

1. O. **Lusitanicum**, *Linn.; Hook. et Grev. Ic. Fil. t.* 80; *Lindl. and Moore's Brit. Ferns, t.* 51 *C*; *Sowerby's Ferns, t.* 47. Ophioglossum vulgatum, *var.* angustifolium, *Hook. Brit. Ferns, t.* 47.—South of Europe.

2. O. **pedunculosum**, *Desv.; Kunze, Fil. t.* 29, *f.* 2.—North America.

3. **O. vulgatum,** *Linn.; Schk. Fil. t.* 153; *Hook. Gen. Fil. t.* 59 *B; Eng. Bot. t.* 108; *Lindl. and Moore's Brit. Ferns, t.* 51 *B; Hook. Brit. Ferns, t.* 46; *Sowerby's Ferns, t.* 46. — Temperate Zone of the Northern Hemisphere.

4. **O. reticulatum,** *Linn.; Hook. et Grev. Ic. Fil. t.* 20 (*Plum. Fil. t.* 164).—Tropics.

** *Fronds pendulous; the sterile segments ribbon-formed, usually dichotomously branched* (Epiphytal). (Ophioderma.)

5. **O. pendulum,** *Linn.; Hook. et Grev. Ic. Fil. t.* 19; *Hook. Gard. Ferns, t.* 33.—Tropics of Eastern Hemisphere.

6. **O. furcatum,** *J. Sm.* Ophioglossum pendulum; β furcatum, *Presl, Tent. Pterid. Supp. p.* 56.—Queensland.

*** *Fronds palmate, pendulous.* (Cheiroglossa.)

7. **O. palmatum,** *Linn.* (*Plum. Fil. t.* 163); *Hook. Ic. Pl. t.* 4. —West Indies, New Granada, Peru, Brazil, Mauritius.

---

156. HELMINTHOSTACHYS, *Kaulf.*

Fronds erect, with dissimilar sterile and fertile segments. *Sterile segments* foliaceous, digitate, pedate. *Veins* forked;

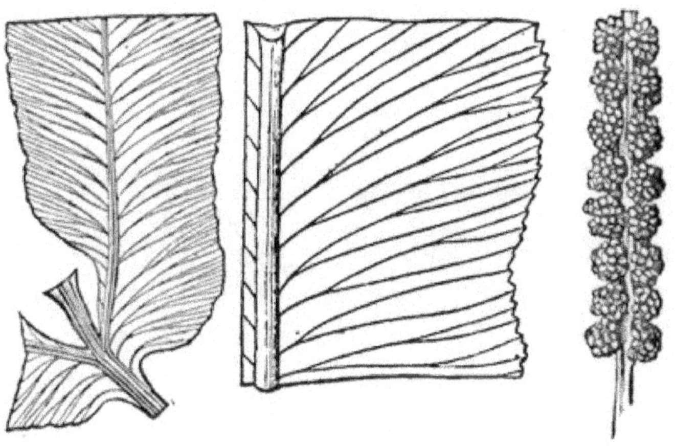

Genus 156.—Portion of sterile pinna, natural size; ditto enlarged; fertile spike, natural size. No. 1.

venules free. *Fertile segment* a rachiform spike. *Sporangia* subglobose, sessile, in pedicellate crested whorls (rarely distinct), forming a clustered simple spike, opening vertically by a slit on the exterior side.

1. **H. Zeylanica,** *Hook. Gen. Fil. t.* 47 *B; Hook.* 2*nd Cent. Ferns, t.* 94; *Hook. Gard. Ferns, t.* 28. Osmunda Zeylanica, *Linn.* Helminthostachys dulcis, *Kaulf.* —Ceylon, Malay, Molucca, Philippine, and other islands.

---

157. BOTRYCHIUM, *Linn.*

Fronds erect, with dissimilar sterile and fertile segments. *Sterile segments* foliaceous, deltoid, bi-tripinnatifidly decom-

Genus 157.—Portion of barren and fertile segments, natural size; fertile enlarged.

pound, rarely pinnate. *Veins* forked; venules free. *Fertile segments* rachiform, compound paniculate. *Sporangia* distinct, in two unilateral rows, opening vertically in two equal valves.

1. **B. simplex,** *Hitchcock; Hook. et Grev. Ic. Fil. t.* 82.—North America.

2. **B. Lunaria,** *Sw.; Schk. Fil. t.* 154; *Hook. Gen. Fil. t.* 47 *A; Lindl. and Moore's Brit. Ferns, t.* 51 *A; Hook. Brit. Ferns, t.* 48; *Sowerby's Ferns, t.* 45. Osmunda Lunaria, *Linn. Eng. Bot. t.* 318;—*β* rutaceum. Botrychium rutaceum, *Sw.; Schk. Fil. t.* 155 *B.*—Temperate Zone of the Northern Hemisphere.

3. **B. lunarioides,** *Sw.* Botrypus lunarioides, *Michx.;*—*β* obliquum, *A. Gray.* Botrychium obliquum, *Muhl.* B. lunarioides, *Schk. Fil. t.* 157;—*γ* dissectum, *A. Gray.* Botrychium dissectum, *Spr.; Schk. Fil. t.* 158.—North America.

4. **B. Virginicum,** *Willd.* Osmunda Virginica, *Linn.* Botrychium Virginianum, *Sw.; Schk. Fil. t.* 156; *Hook. Gard. Ferns, t.* 29.—Temperate Zone of the Northern Hemisphere, Tropical America, East Indies, and Ceylon.

Order IV.—LYCOPODIACEÆ.

Flowerless moss or fern-like plants, consisting of firm, erect, creeping or pendulous, simple or branched, often flagelliform stems, furnished with acerose, rusciform or jungermannia-like, sessile leaves, which are generally imbricate, and often distichous and of two kinds, bearing in their axis, or on contracted terminal spikes, 1–3-celled reniform or globose sessile spore-cases (*sporangia*), of one or of two forms; one called *Antheridangia*, containing numerous spores; the other *Oophoridangia*, containing 1–3–8 large spores.

158. PSILOTUM, *Sw.*

Stems dichotomously forked, compressed or angular, rigid, erect or slender, pendulous; leaves obsolete or small, bract-like. *Sporangia* subglobose, trilobed, vertically trivalved, solitary in the axis of the bract-like leaves.

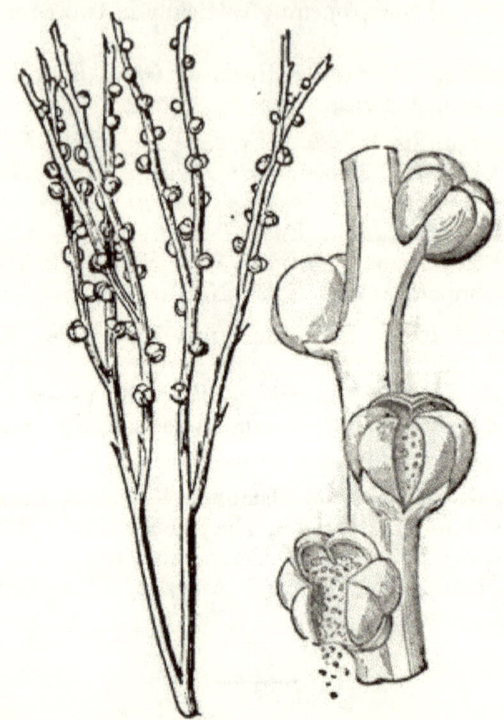

Genus 158.—Portion of fertile frond, natural size; spore-case enlarged. No. 1.

1. **P. triquetrum**, *Sw.; Schk. Fil. t.* 165 *b; Hook. Gen. Fil. t.* 87; *Lodd. Cab. t.* 1916. Lycopodium nudum, *Linn.* — Tropics and sub-tropical regions of both hemispheres.

---

159. LYCOPODIUM, *Linn.*

Stems rigid, erect or slender, flagelliform, pendulous or creeping; leaves acerose, subulate or rusciform, distant or imbricate. *Sporangia* reniform, vertically bivalved, solitary in the axis of the leaves, or in terminal, rarely lateral, contracted, ament-like spikes. *Spores* numerous, small, uniform (*Antheridangia* only ?).

* *Stems creeping.* (Epigeous.)

1. **L. clavatum,** *Linn.; Schk. Fil. t.* 162; *Eng. Bot. t.* 224.—Temperate Zone of the Northern Hemisphere, Britain.
2. **L. annotinum,** *Linn.; Schk. Fil. t.* 160; *Eng. Bot. t.* 239.—Temperate Zone of the Northern Hemisphere, Britain.
3. **L. inundatum,** *Linn.; Schk. Fil. t.* 162; *Eng. Bot. t.* 1727.—Temperate Zone of the Northern Hemisphere, Britain.

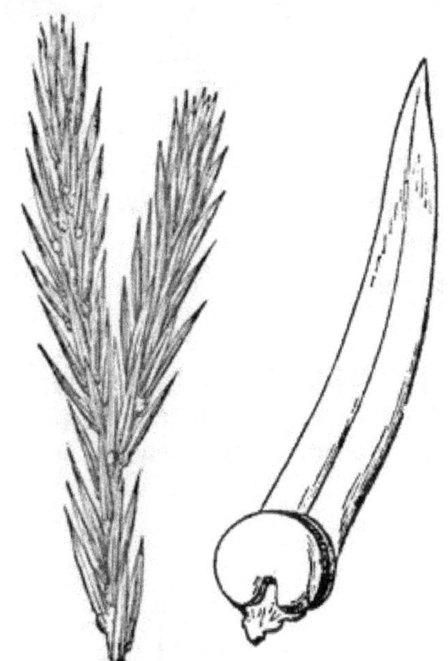

Genus 159.—Portion of plant, natural size; spore-case enlarged.

** *Stems erect.* (Epigeous.)

4. **L. Selago,** *Linn.; Schk. Fil. t.* 159; *Eng. Bot. t.* 233.—Europe, Britain.
5. **L. alpinum,** *Linn.; Schk. Fil. t.* 161; *Eng. Bot. t.* 234.—Temperate Zone of the Northern Hemisphere, Britain.
6. **L. densum,** *Labil. Nov. Holl.* 2, *t.* 251, *f.* 1.—Australia.

7. **L. fastigiatum,** *R. Br.*—New Zealand.

8. **L. complanatum,** *Linn.* (*Plum. Fil. t.* 165, *f. B*); *Schk. Fil. t.* 163.—Temperate Zone of the Northern Hemisphere.

9. **L. dendroideum,** *Michx.; Willd. Sp. Pl.* 5, *p.* 21.—North America.

10. **L. cernuum,** *Linn.* (*Rheed. Mal. t.* 2, *t.* 39); *Burm. Fl. Zey. t.* 66 (*Plum. Fil. t.* 155, *f. A*).—Tropics; very general.

*** *Stems pendulous.* (Epiphytal.)

11. **L. Phlegmaria,** *Linn.* (*Rheed. Mal.* 12, *t.* 14).—Tropics of the Eastern Hemisphere.

12. **L. verticillatum,** *Linn.; Willd. Sp. Pl.* 5, *p.* 48.—Mauritius.

13. **L. ulicifolium,** *Vent.; Willd. Sp. Pl.* 5, *p.* 27.—India.

14. **L. Hookeri,** *Wall.; Hook. et Grev. Ic. Fil. t.* 165.—India.

15. **L. taxifolium,** *Sw.; Willd. Sp. Pl.* 5, *p.* 48.—West Indies.

---

160. SELAGINELLA, *Spring.*

Fern-like plants. *Stems of frondules* creeping, sub-ascending or erect, scandent, of undefined extension (*surculose*), or rising

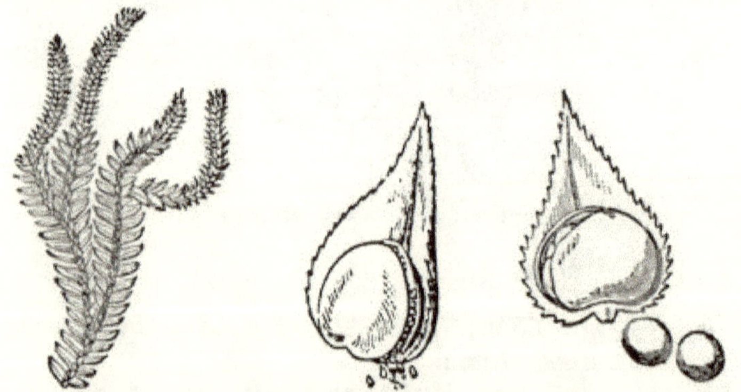

Genus 160.—Portion of plant, and the two kinds of spore-cases enlarged.

singly from an underground elongating stolon (*caulescent*), or in a fasciculate manner from a simple erect centre (*rosulate*);

leaves generally imbricate, distichous, jungermannia-like, of two sizes; the smaller stipulæform. *Sporangia* reniform, vertically bivalved, sessile in the axis of the leaves, or on contracted terminal spikes. *Spores* of two forms: in some, sporangia (*Antheridangia*) small and numerous; in others (*Oophoridangia*), 3–4, large.

A. Plant leafy on all sides.

1. **S. spinulosa,** *Spring.* Lycopodium selaginoides, *Linn.; Schk. Fil. t.* 165; *Eng. Bot. t.* 1148.—Europe, Britain.

2. **S. uliginosa,** *Lab. Nov. Holl.* 2, *p.* 104, *t.* 251, *f.* 2; *Willd. Sp. Pl.* 5, *p.* 32.—Australia and Tasmania.

B. Leaves distichous.

§ 1. *Plants surculose.*

† *Stems decumbent, creeping.*

3. **S. Apus,** *Spring. Monogr. p.* 75. S. densa, *Hort.* Lycopodium Brasiliense, *Radd. Fil. Bras.* 82, *t.* 1.—North America and Brazil.

4. **S. Ludoviciana,** *A. Braun. Revise, No.* 5.—Louisiana.

5. **S. Helvetica,** *Link; Spring. Monogr.* 2, *p.* 83. Lycopodium Helveticum, *Linn.*—Europe, Switzerland.

6. **S. denticulata,** *Link; Spring. Monogr.* 2, *p.* 82. Lycopodium denticulatum, *Linn.*—Central Europe.

7. **S. delicatissima,** *A. Braun. Revise, No.* 8. S. microphylla, *Spring.*—Columbia.

8. **S. serpens,** *Spring. Monogr.* 2, *p.* 102. Lycopodium serpens, *Desv.* S. mutabalis, *Hort.* S. variabilis, *Hort.* S. Jamaicensis, *Hort.*—Jamaica.

9. **S. uncinata,** *Spring. Monogr.* 2, *p.* 109. S. cæsia, *Hort.*—China.

10. **S. hortensis,** *Metten. Fil. Hort. Lips. p.* 128. S. denticulata, *Hort.*—South of Europe.

11. **S. sarmentosa,** *A. Braun.* S. patula, *Spring.* S. apothecia, *Hort.*—West Indies.

12. **S. Pœppigiana,** *Spring.; Hook. Fil. Exot. t.* 56.—Tropical America.

13. **S. stenophylla,** *A. Braun. Revise, No.* 35. S. micro phylla, *Hort.*—Mexico.

†† *Stems sub-erect.*

14. **S. Martensii,** *Spring. Monogr.* 2, *p.* 129. Lycopodium stoloniferum, *Link.* Lycopodium Brasiliense, *Hort.*—Mexico.

15. **S. Breynii,** *Spring. Monogr.* 2, *p.* 119. S. Panamensis, *Hort.* S. Pœppigiana, *Hort.*—Guiana.

16. **S. Galeottii,** *Spring. Monogr.* 2, *p.* 220. Lycopodium stoloniferum, *Mart. et Gal.* S. Schottii, *Hort.*—Mexico.

17. **S. sulcata,** *Spring. Monogr.* 2, *p.* 214. Lycopodium sulcatum, *Desv.*—Columbia.

18. **S. atroviridis,** *Spring. Monogr.* 2, *p.* 124. Lycopodium atroviride, *Wall.; Hook. et Grev. Ic. Fil. t.* 39.—East Indies.

19. **S. inæqualifolia,** *Spring. Monogr.* 2, *p.* 148. Lycopodium inæqualifolium, *Hook. et Grev.*—East Indies and Java.

20. **S. ciliata,** *A. Braun. Revise, No.* 14. Lycopodium ciliatum, *Willd.*—Tropical America.

21. **S. Griffithii,** *Spring.; Veitch. Cat.* (1861).—Borneo.

††† *Stems scandent.*

22. **S. lævigata,** *Spring. Monogr.* 2, *p.* 137. Lycopodium lævigatum, *Willd.* Lycopodium Willdenovii, *Desv.* Selaginella cæsia, *var.* arborea, *Hort.* S. altissima, *Klot.*—East Indies.

§ 2. *Plants caulescent.*

23. **S. caulescens,** *Spring. Monogr.* 2, *p.* 158. Lycopodium caulescens, *Wall. var.* minor, *Veitch. Cat.* (1861).—East Indies.

24. **S. erythropus,** *Spring. Monogr.* 2, *p.* 156. Lycopodium erythropus, *Mart. Ic. Sel. Pl. Crypt. t.* 20, *f.* 3.—Tropical America.

25. **S. viticulosa,** *Klot.; Spring. Monogr.* 2, *p.* 186.—Columbia.

26. **S. flabellata,** *Spring. Monogr.* 2, *p.* 174. Lycopodium flabellatum, *Linn.*—Columbia and Peru.

27. **S. filicina,** *Spring. Monogr.* 2, *p.* 189. S. dichrous, *Hort.* —Columbia and Peru.

28. **S. Africana,** *A. Braun. Revise, No.* 23. S. Vogelii, *Spring.* —Fernando Po.

29. **S. pubescens,** *Spring. Monogr.* 2, *p.* 173. Lycopodium pubescens, *Wall.* S. Willdenovii, *Hort.*—East Indies.

30. **S. Lyallii,** *Spring. Monogr.* 2, *p.* 168. Lycopodium Lyallii, *Hook. et Grev.*—Madagascar.

31. **S. Lobbii,** *Hort.; A. Braun. Revise, App. No.* 26; *Veitch. Cat.* (1861).—Borneo.

32. **S. Wallichii,** *Hort.; Veitch. Cat.* (1861).—Penang.

§ 3. *Plants rosulate.*

33. **S. cuspidata,** *Link; Spring. Monogr.* 2, *p.* 66. S. pallescens, *Klot.* S. circinalis, *Hort.*—Tropical America.

34. **S. convoluta,** *Spring. Monogr.* 2, *p.* 69. Lycopodium convolutum, *Walk. and Arnott.* Lycopodium paradoxa, *Hort.*—Tropical America.

35. **S. involvens,** *Spring. Monogr.* 2, *p.* 63. Lycopodium involvens, *Sw.*—India.

36. **S. lepidophylla,** *Spring. Monogr.* 2, *p.* 27. Lycopodium lepidophyllum, *Hook. et Grev.*—Mexico.

ORDER V.—MARSILEACEÆ.

Plants growing in, or floating in water (rarely not in water), very various in character, consisting of grass or trefoil-like leaves, or branched with imbricated leaves, bearing sporangia (conceptacles*) at their roots, or base, or on the footstalks, or on small branchlets on the under side, which are 1–4, or many-celled.

161. MARSILEA, *Linn.*

Plants with quadrifid (trefoil-like) leaves on a long footstalk, from a rhizome growing under water. *Sporangia* (*conceptacles*) borne on the rhizome at the base of the footstalk, or pedicellate

* A hollow case containing spore-cases; a compound sporangium

on the footstalk, 2-valved, containing numerous obovate receptacles in two longitudiual series, bearing cellular vesicles of

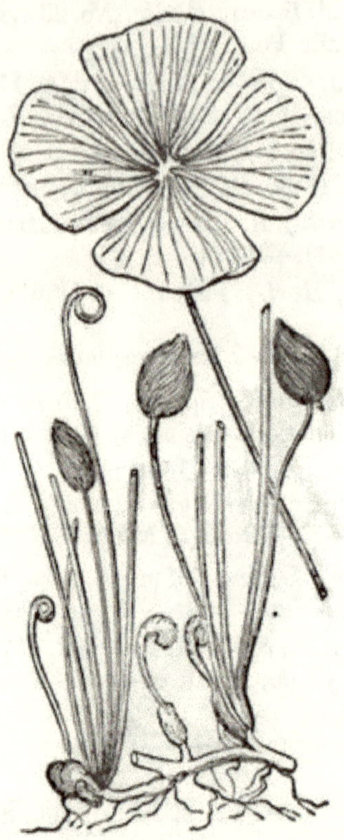

Genus 161.—Plant with spore-cases, natural size. No. 1.

two kinds — 1. (*Antheridangia*), containing numerous small spores; 2. (*Oophoridangia*), containing a single large spore.

1. **M. quadrifolia**, *Linn.; Willd. Sp. Pl.* 5, *p.* 538; *Schk. Crypt. t.* 173.—Germany.

2. **M. macropus**, *Hook. Ic. Pl. t.* 909; *Gard. Ferns, t.* 63; *Seemann, Journ. Bot.* 1, *p.* 6. (The Nardoo plant of Australian explorers.)—Australia.

---

162. PILULARIA, *Linn.*

Plants with filiform leaves, from a creeping rhizome under water. *Sporangia* (*conceptacles*) radical (at the base of the leaves on the rhizome), globose, coriaceous, 2–4-celled, 2-4-valved, each

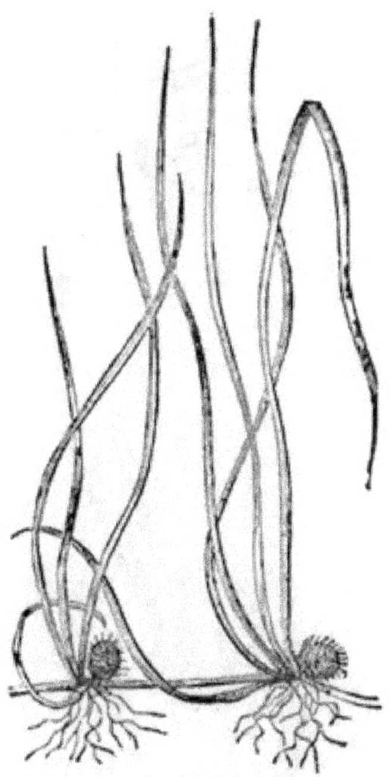

Genus 162.—Plant, natural size. No. 1.

cell containing different kinds of bodies:—1. (*Antheridangia*), consisting of vesicles containing many minute granular spores; 2. (*Oophoridangia*), each containing a single large spore.

1. **P. globulifera**, *Linn.; Willd. Sp. Pl.* 5, *p.* 535; *Bolt. Fil. t.* 40; *Schk. Crypt. t.* 173; *Eng. Bot. t.* 521.—Europe and Britain.

---

163. ISOETES, *Linn.*

Plants with awl-shaped leaves, from a thick creeping rhizome under water or out of water. *Sporangia* (*conceptacles*) borne in the axis of the leaves at their base, globose, 1-celled, traversed by thread-like receptacles of two kinds: — 1. (*Antheridangia*), containing numerous small spores; 2. (*Oophoridangia*), con-atining large 4-sided spores.

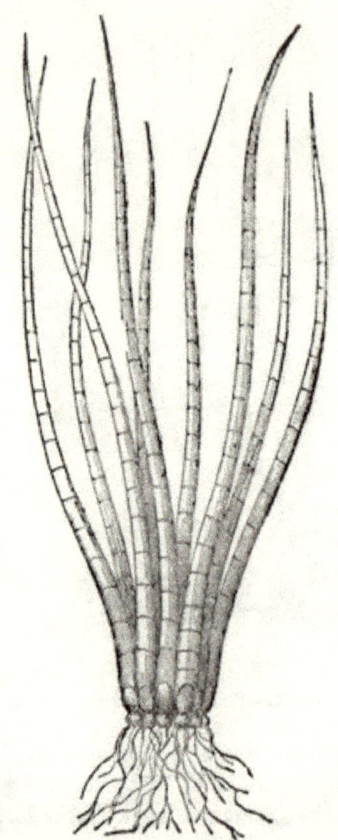

Genus 163.—Plant, natural size. No. 1.

1. **I. lacustris,** *Linn.; Willd. Sp. Pl.* 5, *p.* 534; *Bolt. Fil. t.* 41; *Schk. Crypt. t.* 173; *Eng. Bot.* 1084.— Europe, Britain.

ADDENDA.

A CONSIDERABLE time having elapsed since the preceding Enumeration was prepared and sent to the press, has enabled me to note a number of recently introduced, and a few omitted older species, which I now enumerate in the form of addenda.

N.B.—Those marked thus † after the name of the country are entered on the authority of Mr. T. Moore's notices of them, in the *Proceedings of the Royal Horticultural Society* and the *Gardeners' Chronicle* newspaper; not having myself seen them alive or obtained specimens.

38. LOMARIOPSIS.

2*. **L. fraxinea,** *J. Sm.* Lomaria fraxinea, *Willd.* Acrostichum (Lomariopsis) sorbifolium, *Hook. Sp. Fil.* 5, *p.* 241 (*non J. Sm.*).—Mauritius.

OBS.—This is a very distinct plant from the West Indian type of L. sorbifolia.

---

50. GYMNOGRAMMA.

§ 7. Ampelogramma, *J. Sm. Vernation sarmentose, according to Hook.; fronds indefinite; rachis flexuose; pinnæ bi-tripinnate, refracted; pinnules small, cuneiform.*

18. **G. flexuosa,** *Desv.; Hook. Sp. Fil.* 5, *p.* 192. G. retrofracta, *Hook. et Grev. Bot. Misc.* 3, *t.* 112.—Tropical America.

---

62. DICTYOPTERIS.

2. **D. macrodonta,** *Presl, Tetn. Pterid.; J. Sm. Gen. Fil. Philipp.; Hook. Journ. Bot.* 3, *p.* 396. Polypodium macrodon, *Reinw. in Herb. J. Sm.* Aspidium difforme, *Blume, accord. to Reinw. in Herb. J. Sm.* Polypodium confluens, *Wall.*—East Indies, Malayan Archipelago, Fiji.

Obs.—In Herbariums and books there is great confusion in the synonymy of what, according to Garden plants, seem to be two distinct species; — viz., *Dictyopteris irregularis* and *D. macrodon* of *Presl;* judging from Herbarium specimens, they are difficult to be recognized as distinct; but living plants show the first to have erect vernation, the other decumbent.

The Kew Collection is indebted for this species, as also the beautiful *Microlepia platyphylla*, and others, to Mr. Robert Kennedy, Florist and Fern-dealer in Covent Garden.

3. **D. Cameroonianis,** *J. Sm.* Polypodium (Dictyopteris) Cameroonianum, *Hook. Sp. Fil.* 5, *p.* 104. Dictyopteris varians, *Moore, in Gard. Chron.* (1864). — Tropical West Africa.

Obs.—A plant of this Fern was sent some time ago from Old Calabar to the Royal Botanic Garden, Edinburgh, and I am indebted for a specimen of it to Mr. James McNab, the Curator of that truly scientific garden.

---

63. MENISCIUM.

3*. **M. angustifolium,** *Willd.; Hook. Sp. Fil.* 5, *p.* 164.—Tropical America.

---

65. NEPHRODIUM.

4*. **N. cyatheoides,** *Kaulf.; Hook. Sp. Fil.* 4, *t.* 241. Polystichum Dubreuillianum, *Gaud. in Freyc. Voy. Bot. Crypt. t.* 9.—Sandwich Islands.

Obs.—It is but right to observe that this remarkable species, as also the Sandwich Island Ferns in this addenda, with the previously entered *Phegopteris unidentata*, and the remarkable

Colysis Spectrum, also the beautiful *Cibotium Menziesii*, were sent to the Royal Gardens, Kew, from the Sandwich Islands by Dr. Hillebrand, in 1863.

---

68. CYRTOMIUM.

1*. **C. Fortunei**, *J. Sm.* *Fronds* 1–1½ foot long, pinnate; pinnæ lanceolate, falcate, acuminate, 2–3 inches long, 1 inch wide, entire, the base oblique, the inferior rounded, the superior truncate and subauriculate. *Costa* ebenous, upper surface dull, unreflecting. *Sori* numerous.—Japan.

Obs.—In Herbaria, specimens of this cannot readily be distinguished from *C. falcatum*; but on seeing living plants standing side by side, the difference is evident; *C. Fortunei* being a smaller-growing plant, thinner in texture, and not lucid and reflecting light, as *C. falcatum*.

---

70. ASPIDIUM.

3*. **A. polymorphum**, *Wall.*; *Hook. Sp. Fil.* 4, *p.* 54 (*exclud. syn.*).—India, Ceylon.

3**. **A. Barteri**, *J. Sm.* *Vernation* erect. *Fronds* pinnate, 1–2 feet high; pinnæ linear-lanceolate, 6–7 inches long, 1–1½ broad; the lower pair sometimes binate, or auriculated on the lower margin. *Sori* regularly biserial between the primary veins. *Indusium* small, fugaceous. Aspidium polymorphum, *Wall. according to Hook. Sp. Fil.* 4, *p.* 54.—West Africa, Fernando Po.

Obs.—This is quite distinct from the preceding species. In *A. Barteri* the *sori* are perfectly serial, whereas in *A. polymorphum* the *sori* are scattered.

---

74. POLYSTICHUM.

6*. **P. lepidocaulon**, *J. Sm.* Aspidium lepidocaulon, *Hook. Sp. Fil.* 4, *t.* 217.—Japan.

Obs.—On receiving this plant, it accorded so well with the Indian specimen of *P. obliquum*, which led to that name being inserted at p. 149, but the plant grew out of its imported form; and if the Indian plant (which has not yet been introduced) should assume the character of this by cultivation, then this name must become a synonym.

The introduction of this species is due to Mr. Richard Oldham, the last special collector of plants sent out from Kew, who, after remaining three years in Japan, visited Formosa, where his health failed. He returned to Amoy, and there died in November, 1864.

10*. P. **concavum**, *Moore, Proc. Hort. Soc.* 11, 377. Lastrea Standishii, *Hort.*—Japan.†

10**. P. **ordinatum**, *Fée; Moore, Proc. Hort. Soc.* 11, 367.—Tropical America.†

---

75. LASTREA.

29*. L. **spectabilis**, *J. Sm. in Enum. Fil. Philipp. Hook. Journ. Bot.* 3 (1841). Aspidium spectabile, *Blume.*—Philippine and Malayan Islands, India.

41*. L. sparsa, *Moore.* Nephrodium sparsum, *Don.* Aspidium purpurascens, *Blume, according to Hook. Sp. Fil.* 3, *p.* 133, *t.* 262.—Throughout India, Java, and Ceylon.

41**. L. latifrons, *J. Sm.* Nephrodium (Lastrea) latifrons, *Hook. Sp. Fil.* 4, *p.* 138.—Sandwich Islands.

41***. L. membranifolia, *Presl, Pterid.* Nephrodium membranifolium, *Presl, Reliq. Haenk. t.* 6, *f.* 3; *Hook. Sp. Fil.* 4, *t.* 26.—India, Ceylon.

---

79. NEPHROLEPIS.

3*. N. **falciformis**, *J. Sm.* *Fronds* suberect, linear, pinnate, 1½–2 feet in length; pinnæ numerous; the sterile (lower ones) elliptical, obtuse, base truncate; the superior ones fertile, lanceolate, falcate, acute, subdeflexed, 1½ inch long by ½ inch wide; base truncate, subauriculate on the upper margin. *Sori* uniserial, antimarginal. *Indusium* suborbicular.—Borneo.

Obs.—The affinity of this species is with N. tuberosa; but seeing the two plants together, the difference is readily seen.

79a. ISOLOMA, *J. Sm.*

Vernation fasciculate, erect, stoloniferous. *Fronds* linear, 1–2 feet long, pinnate; pinnæ oblong, elliptical or lanceolate, falcate, coriaceous, the base truncate and subauriculated; petiole short, articulated with the rachis. *Veins* forked; sterile

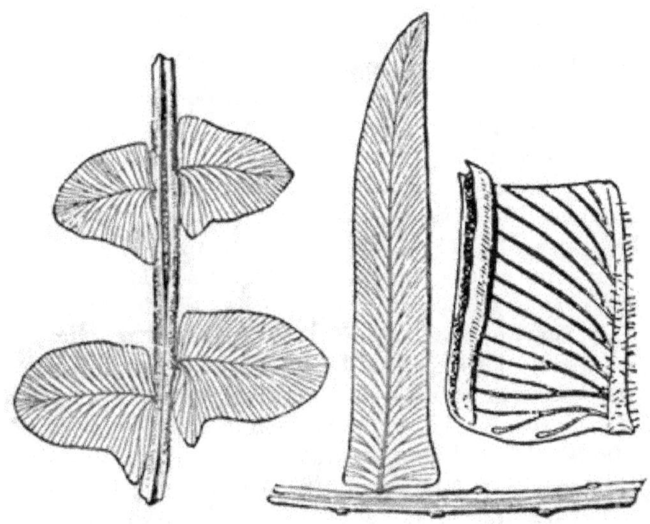

Genus 79a.—Portion of sterile frond and a fertile pinna, natural size ditto enlarged. No. 1.

venules free, the fertile transversely combined at the margin, forming a continuous receptacle. *Indusium* linear, interiorly attached, plane, equal with and conniving with the margin, forming with it a vertical exteriorly open groove, containing the *sporangia*.

1. **I. lanuginosa,** *J. Sm. in Lond. Journ. Bot.* 1, *p.* 420. Lindsæa lanuginosa, *Wall.; Hook. Sp. Fil.* 1, *t.* 69 *B.* —Singapore, Penang, New Guinea, and Seychelles Islands.

Obs.—This remarkable Fern has lately been added to the Kew collection, having been sent from the Mauritius Botanic Garden by Mr. John Horn.* In general habit and appearance this Fern is the prototype of *Nephrolepis;* but is distinguished by the receptacles being laterally confluent, forming a linear marginal sorus, similar to that of *Lindsœa*, from which it is, however, quite distinct in habit.

---

84. PHEGOPTERIS.

2*. **P. plumosa,** *J. Sm.* Asplenium Filix-fœmina, *var.* plumosum, *Moore, Nat. Print. Ferns, oct. ed. p.* 56; *Lowe's New Ferns, t.* 14.—Yorkshire (Mr. Stansfield).

Obs.—A few years ago three plants of this Fern were found wild in Yorkshire. It was soon afterwards described and figured in the works above quoted, under the name of *Asplenium Filix-fœmina, var. plumosum;* but upon what grounds it was referred to *Asplenieœ* I cannot explain, as all the specimens I have examined of it have small, punctiform, naked sori, perfectly characteristic of the genus *Phegopteris*, with which it also agrees in habit. This leaves me no other alternative than to consider it a species of that genus, and consequently a new British species. In doing so, the question arises as to whether it represents an ancient species not before noticed, or the modern result arising from the power of nature to generate new forms, in accordance with the Darwinian Theory of creation of species. It is, however, to be observed that in abnormal or difformed states of *Asplenium* and *Scolopendrium*, the sori are depauperated, in some instances having no vestige of an indusium; but such is not the case with this plant. The fronds are perfect in every respect, and if Herbarium specimens had been received from some foreign country, no Pteridologist, on seeing the naked sori, would refer it to *Asplenieœ*.

---

93. ADIANTUM.

35*. **A. Gheisbeghtii,** *Backhouse, Cat.* (?) **A. tenerum,** *var.* (*J. Sm.*).—Tropical America.†

* Who left Kew in 1861 to be assistant to Mr. Duncan. See p. 15.

39*. **A. colpodes,** *Moore, Gard. Chron.* (1865).—Ecuador.†

44. **A. tinctum,** *Moore, Proc. Hort. Soc.* 11, 369.—Tropical America.†

---

96. PTERIS.

20*. **P. straminea,** *Metten.* P. crispa, *Hort.* (*non Linn.*).—Chili.

---

97. LITOBROCHIA.

14*. **L. areolata,** *Moore.* Pteris areolata, *Lowe's New Ferns t.* 57.—India.†

---

102. LOMARIA.

2*. **L. rigida,** *J. Sm.* *Vernation* fasciculate, erect, becoming cæspitose. *Fronds* (the sterile) lanceolate, 8–10 inches long, 1½–2 inches broad, erect, rigid, pinnatifid to the rachis; sinus acute; segments alternate, contiguous, lanceolate, falcate, finely serrulate; the lower ones decreasing in size and obtuse. *Veins* evident. *Fertile fronds* as broad and rigid as the sterile; segments becoming involute, densely sporangiferous.—Chatham Islands.

12*. **L. Germanii,** *Hook. Sp. Fil.* 3, *t.* 152. L. crenulata, *Hort.*—Chili.

---

110. ASPLENIUM.

33*. **A. Kaulfussii,** *Schlecht. Adum.* (29 *in obs.*). A. protensum, *Kaulf.* (*non Schrad.*).—Sandwich Islands.

19*. **A. tenerum,** *Forst.; Schk. Fil. t.* 69.—Islands of the Pacific, Ceylon.

92*. **A. nigripes,** *Metten.; Hook. Sp. Fil.* 3, *p.* 222.—Ceylon.

---

127. DICKSONIA.

6. **D. Youngii,** *Moore, Proc. Hort. Soc.*—New South Wales.†

---

128. CIBOTIUM.

4. **C. regale,** *Linden; Moore, Gard. Chron.* (1864), 414.—Mexico.†

CYATHEA.

5*. **C. insignis,** *Eat.* Cibotium princeps, *Linden, Cat.*—Mexico.

Obs.—Plants of this Fern were sent to this country under the name of *Cibotium princeps;* but a plant in Messrs. Lee's nursery having produced fructification during 1865 proves it to be a *Cyathea.*

Abstract of the number of species, and date of introduction:—

Exotic species at Kew in 1822		40
Do. My Enumeration of Kew Ferns, 1846 ...		355
Do. My Catalogue of Cultivated Ferns, 1857 ...		559
This Enumeration, Exotic and British ...	1028	
Do. Fern Allies, Exotic and British	56	
Total in 1865	1084	

In closing this enumeration, I deem it necessary to state that a few species recently introduced to the Kew collection have failed to become established, after their names were entered on the list; the principal of which are: *Dicranoglossum furcatum, Aconiopteris nervosa, Pleocnemia Leuziana, Adiantum lunulatum, Lomaria Fraseri, Asplenium lanceum, A. radiatum, Antigramma repanda, Polystichum anomalum, Loxsoma Cunninghamii, Kaulfussia esculæfolia.* Also the following, entered from my catalogue of 1857, were not in the Kew collection in 1864, viz., *Gymnogramma rutæfolia, Hemionites pedata, Notholæna lanuginosa, Myriopteris vestita, Onychium auratum, Blechnum triangulare, Nephrolepis undulata, Lindsæa guianensis, Schizoloma ensifolia, Thyrsopteris elegans, Botrychium virginicum.*

Several special causes which lead to the loss of species have already been noticed. I did not intend to advert to them again; but the recent death of Sir W. J. Hooker * necessitates me to repeat what I have stated at pages 42 and 43, where, in speaking of my having resigned the charge of the Kew collection, I said that "happily it remains under the direction of Sir W. J. Hooker." These words will now apply to Dr. Hooker, the present Director, whose name is sufficient in itself to sustain the scientific reputation of Kew. Let us hope that the general collection of living plants of the Botanic Garden, now famous for more than a century,† will not suffer by the modern taste for showy flowers, and what is now fashionably called "foliage plants."

* Died August 12th, 1865.

† Number of species at Kew in	1768	3,400
Ditto	1786	5,500
Ditto	1813	9,800

Since the latter date no general catalogue has been published, and no public record kept of the plants introduced or lost.

APPENDIX.

Tribe I.—OLEANDREÆ (p. 73).

1. OLEANDRA (p. 74).

5. **O.** musæfolia, *Kze.*; *Hook. Syn. Fil. p.* 46.—Ceylon and Malay Islands. K.*

Tribe II.—DAVALLIEÆ (p. 74).

2. HUMATA (p. 75).

1*a*. **H.** angusta, *J. Sm.* Davallia angustata, *Wall.*; *Hook. Sp. Fil.* 1, *p.* 152; *Hook. et Grev. Ic. Fil. t.* 231.—Malay and Polynesian Islands. K.

1*b*. **H.** pectinata, *J. Sm.* Davallia pectinata, *Sm.*; *Hook. Sp. Fil.* 1, *p.* 153; *Hook. et Grev. Ic. Fil. t.* 139.—Polynesian Islands.

2*a*. **H.** alpina, *J. Sm.* Davallia alpina, *Bl.*; *Hook. Sp. Fil.* 1, *p.* 154.—Java. K.

2*b*. **H.** vestita, *J. Sm.* Davallia vestita, *Bl.*; *Hook. Sp. Fil.* 1, *t.* 41, C.—Java. K.

* All names having the letter K affixed are derived from Mr. Baker's list of the new introductions to the Kew collection.

4. **H. Tyermanii**, *Moore; Gard. Chron.* 1871, *p.* 870, *t.* 178. Davallia Tyermanii, *Baker, in Appendix to Hook. Syn. Fil. p.* 467.—India (non Trop. West Africa as stated in *Gard. Chron.*).

---

3. DAVALLIA (p. 75).

10*a*. **D. Mauritiana**, *Hook. Sp. Fil.* 1, *t.* 55, B.—Mauritius. K.

12*a*. **D. pallida**, *Mett.; Bak. in Hook. Syn. Fil. p.* 469. D. Mooreana, *Masters; Gard. Chron.* 1869, *p.* 964.—Island of Ancitum. K.

---

4. LEUCOSTEGIA (p. 77).

6*a*. **L. membranulosa**, *J. Sm.* Davallia membranulosa, *Wall.; Hook. Sp. Fil.* 1, *t.* 53, A.—India.

Obs.—This genus was founded by Presl in 1836, on the *Davallia immersa* of Wallich, to which additional species have been added, all having articulate vernation, and closely related to true Davallia. Presl also founded another genus on the *Aspidium nodosum* of Blume, which he termed *Acrophorus;* its vernation is fasciculate and adherent, producing large decompound fronds 2–4 feet in height, having the sori seated on the apex of the lacenеæ, with a lateral attached indusium, not technically differing in form from many species of *Lastrea.* While admitting it as a distinct genus, I consider it to be naturally allied to *Lastrea;* nevertheless, in Moore's "Index Filicum" the whole of the species of *Leucostegia* are placed under *Acrophorus;* thus forming an unnatural association, as if *Davallia canariense* was placed in affinity with *Lastrea dilatata* and its allies. *Acrophorus nodosa* is a native of India, and has not yet been introduced in a living state.

TRIBE III.—POLYPODIEÆ (p. 78).

5. POLYPODIUM (p. 78).

4*a*. **P. pollucidum**, *Kaulf.; Hook. Sp. Fil.* 4, *p.* 206; *Hook. Second Century of Ferns, t.* 44, var. bipinnatifida. P. myriocarpum, *Hook. Ic. Pl. t.* 84. — Sandwich Islands.

OBS.—This species produces variously lobed and laciniated fronds, analogous to *Polypodium vulgare* as found in this country, of which, in the Catalogue of the Todmorden Nursery, no less than twenty varieties are recorded.

---

7. GONIOPHLEBIUM (p. 80).

a. Fronds simple.*

1*a*. **G. glaucophyllum**, *Kze.* Polypodium glaucophyllum, *Hook. Sp. Fil.* 4, *p.* 18.—Trop. America.

1*b*. **G. plesiosorum**, *Kze.*—Trop. America.

2*a*. **G. Scouleri**, *J. Sm.* Polypodium Scouleri, *Hook. et Grev. t.* 56.—N. W. America.

2*b*. **G. californicum**, *Kaulf.* Polypodium, *Hook. Sp. Fil.* 5, *p.* 18, *not Mett.*—California. K.

2*c*. **G. surrucuchense**, *Hook.* Polypodium, *Sp. Fil.* 5, *p.* 30, *Ic. Pl. t.* 69.—West Indies to Ecuador. K.

7*a*. **G. adnatum**, *Kze.* Polypodium, *Hook. Sp. Fil.* 5, *p.* 27. —Guatemala to Guiana and Galapagos. K.

---

9. PHLEBODIUM (p. 83).

1*a*. **P. nigripes**, *J. Sm.* Polypodium nigripes, *Hook. Sp. Fil.* 5, *p.* 17.—Venezuela.

---

21. PHYMATODES (p. 93).

1*a*. **P. sinuosum**, *J. Sm.* Polypodium, *Wall.*—India, Malay, and Philippines.

OBS.—The genus *Lecanopteris* of Blume is founded on an abnormal state of this. *See* "Historia Filicum," p. 105.

---

23. SELLIGUEA (p. 96).

*a*1. **S. Hamiltoniana**, *Pr.* Ceterach pedunculata, *Hook. et Grev. Ic. Fil. t.* 5.—E. Indies.

---

24. COLYSIS (p. 97).

3. **C. dilatata**, *J. Sm.* Polypodium dilatatum, *Wall.; Hook Sp. Fil.* 5, *p.* 8.—N. India. K.

---

26. NIPHOBOLUS (p. 99).

4*a*. **N. Heteractes**, *J. Sm.* Polypodium Heteractes, *Mett.; Kuhn. Linnæa*, 36, *p.* 140.—Upper India.

5*a*. **N. africanum**, *Kunze.*—E. Africa.

28*a*. AGLAOMORPHIA*, *Schott.*

Rhizome short, thick. *Fronds* sessile, of two forms, the sterile rigid, querciform; fertile fronds 2–4 feet in length, pinnatifid and sterile below, pinnate and fertile above. *Veins* of sterile segments costæform; venules and veinlets compound, anastomosing, forming nearly equal quadrangular areoles, containing free veinlets. *Fertile pinnæ* contracted, linear, sinuosely moniliform, 8–10 inches long. *Receptacles* compital, solitary on each lobule, forming a row of punctiform sori on each side of the midrib.

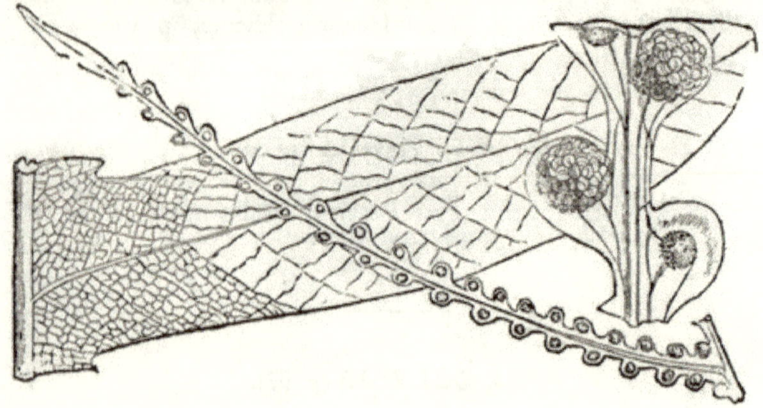

Genus 28.—Portion of sterile and fertile frond, and portion of the latter magnified.

Illust. *Schott, Gen. Fil. t.* 19; *Hook. et Bauer, Gen. Fil. t.* 91; *Moore's Index Fil. pl.* 63 *B; Presl, Tent. Pterid. t.* 8, *f.* 21, 22 (Psygmium, *Presl*).

1. **A. Meyenianum**, *Schott, Gen. Fil.* Polypodium Meyenianum, *Hook. Sp. Fil.* 5, *p.* 94.—Philippine Islands. K.

Obs.—This, as its name imports, is a splendid fern, having the habit of *Drynaria coronans*, but differing in the upper pinnæ being contracted and fertile.

---

* *Aglao*, splendid, *morpha*, form; splendid form.

TRIBE IV.—ACROSTICHEÆ (p. 104).

29. ELAPHOGLOSSUM (p. 104).

2*a*. **E. palustre,** *J. Sm.* Acrostichum palustre, *Hook. Sp. Fil.* 5, *p.* 214.—Guinea Coast. K.

16*a*. **E. perelegans,** *J. Sm.* Acrostichum perelegans, *Fée, Acrost. t.* 23; *Hook. Sp. Fil.* 5, *p.* 232.—Dominica. K.

16*b*. **E. melanopus,** *Kze. Mett. Fil. H. Lips. p.* 19, *t.* 1, *Hook. Syn. Fil. p.* 403.—Venezuela.

16*c*. **E. villosum,** *J. Sm.* Acrostichum villosum, *Sw. Hook. et Grev. t.* 95.—Trop. America.

20. **E. Prestoni,** *J. Sm.* Acrostichum Prestoni, *Bak. Sym. Fil. App.*—Rio Janeiro.

---

46. PLATYCERIUM (p. 120).

OBS.—Consequent on this genus having amorphous sori, it has always been placed in the tribe *Acrosticheæ;* it, however, in habit differs from any other genus in that tribe. Its vernation is peculiar; and although the axis of development is not very evident, I have, nevertheless, satisfied myself that it is articulate, which, with the coriaceous texture and stellated pubescence of the fronds, being analogous to *Niphobolus;* I therefore consider the division *Eremobrya* the natural place for the tribe *Platycereæ.*

6. **P. Willinckii,** *Moore; Gard. Chron.* (Jan. 1876).—Java.

TRIBE V.—GRAMMITIDEÆ (p. 122).

50. GYMNOGRAMMA (p. 125).

3a. **G. triangularis**, *Kaulf.; Hook. et Grev. Ic. Fil. t.* 153; *Hook. Sp. Fil.* 5, *p.* 146; *Hook. Fil. Exot. t.* 153.—California.

15a. **G. decomposita**, *Bak.; Gard. Chron.* (1872) *p.* 1587.—Andes of South America. K.

---

54. HEMIONITIS (p. 128).

4. **H. Muelleri**, Sect. Sericonitis, *J. Sm, Hist. Fil.* p. 151. Gymnogramma Muelleri, *Hook. Sp. Fil.* 5, *p.* 295.—Queensland. K.

---

55. ANTROPHYUM (p. 129).

5. **A. latifolium**, *Bl.; Hook. Sp. Fil.* 5, *p.* 172.—Java. K.

55a. TÆNITIS, *Willd.*

Vernation uniserial; sarmentum short, sub-cæspitose, naked, *Fronds* pinnate, long-stipate, smooth, 1–2 feet in length; pinnæ 4–6 pair, linear - lanceolate or elliptical - acuminate entire, 6–10 inches long, by 1–2 inches broad. *Veins* uniform, reticulated, areoles oblong-hexagonal. *Receptacles* compital, obliquely-oblong and coalescent, forming a broad, transverse, linear, medial, compound sorus.

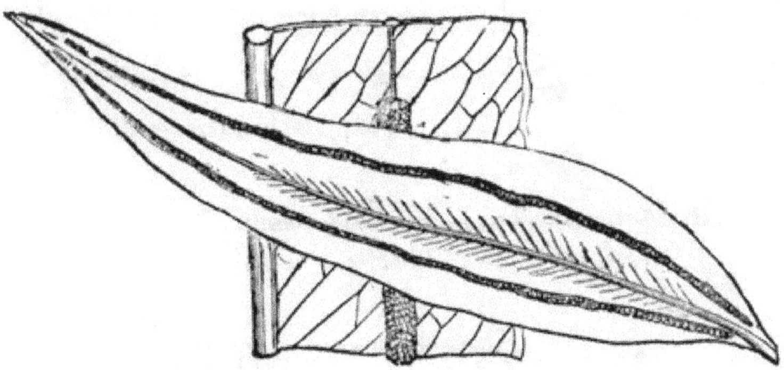

Genus 55a.—Portion of pinnæ; ditto magnified, with part of the sporangia removed.

Illust. *Hook. et Grev. Ic. Fil. t.* 63; *Hook. et Bauer, Gen. Fil. t.* 77 *B*; *Schott., Gen. Fil. t.* 20; *Moore, Ind. Fil. pl.* 17 *A*.

1. T. **blechnoides**, *Willd.*; *Sw. Synop. Fil.* Tænitis pteroides, *Schk. Fil. t.* 6 *B*.—India, Malay, and Philippine Islands. K.

Obs.—This genus probably consists of only a single species, and in affinity it may be ranked with *Syngramma* (not yet in cultivation), and is readily distinguished by its continual linear sorus, which is formed by the confluence of numerous short, oblique, sporangiferous receptacles. Specimens from the Fiji Islands exhibit fructiform or ovate scattered sori, which I consider to be only abnormal conditions, consequent on local influences affecting development. The name is derived from *Tænia*, a tape or riband, in allusion to the narrow pinnæ.

---

56. VITTARIA (p. 130).

2. **V. remota,** *Fée; Hook. Sp. Fil. t. p* 185.—W. Indies. K.

3. **V. stepitata,** *Kze.; Hook. Sp. Fil. p.* 179.—Columbia to Peru. K.

Tribe V*a.*—CTENOPTERIDEÆ.

J. Sm. Hist. Fil. p. 183.

Fronds linear, repand, sinuose, moniliform or pinnatifid, rarely pinnate, or more compound, from an inch to a foot or more in length. *Veins* free; *sori* punctiform, naked.

Obs.—This tribe embraces about 50 or 60 small, neat-growing ferns, none of which have as yet been introduced except the following.

60*a*. CTENOPTERIS, *Bl.*

(Polypodium sp., *Auct. Hook. Sp. Fil.*)

Vernation fasciculate or sarmentose. *Fronds* linear lanceolate, sinuose, or more or less deeply pinnatifid or pinnate, rigid, erect: or thin and pendulous; *segments* entire, dentate or laciniated. *Veins* once forked, rising from the midrib of each segment or laciniæ, generally obscure. *Sori* punctiform, terminal, solitary, or few in each laciniæ, or uniserial, often becoming confluent, naked, or beset with rigid hairs.

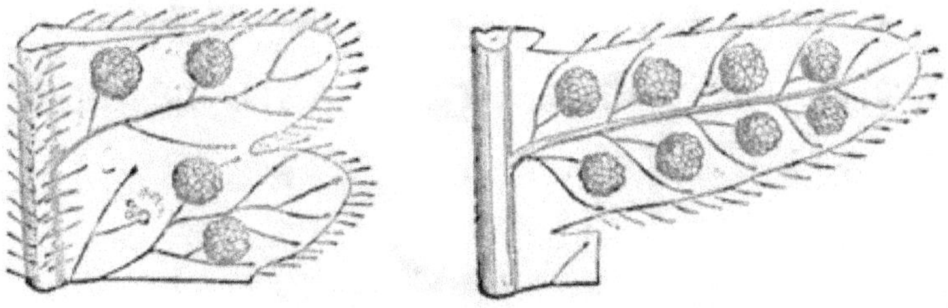

Genus 60*a*.—Portion of a frond of C. *trichosorus*.

Genus 60*a*.—Portion of a frond of C. *rigescens*.

C. trichomanoides, *Sw.* Polypodium trichomanoides, *Bedd. Ferns Brit. Ind. pl.* 2.—West Indies and Trop. America, Himalaya, Malay islands, and other localities.

Tribe VI.—PHEGOPTERIDEÆ (p. 136).

63. MENISCIUM (p. 136).

1*a*. **M. Thwaitesii**, *Hook. Syn. Fil. p.* 391.—Ceylon. K.

63*a*. DICTYOCLINE, *Moore*.

Corm decumbent, sub-sarmentose. *Fronds* long, stipate, pinnatifid or pinnate, 1½–2½ feet high, pinnæ 3–4 pairs, short petiolate, broad lanceolate, accuminate, falcate, 5 – 6 inches long, 1 –1½ inch broad. *Primary veins* costæform, combined by transverse, arcuate or zigzag, anastomosing venules, the lower forming one oblong costal areole between each pair of primary veins, the exterior ones unequally hexagonal. *Venules* sporangiferous, forming reticulated sori.

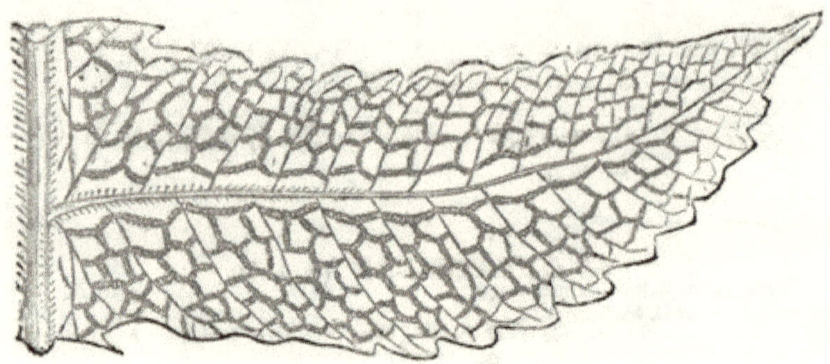

Genus 63*a*.—Portion of fertile frond slightly magnified.

Illust. *Moore, Ind. Fil. pl.* 46 *A; Hook. Fil. Exot. t.* 93.

1. **D. Wilfordii,** *J. Sm.* Hemionitis Wilfordii, *Hook. Fil. Exot. t.* 93. Sub Hemionitis Griffithii, *Hook. Syn. Fil. p.* 399.—Formosa. K.

Obs.—The present genus was founded by Mr. Moore upon a pinnate fronded fern from Assam, and placed by him, on account of its sori being reticulated, next to *Hemionitis*, under which genus it is also placed in the "Species Filicum;" but it possesses no other point of affinity with *Hemionitis*, its whole habit and nature of venation agreeing with the meniscioid group of *Phegopterideæ*, differing only in the venation and sori being more decidedly reticulated. The name is derived from *dictyon*, a net, *cline*, a bed, in allusion to the reticulated fructification.

---

64. GONIOPTERIS (p. 137).

14. **G. stegnogrammoides,** *J. Sm.* Polypodium stegnogrammoides, *Bak.; Hook. Syn. Fil. p.* 317. Polypodium sandwicense, *Hook. Sp. Fil.* 5, *p.* 5 (*non Hook. Sp. Fil.* 4, *p.* 267).—Sandwich Islands. K.

65. NEPHRODIUM (p. 138).

1*a*. **N. sophoroides**, *Desv.; Hook. Syn. Fil. p.* 289.—Japan to Hong-Kong and Formosa. K.

2*a*. **N. extensum**, *Hook. Sp. Fil.* 4, *t.* 240.—Ceylon, India, and Malayan Islands. K.

11*a*. **N. amboinense**, *Pr.; Hook. Sp. Fil.* 4, *p.* 75.—E. Indies and Malay.

---

68. CYRTOMIUM (p. 141).

4. **C. abbreviatum**, *J. Sm.* Aspidium abbreviatum, *Schrad.; Hook. Sp. Fil.* 4, *t.* 234.—West Indies. K.

---

70. ASPIDIUM (p. 143).

4*a*. **A. elatum**, *J. Sm.* (*non Bory*). Nephrodium elatum, *Bak.; Hook. Syn. Fil. p.* 298.—Mount Chimborazo. K.

5*a*. **A. membranaceum**, *Hook. Sp. Fil.* 5, *p.* 105.—India, China, and Philippine Islands. K.

---

74. POLYSTICHUM (p. 148).

2*a*. **P. munitum**, *J. Sm.* Aspidium munitum, *Kaulf.; Hook. Sp. Fil.* 4, *t.* 219.—California. K.

3*a*. **P. Richardii**, *J. Sm.* Aspidium Richardii, *Hook. Sp. Fil.* 4, *t.* 222.—New Zealand. K.

15*a*. **P. laserpitiifolium**, *Mett.; Hook. Syn. Fil. p.* 254.—Japan.

---

75. LASTREA (p. 152).

2*a*. L. **Beddomei,** *J. Sm.* Nephrodium Beddomei, *Bak.*; *Hook. Syn. Fil. p.* 267. Lastrea gracilescens, *Beddome, Fil. t.* 110 (*non Hook. Sp. Fil.*).—Ceylon. K.

6*a*. L. **lanciloba,** *J. Sm.* Nephrodium lancilobum, *Bak.*; *Hook. Syn. Fil. edit.* 2, *p.* 499.—North Australia.

16*a*. L. **Bergeanum,** *J. Sm.* Nephrodium Bergeanum, *Bak.*; *Hook. Syn. Fil. p.* 269.—South Africa. K.

26*a*. L. **crinitum,** *J. Sm.* Nephrodium crinitum, *Desv.*; *Hook. Sp. Fil.* 4, *p.* 111 (*in part*).—Mauritius and Bourbon. K.

28*a*. L. **prolixa,** *J. Sm.* Aspidium prolixum, *Willd.* Nephrodium prolixum, *Bak.*; *Hook. Syn. Fil. p.* 268.—Ceylon and India. K.

30. L. **podophylla,** Hong-kong, exclude *Aspidium Sieboldi, &c.*

30*a*. L. **Sieboldi,** *J. Sm.* Aspidium Sieboldi, *Van Houtte, Cat. Mett. Fil. Hort. Lips. t.* 20, *f.* 1–4.—Japan.

Obs.—I originally considered *L. Sieboldi* of Japan to be the same as *L. podophylla*, Hook., of Hong-kong; but plants of the latter having been recently received at Kew from Hong-kong, show that the two are distinct species.

30*b*. L. **cuspidata,** *J. Sm.* Nephrodium cuspidatum, *Bak.*; *Hook. Syn. Fil. p.* 260. Polypodium elongatum, *Wall.*—Ceylon and India. K.

34*a*. L. **sageniodes,** *J. Sm.* Aspidium sageniodes, *Mett.* Nephrodium sageniodes, *Bak.*; *Hook. Syn. Fil. p.* 271. Nephrodium melanopus, *Hook. Sp. Fil.* 4, *p.* 110.—Malayan Peninsula and Islands. K.

38*a*. L. **cognata,** *J. Sm.* Nephrodium cognatum, *Hook. Sp. Fil.* 4, *t.* 256.—St. Helena. K.

42*a*. **L. obtusiloba,** *J. Sm.* Nephrodium obtusilobum, *Bak.; Hook. Syn. Fil. p.* 284.—Ceylon. K.

42*b*. **L. Floridanum,** *J. Sm.* Nephrodium Floridanum, *Hook. Fil. Exot. t.* 99.—Florida. K.

42*c*. **L. catopteron,** *J. Sm.* Nephrodium catopteron, *Hook. Sp. Fil.* 4, *p.* 137.—Venezuela. K.

42*d*. **L. Blumei,** *J. Sm.* Aspidium intermedium, *Bl.* (*non Willd.*). Nephrodium intermedium, *Bak.; Hook. Syn. Fil. p.* 283.—Ceylon. K.

42*e*. **L. hirsuta,** *J. Sm.* Nephrodium hirsutum, *Don.* Aspidium eriocarpum, *Wall.* Nephrodium eriocarpum, *Hook. Sp. Fil.* 4, *p.* 141. Hypodematium Ruppellianum, *Kze. Fil. t.* 21. H. onustum, *Kze. Analect. t.* 28.—Cape Verdes, India, Malayan Peninsula, and China. K.

45*a*. **L. flaccidum,** *Hook. Sp. Fil.* 4, *t.* 263.—Ceylon, India, and Java K.

46*a*. **L. setosa,** *Bl.* Aspidium, *Bl.;* Nephrodium setosum, *Bak. Syn. Fil. p.* 274.—Java. K.

47*a*. **L. inequale,** *J. Sm.* Nephrodium inequale, *Hook. Sp. Fil.* 4, *p.* 125.—South Africa. K.

48*a*. **L. fragrans,** *J. Sm.* Nephrodium fragrans, *Hook. Sp. Fil.* 4, *p.* 122; *Hook. et Grev. Ic. Fil. t.* 70.—General in the cold regions of the North Temperate zone.

53*a*. **L. undulatum,** *J. Sm.* Nephrodium undulatum, *Bak.; Hook. Syn. Fil. p.* 276.—Ceylon. K.

53*b*. **L. Thwaitesii,** *J. Sm.* Nephrodium Thwaitesii, *Bak.; Hook. Syn. Fil. p.* 277. Aspidium concinnum, *Thwaites* (*non Willd.*).—Ceylon. K.

---

77. WOODSIA (p. 161).

2*a*. **W. glabella**, *R. Br.; Hook. Sp. Fil.* 1, *p.* 64; and in *Fl. Boreal. Americ.* 2, *t.* 237.—Pyrenees and Arctic Regions.

4*a*. **W. oregana**, *Eat.; Hook. Syn. Fil. p.* 49.—Rocky Mountains. K.

77*a*. DIACALPE, *Blume.*

Vernation fasciculate, erect. *Fronds* stipate, 2–3 feet high, deltoid, decompound-multifid; primary pinnæ alternate; ultimate pinnules small ($\frac{1}{4}$-inch in length), linear, obtuse, oblique-cuneate at the base. *Veins* simple or forked, clavate. *Receptacles* medial, small, punctiform. *Indusium* superficial, at first globose, entire, ultimately opening irregularly (calyciform), including the sporangia.

Genus 77*a*.—Portion of frond slightly magnified.

Illust. *Hook. et Bauer, Gen. Fil. t.* 99; *Moore, Ind Fil. pl.* 81.

1. **D. aspidioides**, *Bl.; Hook. Sp. Fil.* 1, *p.* 59.—Java, Assam, and Moulmein. K.

Obs.—This genus consists of one species only, which in habit is similar to the multifid fronded species of *Lastrea*, but differs in the sporangia being enclosed in a globose indusium, which ultimately opens, becoming calyciform, similar to the *Physematium* section of *Woodsia*.

The name *Diacalpe* is derived from *dia* and *kalpe*, the Greek for a vessel, in allusion to the cup form of the indusium.

---

82. STRUTHIOPTERIS (p. 166).

1*a*. **S. orientalis**, *Hook. Second Century of Ferns, t.* 4. Onoclea orientalis, *Hook. Sp. Fil.* 4, *p.* 161.—Sikhim, Assam, and Japan.

---

84. PHEGOPTERIS (p. 168).

5*a*. **P. molle**, *J. Sm.* Polypodium molle, *Roxb.; Hook. Syn. Fil. p.* 309. Polypodium Dianæ, *Hook. Sp. Fil.* 4, *p.* 234.—St. Helena. K.

12*a*. **P. platyphylla**, *J. Sm.* Polypodium platyphyllum, *Hook. Sp. Fil.* 4, *p.* 248.—Cuba. K.

---

85. HYPOLEPIS (p. 171).

5. **H. millefolia**, *Hook. Sp. Fil.* 2, *t.* 95 *B.*—New Zealand.

6. **H. Bergiana**, *Hook. Sp. Fil.* 2, *p.* 67.—South Africa. K.

Tribe VII.—PTERIDEÆ (p. 172).

86. NOTHOLÆNA (p. 172).

13. **N. canescens**, *Kze.; Hook. Sp. Fil.* 2, *p.* 110.—Mexico.

---

87. MYRIOPTERIS (p. 173).

1*a*. **M. Fendleri**, *Hook.* Cheilanthes Fendleri, *Hook. Sp. Fil.* 2, *t.* 107, *B.*—New Mexico, California. K.

1*b*. **M. gracillima**, *Eat.* Cheilanthes gracillima, *Eat.; Hook. Syn. Fil. p.* 139.—Brit. Columbia to Guatemala. K.

88. CHEILANTHES (p. 174).

1*a*. **C. Mathewsii**, *Kze.; Hook. Sp. Fil.* 2, *p.* 91.—Peru. K

5*a*. **C. pulchella**, *Bory; Hook. Sp. Fil.* 2, *p.* 94.—Madeira.

90. CINCINALIS (p. 178).

4*a*. **C. Fendleri**, *Kze.* Nothochlæna Fendleri, *Hook. Syn. Fil. p.* 374.—New Mexico. K.

91. PELLÆA (p. 179).

1*a*. **P. Stilleri**, *Bak.; Hook. Syn. p.* 453. Pellea gracilis, *Hook. Sp. Fil.* 2, *t.* 133 *B.* Pteris gracilis, *Michx.* Pteris Stilleri, *Gmelin.*—North America and North Asia. K.

10*a*. **P. ornithopus**, *Hook. Sp. Fil.* 2, *t.* 116 *A.*—California. K.

10*b*. **P. mucronata**, *Eat.; Hook. Syn. Fil. p.* 148. P. Wrightiana, *Hook. Sp. Fil.* 2, *t.* 115 *B.* P. longimucronata, *Hook. Sp. Fil.* 2, *t.* 115 *A.*—California, New Mexico. K.

16. **P. densa**, *Hook. Sp. Fil.* 2, *t.* 125 *B.*—North America.

92. PLATYLOMA (p. 181).

2*a*. **P. Bridgesii,** *J. Sm.* Pellea Bridgesii, *Hook. Sp. Fil. t.* 142 *B.*—California. K.

4. **P. bellum,** *Moore; Gard. Chron.* (1873), *p.* 213.—California.

5. **P. brachypetrum,** *Moore; Gard. Chron.* (1873), *p.* 141.—California.

6. **P. andromedæfolia,** *Kaulf.* Pellæa andromedæfolia, *Fée; Hook. Sp. Fil.* 2, *p.* 149.—California.

---

93. ADIANTUM (p. 182).

1*a*. **A. asarifolium,** *Willd.; Hook. Fil. Exot. t.* 11; *Hook. Sp. Fil.* 2, *t.* 71 *B.*—Mauritius.

9*a*. **A. Seemanii,** *Hook. Sp. Fil.* 2, *t.* 81 *A.*—Central America.

13*a*. **A. Cayenense,** *Willd.; Hook. Sp. Fil.* 2, *t.* 61.—Guiana.

13*b*. **A. Lindenii,** *Moore; Gard. Chron.* (1866), *p.* 778; *Appendix, Hook. Syn. Fil. p.* 473.—Brazil.

13*c*. **A. velutinum,** *Moore; Gard. Chron.* (1866), *p.* 777; *Appendix, Hook. Syn. Fil. p.* 473.—South America.

15*a*. **A. peruvianum,** *Klot.; Hook. Sp. Fil.* 2, *t.* 81 *C*; *Gard. Chron.* (1870), *p.* 457, *with a fig.*—Peru.

17*a*. **A. Sanctæ-Catherinæ,** *Hort.*—Brazil.

Obs.—Resembles *A. pentadactylon,* but differs in having a strong feline scent, similar to that of *Pteris felosma.*

23*a*. **A. speciosum,** *Hook. Sp. Fil.* 2, *t.* 85 *C.*—Peru.

28*a*. **A. hirtum,** *Klot.; Hook. Sp. Fil.* 2, *t.* 82 *A.*—Brazil.

28*b*. **A. cubense,** *Hook. Sp. Fil.* 2, *E.* 73 *A.*—Jamaica. Cuba. K.

35*a*. **A. Moorei**, *Bak.; Appendix, Hook. Syn. Fil. p.* 474, A. amabile, *Moore; Gard. Chron.* (1868), *p.* 1090.—Peru.

35*b*. **A. decorum**, *Moore; Gard. Chron.* (1869), *p.* 582. A. Wagnerii, *Mett.; Appendix, Hook. Syn. Fil. p.* 473.—Peru.

35*c*. **A. rubellum**, *Moore; Gard. Chron.* (1868), *p.* 865; *Appendix, Hook. Syn. Fil. p.* 474.—Bolivia. K.

35*d*. **A. Veitchianum**, *Moore; Gard. Chron.* (1868), *p.* 1090; *Appendix, Hook. Syn. Fil. p.* 473.—Peru.

35*e*. **A. venustum**, *Don; Hook. Sp. Fil.* 2, *t.* 96 *B.*—Himalayas. K.

35*f*. **A. Henslovianum**, *Hook. fil.; Hook. Sp. Fil.* 2, *p.* 43; *Moore, in Flo. et Pomo.* (1873), *p.* 277, *with fig.*—Peru.

35*g*. **A. Farleyense**, *Moore; Gard. Chron.* (1866), *p.* 6.—Barbadoes.

Obs.—This has the general habit of *A. tenerum*, but is a much larger-growing plant, and, as it does not produce fructification, it is supposed to be a hybrid originated in a garden in the island of Barbadoes.

35*h*. **A. princeps**, *Moore; Gard. Chron.* (Jan. 22, 1876).—New Grenada.

39*a*. **A. excisum**, *Kze.; Hook. Sp. Fil.* 2, *p.* 41.—Chili.

42*a*. **A. glaucophyllum**, *Hook. Sp. Fil.* 2, *p.* 40.—Mexico.

Hybrid. **A. gracillimum**, *Moore; Gard. Chron* (1874), *p.* 14.

---

96. PTERIS (p. 188).

2*a*. **P. Hookeriana**, *Ag.; Hook. Sp. Fil.* 2, *p.* 165.—Adam's Peak, Ceylon. K.

25a. P. **pellucens**, *Ag.; Hook. Sp. Fil.* 2, *p.* 201.—Himalayas. K.

---

97. LITOBROCHIA (p. 192).

1a. L. **Currori**, *J. Sm.* Pteris Currori, *Hook. Sp. Fil.* 2, *t.* 190.—Fernando Po. K.

11a. L. **gigantea**, *Willd.; Hook. Sp. Fil.* 3, *t.* 217.—West Indies and Tropical America. K.

TRIBE VIII.—BLECHNEÆ (p. 196).

100. BLECHNUM (p. 196).

16a. B. **nitidum**, *Pr.; Hook. Sp. Fil.* 3, *t.* 55.—Tropical America and Asia. K.

---

101. DOODIA (p. 199).

7. D. **duriuscula**, *Moore; Gard. Chron.* (1868), *p.* 114.

---

102. LOMARIA (p. 199).

11. L. **gibba**, *add. var. Belli.*—Chatham Islands.

24a. L. **ciliata**, *Moore; Gard. Chron.* (1866), *p.* 290; *Hook. Syn. Fil. p.* 175.—New Caledonia.

26a. L. **cycadoides**, *Pappe and Rawson, Syn. Fil. Af. Aust. p.* 28.—Natal.

OBS.—This plant of Natal is represented by allied forms found in the West Indies, Brazil, southward to the Falkland Islands and Straits of Magellan; all of which in the "Species

Filicum" are united under *L. Boryana* (as quoted at p. 202), but from an examination of them, I am led to believe that the plants from the different localities represent different species.

102*a*. SADLERIA,* *Kaulf.*

Hook. Sp. Fil.

Vernation fasciculate, erect, arborescent, 2 to 3 feet high, stout. *Fronds* 3 to 4 feet long, rigid, bipinnatifid; pinnæ linear lanceolate, 10 to 12 inches long, contiguous, sessile, articulated with the rachis, deeply pinnatifid, coriaceous, opaque; laciniæ ¾ to 1 inch long, linear falcate. *Veins* obscure, arcuately anastomosing, forming costal areoles; venules simple or forked, their apices terminating in a thickened margin. *Sporangiferous receptacle* transverse, linear, continuous on the costal anastomose, elevated in the form of a ridge. *Indusium* linear, laterally attached on the exterior side of the receptacle, its inner margin free, becoming reflexed, coriaceous, persistent.

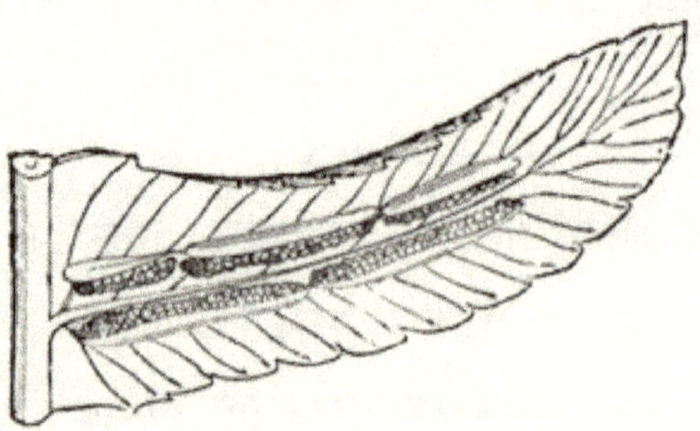

Genus 102*a*.—Fertile pinnule.

S. cyatheoides, *Kaulf.*; *Moore. Ind. Fil. p.* 12 *B.*; *Hook. Syn. Fil. t.* 4, *fig.* 35.—Sandwich Island.

107. LORINSERIA (p. 206).

1*a*. **L. Harlandii**, *J. Sm.* Woodwardia Harlandii, *Hook. Sp. Fil.* 3, *p.* 70; *Hook. Fil. Exot. t.* 7.—Hong-Kong. K.

---

* In honour of Josephus Sadler, an Hungarian botanist.

TRIBE IX.—ASPLENIEÆ (p. 209).

110. ASPLENIUM (p. 209).

10*a*. **A. normale**, *Don, Prod. Fl. Nep. p.* 7; *Hook. Syn. Fil. p.* 197. A. multijugum, *Wall.; Hook. Sp. Fil.* 3, *t.* 188.—India and Ceylon.

12*a*. **A. heterocarpum**, *Wall.; Hook. Sp. Fil.* 3, *t.* 173.—Java.

15*a*. **A. Fernandezianum**, *Kze.; Hook. Syn. Fil.*, sub-erectum, *p.* 202.—Juan Fernandez.

17*a*. **A. angustifolium**, *Mich.; Hook. Sp. Fil.* 3, *p.* 115.—North America.

17*b*. **A. Wightianum**, *Wall.; Hook. Sp. Fil.* 3, *t.* 167.—Madras, Ceylon. K.

18*a*. **A. auriculatum**, *Sw.; Hook. Sp. Fil.* 3, *t.* 171.—Tropical America, West Indies. K.

22*a*. **A. anisophyllum**, *Kze.; Hook. Sp. Fil.* 3, *t.* 136.—South Africa. K.

25*a*. **A. resectum**, *Sm.; Hook. Sp. Fil.* 3, *p.* 130.—Java. K.

27*a*. **A. Schizoden**, *Moore; Gard. Chron.* (1871), *p.* 1004.—New Zealand.

34*a*. **A. Dregianum**, *Kze.; Hook. Sp. Fil.* 3, *p.* 214.—Natal, K.

45*a*. **A. Nova-Caledoniæ**, *Hook. Sp. Fil.* 2, *p.* 213; *Hook. Ic. Pl. t.* 911.—New Caledonia.

8*a*. *Gibberosa Group.*

Vernation fasciculate, erect; caudex small, undefined. *Fronds* bi-tripinnatifid, firm, smooth, from 6–18 inches high; ultimate pinnæ pinnatifidly laciniate. *Segments* linear, widening upwards, generally with a projecting dent. *Veins* forked, sporangiferous receptacle oblong, sub-terminal. *Indusium* short, conniving with the margin, forming a round or oblong open cyst, containing the sporangia. (Loxoscaphe, *Moore*.)

57*a*. **A. concinnum**, *J. Sm.* Davallia concinnum, *Schred.; Hook. Sp. Fil.* 1, *p.* 193. Loxocaphe concinna, *Moore*.—Trop. America. K.

57*b*. **A. davallioides**, *Hook. Sp. Fil.* 3, *p.* 212; *Second Cent. Ferns, t.* 40.—Japan, Formosa.

57*c*. **A. ferulaceum**, *Moore; Hook. Second Cent. Ferns, t.* 38.—New Grenada.

59*a*. **A. elegantulum**, *Hook. Sp. Fil.* 3, *p.* 190; *Hook. Second Century of Ferns, t.* 38.—Japan.

68*a*. **A Gardneri**, *Baker; Gard. Chron.* (1873), *p.* 712.—Ceylon.

91*a*. **A. Goringianum**, *Mett;. Hook. Sp. Fil.* 3, *p.* 224.—Japan.

92*a*. **A. Japonicum**, *Mett. Fil. Ind.* 2, *p.* 240; *Hook. Syn. Fil. p.* 227.—Japan and China.

95*a*. **A. aspidioides**, *Schlecht; Hook. Syn. Fil. p.* 228.—Tropics and sub-tropical regions generally. K.

---

111. DIPLAZIUM (p. 221).

5*a*. **D. fraxinifolium**, *Wall; Hook. Second Cent. Ferns, t.* 19.—E. Indies.

5*b*. **D. Pullingeri**, *Bak.; Gard. Chron.* (Oct. 1875).—Hong-Kong.

---

112. SCOLOPENDRIUM (p. 223).

2. **S. Hemionites**, *Sw. Schk. Fil. t.* 84; *Hook. Sp. Fil.* 4, p. 2—South Europe.

114. ANTIGRAMMA (p. 226).

3. **A. plantaginea**, *Presl.* Scolopendrium Douglasii, *Hook. Sp. Fil.* 4, *p.* 3. Asplenium Douglasii, *Hook. et Grev. t.* 150.—Brazil. K.

115*a*. ALLANTODIA, *R. Br.* (*in part*).

Vernation fasciculate, corm decumbent. *Fronds* pinnate, 2–3 feet long, sub-erect; pinnæ linear, membranaceous, 4–6 inches long, 1 inch broad. *Veins* forked near the midrib, anastomosing towards the margin, forming two rows of hexagonoid areolæ bounded by a continuous marginal vein. Sporangiferous *receptacles* unilateral on the lower part of the veins. *Indusium* vaulted, cylindrical, forming linear sori.

Obs.—This genus is now restricted to a single species; it originally contained a group of species, which now form the section *Athyrium* of the genus *Asplenium* (p. 219): they differ in having large decompound fronds and free venation, agreeing only with the present genus in having turgid indusia, which is not uncommon to other species of *Asplenium*.

The name *Allantodia* is derived from the Greek *allantos*, a sausage, the indusium before bursting resembling a miniature sausage in shape and colour.

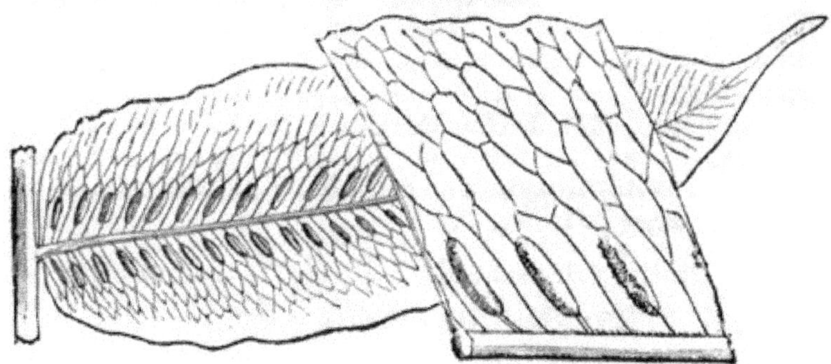

Genus 115*a*.—Portion of fertile frond, natural size, and ditto magnified.

1. **A. Brunoniana**, *Wall. Pl. Asiat. Rar.* i. 44, *t.* 52; *Hook. et Bauer, Gen. Fil. t.* 120 *A*; *Hook. Sp. Fil.* 3, *p.* 275.—Himalayas, Ceylon, Java, Tahiti. K.

117. CETERACH (p. 228).

2. C. **aureum**, *Desv.* Asplenium Ceterach, *var.* aureum, *Hook. Sp. Fil.* 3, *p.* 273.—Canaries and Madeira.

3. C. **cordatum**, *Kaulf.; J. Sm. En. Fil.* (1841). Gymnogramma cordata, *Hook. et Grev. Ic. Fil. t.* 156.—S. Africa. K.

Obs.—The general appearance of this species seems to indicate its affinity to be with *C. officinarum* rather than with any species of *Gymnogramma*, differing only from *C. officinarum* in the veins being generally free, rarely anastomosing.

Tribe X. DICKSONIEÆ (p. 229).

Sect. Lindsœeœ.

118. LINDSÆA (p. 230).

8*a*. L. **parvula**, *Fée; Hook. Syn. Fil. p.* 452.—Trinidad. K.

Obs.—With regard to *L. reniformis*, *L. sagittata*, *L. Leprieusii*, *L. falcata*, *L. trapeziformis*, *L. stricta*, *L. crenata*, and *L. dubia*, entered at p. 230, it is proper to state that they were entered on the evidence of living specimens sent to me from Mr. Backhouse, of York, who had recently imported them from their native country, and I expected that they would soon have been added to the Kew Collection; but they proved to be of difficult cultivation; and Mr. Backhouse writes me that he has not succeeded in establishing any of them. This is much to be regretted, as *Lindsœa* is a very interesting genus of Ferns.

---

120. ODONTOSORIA (p. 232).

*a*1. O. **clavata**, *J. Sm.* Davallia clavata, *Sw.* (*Plum. t.* 101 *B*), *Hook. Sp. Fil.* 1, *p.* 187.—West India.

120*a*. ODONTOLOMA, *J. Sm. Gen. Fil.* 1841.

Vernation uniserial, sarmentose. *Fronds* slender, 1–2 feet high, pinnate or bipinnate; pinnæ and pinnules oblong-dimidiate, the upper margin entire, dentate or incise-lobed. *Costæ* excentric. *Veins* unilateral; venules direct, free. *Receptacles* terminal, punctiform. *Indusium* subreniform, plane, its sides free, shorter than the subindusæform margin.

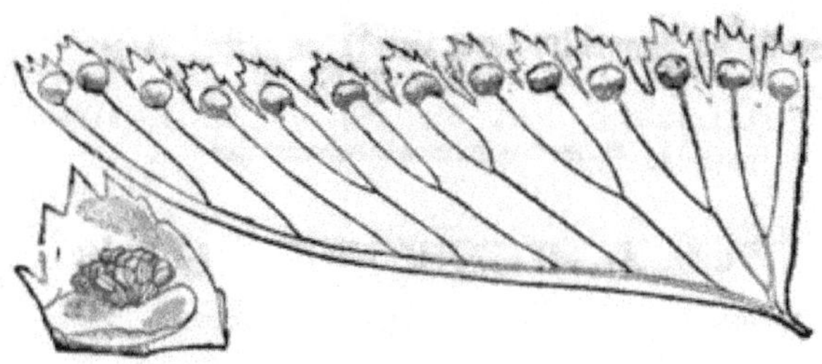

Genus 120*a*.—Pinnæ, and portion magnified.

1. O. repens, *Pr. Epim. Bot. p.* 97. Dicksonia repens, *Bory,* Davallia repens, *Desv.* Davallia hemiptera, *Bory; Hook. Sp. Fil.* 1, *p.* 176. Davallia Boryana, *Hook. et Grev. Ic. Fil. t.* 143; *Hook. Sp. Fil.* 1, *p.* 175. Odontoloma Boryanum, *J. Sm. Gen. Fil.* 1841; *Hook. et Bauer, Gen. Fil. t.* 114 *B.*—Mauritius, India, Malayan and Polynesian Islands. K.

Obs.—I may here remark that Mr. Moore has united the species of *Odontoloma* and *Leucostegia* under *Acrophorus* of *Presl.* (See *Leucostegia.*)

From *odontos*, gen. of *odous*, a tooth, *loma*, a margin; the sori being seated on marginal lobes.

---

121. MICROLEPIA (p. 233).

*a*1. **M. pinnata,** *J. Sm.* Davallia pinnata, *Cav.; Hook. Sp. Fil.* 1, *t.* 60, *f.* 1 and 4. D. flagellifera, *Wall.; Hook. et Grev. Ic. Fil. t.* 183.—Malay and Philippine Islands.

4*a*. **M. Thwaitesii**, *J. Sm.* Davallia Thwaitesii, *Bak.; Hook. Syn. Fil. p.* 99.—Ceylon. K.

5*a*. **M. hirta**, *J. Sm.* Davallia hirta, *Kaulf.; Hook. Sp. Fil.* 1, *p.* 181.—Sandwich Islands. K.

---

125. SITOLOBIUM (p. 236).

3*a*. **S. Plumieri**, *J. Sm.* Dicksonia Plumieri, *Hook. Sp. Fil.* 1, *p.* 72. Davallia adiantoides, *Sw. Plum. Fil. t.* 7. D. Lindenii, *Hook. Sp. Fil.* 1, *t.* 25 *B.*—West Indian Islands. K.

7*a*. **S. Smithii**, *J. Sm.* Dicksonia Smithii, *Hook. Sp. Fil.* 1, *t.* 28 *D.* Sitolobium flaccidum, *J. Sm.; Hook. Journ. Bot.* 3, *p.* 418 (*non* Dicksonia flaccida, *Sw.*).

---

127. DICKSONIA (p. 238).

6. **D. fibrosa**, *Col.; Hook. Sp. Fil.* 1, *t.* 23 *B.*—New Zealand.

7. **D. chrysotricha**, *Moore; Hook. Syn. Fil. p.* 50. Balantium chrysotrichum, *Hassk. Fil. Jav. p.* 53 (*Lind. Cat.* 1871).—Java.

8. **D. Deplanchei**, *Vieill.; Hook. Syn. Fil. App. p.* 462.—New Caledonia.

---

128. CIBOTIUM (p. 239).

3*a*. **C. Wendlandi**, *Mett.* Dicksonia Wendlandii, *Bak.; Appendix Hook. Syn. Fil. p.* 460. Cibotium spectabile, *Lind. Cat.* 1871.—Guatemala.

TRIBE XI.—CYATHEÆ (p. 240).

131. CYATHEA (p. 241).

2*a*. C. **Schanshin**, *Mart. l. c. t.* 54; *Hook. Sp. Fil.* 1, *p.* 20.—Tropical America. K.

2*b*. C. **Imrayana**, *Hook. Sp. Fil.* 1, *t.* 9 *B.*—West Indies. K.

5. C. **princeps**, *p.* 291, *add.* C. **insignis**, *Cat. Gard. Chron* 1873, p. 776.—Cuba.

8*a*. C. **Burkei**, *Hook. Sp. Fil.* 1, *t.* 17 *B.*—Natal.

11*a*. C. **spinulosa**, *Wall.; Hook. Sp. Fil.* 1, *t.* 12 *C.*—India.

16. C. **funebris**, *Linden's Cat.* 1871, *with a fig.*—New Caledonia. K.

---

133. ALSOPHILA (p. 243).

5*a*. A. **sagittifolia**, *Hook. Syn. Fil. p.* 37; *Gard. Chron. p.* 321, *f.* 112.—Trinidad. K.

8*a*. A. **atrovirens**, *Presl; Hook. Sp. Fil.* 1, *p.* 46.—Brazil. K.

8*b*. A. **infesta**, *Kze.; Hook. Sp. Fil.* 1, *p.* 42.—Peru. K.

8*c*. A. **phalerata**, *Mart. Crypt. Bras. t.* 42; *Hook. Sp. Fil.* 1, *p.* 42.—South America. K.

13*a*. A. **oblonga**, *Klot.; Hook. Syn. Fil. App.*—Brit. Guiana.

14*a*. A. **tomentosa**, *Hook. Sp. Fil.* 1, *pp.* 54, 55.—Java.

15*a*. A. **Scottiana**, *Bak.; Gard. Chron.* 1869, *p.* 699; *Appendix Hook. Syn. Fil. p.* 460.—Sikhim, Himalaya. K.

15*b*. A. **Walkeræ**, *J. Sm.* Cyathea Walkeræ, *Hook. Sp. Fil.* 1, *p.* 24; *Hook. Ic. Pl. t.* 647. Hemitelia Walkeræ, *Hook. Syn. Fil. p.* 30.—Ceylon.

15c. **A. ornata**, *Scott. Bedd. Ferns Brit. Ind. t.* 342; *Bak. Syn. Fil. App.*—Sikkim.

18. **A. Moorei**, *add. syn.* A. Macarthurii, *Hook. Syn. Fil. p.* 40, *and* A. Leichardtiana, *Muell; Hook. Syn. Fil. p.* 450.

Obs.—This species was first discovered by Mr. Thomas Moore, the Director of the Botanic Garden, Sydney.

18a. **A. Cooperi**, *Hook.; Appendix Hook. Syn. Fil. p.* 459.—Queensland, New South Wales.

The following names appear in Nurserymen's Catalogues, but their identification as distinct and new species has not yet been botanically ascertained:—A. Shepherdii, *Bull. Cat.* 1871. A. denticulata, *Lind. Cat.* and *Veitch Cat.* A. Amazonnica, *Lind. Cat.* 1871. A. Van Geertii, *Van Geert's Cat.*

Sub-Order II.—GLEICHENIACEÆ (p. 247).

136. GLEICHENIA (p. 247).

9a. **G. glauca**, *Hook. Sp. Fil. t.* 3 *B* (*non Sw.*). Polypodium glaucum, *Thunb. Fil. Jap.* G. gigantea, *Wall. in Hook. Sp. Fil.* 1, *t.* 3 *A.* G. excelsa, *J. Sm. Hook. Sp. Fil.* 1, *t.* 4 *B.*—General throughout the tropics and sub-tropics of both the Old and New World.

Sub-Order III.—HYMENOPHYLLACEÆ (p.249).

137. HYMENOPHYLLUM (p. 249).

a1. **H. cruentum**, *Cav.; Hook. Sp. Fil.* 1, *t.* 31 *A.*—S. Chili. K.

5*a*. H. **Javanicum**, *Spreng.; Hook. Sp. Fil.* 1, *p.* 106.—Java.

17*a*. H. **Neesii**, *Hook. Sp. Fil.* 1, *p.* 99.—Java.

17*b*. H. **sabinæfolium**, *Bak.; Hook. Syn. Fil. p.* 71.—Java.

18*a*. H. **Magellanicum**, *Willd.* H. attenuatum, *Hook. Sp Fil.* 1, *t.* 36 *B.*—Chili.

18*b*. H. **pectinatum**, *Cav.; Hook. Sp.* 1, *Fil. t.* 34 *D.*—Southern Chili.

20*a*. H. **Plumieri**, *Hook. et Grev. Ic. Pl. t.* 123.—W. Indies.

21*a*. H. **elasticum**, *Bory; Hook. Sp. Fil.* 1, *p.* 93.—Mauritius.

25*a*. H. **Catherinæ**, *Hook. Syn. Fil. p.* 67.—Jamaica. K.

25*b*. H. **interruptum**, *Kze.; Hook. Sp. Fil.* 1, *t.* 33 *B.*—W. Indies.

---

138. TRICHOMANES (p. 252).

6*a*. T. **Petersii**, *A. Gray; Hook. Ic. Pl. t.* 986.—Alabama.

7*a*. T. **proliferum**, *Bl.; Hook. Sp. Fil.* 1, *t.* 39 *B.*—Java.

11. T. **pyxidiferum**, *Linn. add. var.* T. olivaceum, *Kze.; Hook. Syn. Fil. p.* 81.—Venezuela.

14*a*. T. **Colensoi**, *Hook. Second Century of Ferns, t.* 79; *Hook. Syn. Fil. p.* 85.—New Zealand.

15*a*. T. **auriculatum**, *Bl.; Hook. Sp. Fil.* 1, *p.* 133.—Java. K.

18*a*. T. **Ankersii**, *Parker; Hook. Sp. Fil.* 1, *p.* 121; *Hook. et Grev. t.* 201.—Trinidad.

18*b*. T. **brachypus**, *Kze.; Hook. Sp. Fil.* 1, 121.—Trinidad.

18c. **T. humile**, *Forst.; Hook. Sp. Fil.* 1, *p.* 123.—New Zealand.

20a. **T. plumosum**, *Kze.; in Linn.* 9, *p.* 104.—Peru.

20b. **T. Sellowianum**, *Presl, Hymen. p.* 37; *Hook. Sp. Fil., noticed at p.* 145.—Brazil.

32a. **T. longisetum**, *Bory; Hook. Sp. Fil.* 1. *p.* 137.—Java.

33a. **T. maximum**, *Bl.; Hook. Sp. Fil.* 1, *p.* 137.—Java.

Obs.—I deem it here proper to state that nearly the whole of the above species of *Hymenophyllum* and *Trichomanes* have been introduced from their native countries by Mr. J. Backhouse, of York, who informs me that he has recently introduced five or six other beautiful species from Borneo and New Granada, not yet named.

Sub-Order IV.—OSMUNDACEÆ (p. 257).

Tribe I.—SCHIZÆÆ (p. 257).

141. LYGODIUM (p. 257).

2a. **L. venustum**, *Sw.; Hook. Syn. Fil. p.* 438.—West Indies Tropical America. K.

---

142. LYGODICTYON (p. 258).

3. **L. lanceolatum**, *J. Sm.* Lygodium lanceolatum, *Desv.; Hook. Syn. Fil. p.* 439.—Madagascar. K.

---

143. ANEMIA (p. 259).

1a. **A. Gardneri**, *Hook. Ic. t.* 190; *Hook. Syn. Fil. p.* 431.—S. Brazil. K.

1*b*. **A. rotundifolia**, *Schrad.; Hook. Syn. Fil. p.* 432.—S. Brazil. K.

---

146. SCHIZÆA (p. 261).

1*a*. **S. flabellum**, *Mart. Crypt. Bras. t.* 55.—Brazil.

Tribe II.—OSMUNDEÆ (p. 263).

148. OSMUNDA (p. 263).

2*a*. **O. lancea**, *Thunb.; Hook. Syn. Fil. p.* 427.—Japan. K.

---

149. TODEA (p. 264).

5. **T. Wilkesiana**, *Brack. t.* 43.—New Caledonia.

Order II.—MARATTIACEÆ (p. 265).

150. MARATTIA (p. 266).

8. **M. attenuata**, *Lab. Nov. Caledonia, t.* 13, 14. M. Cooperi, *Veitch Cat.* (1873–74).—New Caledonia.

---

153. DANÆA (p. 269).

*a*1. **D. simplicifolia**, *Rudge, Pl. t.* 36; *Hook. Syn. Fil. p.* 442. —Guiana and North Brazil. K.

Order III.—OPHIOGLOSSACEÆ (p. 270).

157. BOTRYCHIUM (p. 273).

5. **B. daucifolium**, *Wall.; Hook. et Grev. t.* 161; *Hook. Syn. Fil. p.* 448.—India. K.

Order IV.—LYCOPODIACEÆ (p. 274).

LYCOPODIUM (p. 275).

5*a*. **L. rupestre**, *Linn.*—N. America.

16. **L. lucidulum**, *Mich.*

17. **L. squarrosum**, *Forst. Prod. No.* 479.—Polynesia. K.

18. **L. dichotomum**, *Jacq. Hort. Vind.* 3, *p.* 26, *t.* 45. —Tropical America. K.

19. **L. carinatum**, *Desv. Enc. Bot. Suppl.* 3, *p.* 559. —Malay. K.

20. **L. gnidioides**, *Linn.; Hook. et Grev. Icon. Fil. t.* 50. —Cape and Mauritius. K.

21. **L. linefolium**, *Linn.*—Tropical America. K.

SELAGINELLA (p. 277).

S. albonitens, *Spr. Mon. Lycopod.* 2, *p.* 80. — West Indies. K.

S. Braunii, *Baker; Gard. Chron.* (1867), *p.* 1120.—China. K.

S. bulbillifera, *Baker; Gard. Chron.* (1867), *p.* 950. S. increscentifolia, *Stansfield's Cat.*—Venezuela. K.

S. conferta, *McNab, Monog. Selag. Hort. Edinb., p.* 8.—Borneo. K.

S. hæmatocles, *Spring, Mon. Lycopod.* 2, *p.* 156.—Trop. America. S. Karsteniana, *Veitch Cat.* K.

S. Kraussiana, *A. Braun; Ind. Sem. Berol.* (1859), *p.* 22.—South Africa. K.

S. pilifera, *A. Br.; Ind. Sem. Berol.* (1857), *p.* 20.—Texas, Mexico. K.

S. rubella, *Moore; Gard. Chron.* (1871), *p.* 902. S. divaricata (?), *Backhouse Cat., is the same.* K.

S. rubricaulis, *A. Braun, in Fil. Afric. p.* 211. K.

S. suberosa, *Spring, Mon.* 2, *p.* 252.—East Indies. K.

S. argentea, *Veitch Cat., syn.* S. serpens. No. 8.

S. flexuosa, *Williams Cat., var.* S. Martensii. No. 14.

S. formosa, *Veitch Cat., var.* S. Martensii. No. 14.

S. Poulterii, *Veitch Cat., var.* S. denticulata. No. 6.

S. triangularis, *Williams Cat., syn. of* S. Vogelii. No. 28.

S. Warsæwiczii, *Backhouse Cat., Syn. of* S. erythropus. No. 24.

ETYMOLOGY.

MANY cultivators and plant amateurs, not versed in scientific literature, complain of the (to them) difficulty to pronounce names given to plants by botanists. This complaint is, however, more imaginary than real. They have only to recollect that time and use has made them perfectly familiar with such names as *Pelargonium, Geranium, Hydrangea, Calceolaria, Rhododendron, Chrysanthemum, Mesembryanthemum, Elscholtzia, Fuchsia,* and many others now fluently spoken as if they were original words of their mother-tongue. Others say, these long-sounding names would be more readily reconciled to them if the meaning and bearing upon the plant were explained. To meet this desire, I have drawn up the following table, showing the derivation of the genera of Ferns characterized in the preceding pages.

As a general rule, it may be admitted that names of plants are derived from three principal sources. The first, and greater number, being generally a compound of two Greek words. The second, names of persons, with the addition of a Latin termination. The third source is various; such as the adoption of local aboriginal names, names from Heathen Mythology, and often words of doubtful origin, and unmeaning application to the plants, have been given as generic names by even the most learned botanists.

N.B.—For all names ending and beginning with *Pteris,* see Pteris.

Aconiopteris, *akōn*, a point, and *pteris;* the veins forming sharp angles close to the margin of the frond in the typical species.

Acrostichum, *akros*, highest, *stichos*, order; the fructification at the top of the fronds.

Adiantum, *adiantos*, dry; when plunged in water comes out dry.

Alsophila, *alsos*, grove, *phileo*, to love; grows in groves and shady places.

Amphidesmium, *amphi*, around, or on both sides, *desmos*, a band; relating to the position of the sori to one another.

Anapeltis, *ano*, without, *peltis*, shield; the sori naked, in opposition to *Pleopeltis.*

Anchistea, *agnati*, kindred, intermediate between *Woodwardia* and *Doodia.*

Anemia, *aneimon*, naked; naked spikes of fructification.

Anemidictyon, veins reticulated; not free as in Anemia.

Anetium, probably from the Greek *anaitios*, guiltless. This is perhaps meant to indicate that *A. citrifolium* had been put wrongfully to *Acrostichum.*

Angiopteris, *angio*, open, *pteris*; the open sporangia.

Antigramma, *anti*, opposite, *gramma*, a line; the linear sori being in pairs opposite to each other.

Antrophyum, *antrum*, a den or hollow; the receptacles of sporangia being in a grove or channel.

Arthropteris, *arthron*, joint, *pteris;* the frond and pinnæ being articulate.

Aspidium, *aspidos;* the indusium being like a buckler or shield.

Asplenium, *a*, privative, *splen*, the spleen; medical qualities.

Balantium, *balantion*, a purse or bag; the form of the indusium.

Blechnum, one of the Greek names for a fern.

Botrychium, *botrys*, bunch or raceme; the fructification like a bunch of grapes.

Brainea, in honour of C. Braine, Esq., who introduced the first plant.

Callipteris, *kallos*, beautiful, and *pteris*, beautiful fern.
Campyloneurum, *kamptos*, arched, *neuron*, a nerve; the veins forming arches.
Ceratopteris, *keras*, *keratos*, a horn, *pteris*; horned fern.
Ceterach, a name given by Persian physicians.
Cheilanthes, *cheilos*, lip or margin, *anthos*, flower; the fructification on the margin.
Cibotium, *kibotos*, a casket or coffer; the form of the indusium.
Cincinalis, meaning unknown.
Colysis, *kolysis*, separation, a species separated from other genera.
Coniogramma, *konis*, dust, *gramma*, a line; imperfect sporangia, but not applicable in perfect specimens.
Cryptogramma, *kryptos*, hidden, *gramma*, a line or writing; the lines of fructification being hidden.
Cyathea, *kyathos*, a cup; the form of indusium.
Cyclodium, *kyklos*, circle; the form of indusium.
Cyclopeltis, *kyklos*, circular, *peltis*, shield; the form of indusium.
Cystopteris, *kystos*, a bladder; the inflated indusium.

Danæa, in honour of Pierre Martin Dana, a Piedmont botanist.
Davallia, in honour of Edmond Davall, a Swiss botanist.
Dicksonia, in honour of James Dickson, a British cryptogamist.
Dicranoglossum, *dikranos*, forked, *glossa*, tongue; the fronds forked and pendulous.
Dictymia, *diktyon*, a net; the netted venation.
Dictyogramma, *diktyon*, *gramma*, a line; the sori reticulated.
Dictyopteris, *diktyon*, *pteris*; the veins reticulated.
Dictyoxiphium, *diktyon*, *xiphion*, sword; the veins reticulated, and the form of the frond.
Didymochlæna, *didymos*, two or double, *chlæna*, a cloak; the indusium being double.
Diplazium, *diplazo*, to be double; two indusia on the same receptacle.
Doodia, in honour of Samuel Doody, a London apothecary, and British cryptogamist.

Doryopteris, *dory*, spear or halbert, *pteris;* form of the fronds.

Drymoglossum, *drymos*, wood, *glossa*, tongue; the fronds like tongues on trees.

Drynaria, *dryads;* the sterile fronds being like oak leaves, which tree was sacred to the Dryads.

Dryomenis, *dryos*, *dryads* (as above), *meniskos*, a crescent; shape of the sori.

Egenolfia, in honour of Christian Egenolph, a German author of a book on herbs.

Elaphoglossum, *elaphos*, a stag, *glossa*, tongue; the fronds being like the stag's tongue.

Eupodium, *eu*, good, *pous*, *podos*, a foot; the sorus having a foot-stalk.

Fadyenia, in honour of Dr. M'Fadyen, an eminent physician in Jamaica.

Feea, in honour of Mons. A. L. A. Fée, Professor of Botany at Strasburg, a celebrated writer on ferns.

Gleichenia, in honour of Baron P. F. von Gleichen, a German botanist.

Goniophlebium, *gonia*, angle, *phlebes*, veins; the veins meeting, forming angles.

Goniopteris, *gonia*, angle, *pteris;* the veins meeting, forming angles.

Grammitis, *gramma*, a line; the sori being linear.

Gymnogramma, *gymnos*, naked, *gramma*, a line; the sori on the veins in lines, and naked.

Gymnopteris, *gymnos*, naked, *pteris;* the fructification naked.

Haplopteris, *haploos*, simple, *pteris;* simple venation.

Helminthostachys, *helminthos*, worm, *stachys*, a spike; the fructification in compact spikes, worm-like.

Hemidictyum, *hemi*, half, *diktyon*, a net; the outer portion only of the veins being reticulated.

Hemionites, a name given by ancient botanists to a plant now called *Asplenium hemionites*, but retained as a generic name for a West Indian fern by Linnæus.

Humata, *humatus*, humid; in opposition to *Adiantum*.

Hymenodium, *hymen*, membrane; the character of the frond.

Hymenolepis, *hymen*, membrane, *lepis*, a scale; membraneous scales covering the fructification.

Hymenophyllum, *hymen*, membrane, *phyllon*, leaf; "Filmy-leaf Fern."

Hymenostachys, *hymen*, membrane, *stachys*, a spike; the fructification on spikes.

Hypoderris, *hypo*, under, *derma*, skin; the indusium under the sori, like a thin skin.

Hypolepis, *hypo*, under, *lepis*, a scale; the sporangia under the indusium.

Isoetes, *isos*, equal, *etes*, the year; remaining the same throughout the year.

Isoloma, *isos*, equal, *loma*, a border or margin; the indusium and margin equal.

Kaulfussia, in honour of D. G. F. Kaulfuss, of Halle, a celebrated writer on ferns.

Lastrea, in compliment to Chev. de Lastre, a French nobleman.

Lepicistis, *lepis*, scale, *kistis*, a cyst or cell; the sori being immersed in scales, which form a cyst.

Leptochilus, *leptos*, slender, *cheilos*, lip; narrow indusium.

Leptogramma, *leptos*, slender, *gramma*, a line; short linear sori.

Leucostegia, *leukos*, white, *stegos*, a cover; the indusium being pale-coloured, approaching white.

Lindsæa, in honour of Dr. Lindsay, of Jamaica, a writer on the germination of mosses and ferns.

Lithobrochia, *lithos*, a stone, *brocha*, spots; the areoles of the reticulated veins like pavement.

Llavea, named by Lagasca after a Mexican traveller.

Lomaria, *loma*, a fringe or border; relating to the indusium.

Lomariopsis, *lomaria*, *opsis*, like; like Lomaria.

Lonchitis, *logche*, a lance; form of pinnæ.

Lopholepis, *lophos*, a crest, *lepis*, scale; the sori being furnished with a tuft of slender scales.

Lophosoria, *lophos*, crest or tuft; sori furnished with hairs like a tuft.

Lorinseria, in honour of Gustave Lorinser, a Bohemian physician.

Loxsoma, *loxos*, oblique, *soma*, band; the oblique broad ring of the sporangium.

Lycopodium, *lykos*, wolf, *pous*; foot-resemblance.

Lygodium, *lygodes*, flexible; climbing plant.

Marattia, in honour of J. F. Maratti, of Tuscany, who wrote on ferns.

Marsilea, in honour of Count L. F. Marsigli, of Bologna.

Meniscium, *meniskos*, crescent; shape of the fructification.

Mesochlæna, *mesos*, middle, *chlæna*, a cloak; attachment of the indusium.

Microlepia, *mikros*, small, *lepis*, scale; the indusium small.

Microsorum, *mikros*, small; small sori.

Microstaphyla, *mikros*, small, *staphyle*, a bunch; the fertile frond being like a little bunch.

Mohria, in honour of D. D. Mohr, a German botanist.

Myriopteris, *myrios*, myriad, *pteris;* the frond being divided into a number of small parts.

Neottopteris, *neottia*, nest, *pteris;* bird's-nest fern.

Nephrodium, *nephros*, kidney; form of the indusium.

Nephrolepis, *nephros*, kidney, *lepis*, scale; the indusium being kidney-shaped and scale-like.

Neurocallis, *neuron*, a nerve, *kallos*, beautiful; the venation.

Neurodium, *neuron*, nerve; the venation.

Niphobolus, *niphos*, of snow, *bolos*, a large pill; the under side of the frond densely covered with white scales, snow-like and the round sori.

Niphopsis, *niphos*, of snow, *opsis*, like; like niphobolus.

Notholæna, *nothos*, spurious, *chlæna*, cloak; the imperfect indusium.

Ochropteris, *ochros*, pale; pale fern.

Odontosoria, *odontos*, of a tooth, *sori;* the sori-like teeth.

Oleandra, resemblance to *Nerium Oleander;* the *Oleander*.

Olfersia, in honour of Professor Olfers, a celebrated astronomer.

Onoclea, *onos*, a vessel, *kleio*, to enclose; the rolled-up fertile segments of the frond.

Onychium, *onychion*, a little nail; resemblance to the fertile segments of the fronds.

Ophioglossom, *ophios*, of a serpent, *glossa*, tongue; the spikes of fructification.

Osmunda, *Osmunder*, one of the names of Thor, a Celtic divinity.

Paragramma, *para*, near to, *gramma*, a line; the sori in short lines parallel with and close to the margin.

Pellæa, *pellos*, dark-coloured; the dusky colour of the fronds.

Phegopteris, *phegos*, beech; beech fern, by some called sun fern.

Phlebodium, *phlebes*, veins, *odous*, a tooth; the joining of the veins in the areoles being like teeth.

Phymatodes, *phymata*, tubercles; the impressed sori having the appearance of tubercles on the upper side of the frond.

Pilularia, *pilula*, a pill; the form of the spore-cases.

Platycerium, *platys*, broad, *keras*, horn; the fronds divided in broad segments like stags' horns. (The stags'-horn fern.)

Platyloma, *platys*, broad, *loma*, border; having broad sori close to the margin.

Pleocnemia, *pleos*, full, *knemia*, rays; full of rays; the venation.

Pleopeltis, *pleos* full, *peltis*, shield; the sori being furnished with numerous round scales.

Pleuridium, *pleura*, side (ribs); the primary veins being costæform, like ribs.

Pœcilopteris, *pœkilos*, spotted, *pteris;* the appearance of the venation.

Polybotrya, *polys*, many, *botrys*, bunch; the fructification being in bunch-like racemes.

Polypodium, *polys*, many, *pous*, foot; polypus; the rhizome, when destitute of the fronds, having the appearance of some kind of sea-polypus.

Polystichum, *polys*, many, *stichos*, order; not specially applicable to the genus *Polystichum*, as now defined.

Psilotum, *psilos*, naked; destitute of leaves.

Psomiocarpa, *psomion*, a small pellet, *karpa*, fruit; the sporangia being in small round patches, in spikes.

Pteris, *pteryx*, wing; the scientific name given to the plant known by the names of *Brake, Bracken,* and *Fern;* on account of the supposed likeness of the branching of its fronds to wings. This, being the commonest of all Ferns, has become the type of the whole race; hence *Pteris* means *Fern*, all generic names of ferns ending in pteris; such as *Ceratopteris* means horned *Fern*, *Dictyopteris* netted *Fern*; also, if at the beginning of a word, as *Pteridography*, a book or writing on Ferns; *Pteridologist*, a studier and writer on Ferns; *Pteridophilist*, a lover of Ferns.

Rhipidopteris, *rhipis*, fan; shape of the fronds like a fan.

Saccoloma, *sakkos*, a bag or sack, *loma*, margin; the union of the indusium with the margin, forming a hood or cyst.

Salpichlæna, *salpi*, pipe, *chlæna*, a cloak; the indusium being like a pipe.

Schellolepis, *schello*, skeleton, *lepis*, scale; the character of the scale surrounding the sori.

Schizoloma, *schizo*, I cut, *loma*, a border or margin; the sori in the form of a slit on the margin of the frond.

Scolopendrium, *skolopendra*, a centipede; the appearance of fructification on the under side of the frond.

Selaginella, the diminutive of *selago* (*Lycopodium Selago*).

Selliguea, in memory of M. Selligue, a French optician.

Sitibolium,* *sitos*, food, *bolos*, a large pill; the appearance of fructification.

Soromanes, *soros*, heap, *mania*, fancy; the sporangia in fanciful and irregular heaps.

Stenochlæna, *stenos*, narrow, *chlæna*, cloak; narrow indusium.

Stenosemia, *stenos*, narrow, *sema*, standard; the fertile frond with its narrow segments being like a flag.

Struthiopteris, *struthios*, ostrich; the fronds being like the feathers of an ostrich.

Thyrsopteris, *thyrsos*, bunch or raceme, *pteris*; the fructification in racemose bunches.

Todea, in honour of Henry Julius Tode, of Mecklenburg, an experienced mycologist.

Trichocarpa, *thrix*, *trichos*, a hair, *karpa*, fruit; the fructification borne on a hair-like stalk.

Trichomanes, *thrix*, hair, *manos*, soft; the delicate nature of the fronds.

Trichopteris, *thrix*, hair, *pteris*; the sori being furnished with hairs.

Vittaria, *vitta*, riband; shape of narrow fronds.

Woodsia, in honour of Joseph Woods, a celebrated British botanist (died 1864).

Woodwardia, in honour of Thomas Jenkinson Woodward, an English botanist.

Xiphopteris, *xiphos*, sword, *pteris*; form of the frond.

---

* For *Sitilobium*, at page 236 and elsewhere in this work, read *Sitibolium*.

CULTIVATION.

1. Preliminary Remarks.

HAVING given an account of how the Exotic Ferns of the preceding enumeration have been introduced to this country, I shall now proceed to state the best means for growing and preserving them in our collections; and as the successful cultivation of Ferns depends much upon a knowledge of the conditions under which they grow in their native country, I begin with a few observations on that point.

Ferns have already been spoken of as favourites with the plant-loving public; but it is not simply on their merits as pretty and interesting objects that they claim attention. Geology reveals to us that Coal—that source of our domestic comforts and national greatness—is formed chiefly of Ferns, which at some remote period grew upon the earth. However, this is not the place to discuss the views and speculative theories regarding the thermal and gaseous condition of the earth and atmosphere, under which Ferns then flourished, or to reason on the manner by which they were converted into coal. I may simply remark that fossil remains show, that contemporaneous with Ferns grew plants of remarkable character, quite distinct from those of the present era. Not so the Ferns, for their beautiful and well-preserved remains show that their fronds were of various sizes and forms, in every way analogous to the present race; having free and anastomosing venation, round and linear fructification, and, in some instances, almost identical with species now living. At present, Ferns rank amongst the widest spread of all the orders of the vegetable kingdom, being found in more or less number in all climates, between the most northern and southern limits of vegetable life, and at elevations ranging from the sea-level to 14-15,000 feet within the tropics, their number in any localities being generally in proportion to the degree of atmospheric moisture in conjunction with elevation, the latter applying specially to the interior of continents. Comparatively few species are found in open, grassy, thinly-wooded countries, whether it be the plain or mountain-slope; such districts are often in full possession of the most gregarious and abundant of all Ferns, the common Brake (*Pteris aquilina*), which, under slightly different forms, and in some countries accompanied by different species of *Gleichenia*, occupy vast tracts of the earth's surface. In hot and moist plains, in valleys of great extent, the number of different species are few; even in the valley of the Amazon, teeming as it does with vegetable life, the number of Ferns found by Dr. Spruce after he left the coast Flora, at Para, in his journey of 2,000 miles, were very few. They became more numerous on attaining an elevation of 1,500 feet, and in one locality, at a higher elevation, he found 250 species in a diameter of fifty miles. Another extensive tract with but few Ferns is the dry zone of Northern Africa, and few have as yet been recorded from the interior of Australia; while, on the contrary, on elevated coast-ridges and islands, they form a large proportion of the entire Flora. They are the most numerous, both in regard to genera and species, in the tropical regions, where, too, a greater number of individual plants are to be found than in temperate regions.*

On reviewing the above, it may readily be supposed that the varied influences under which Exotic Ferns naturally grow, necessitate various kinds of treatment for cultivating and preserving them in this country; but such is not the case. As might be expected, our native representatives (although only forty-two in number) have received special attention. We have Fern Tourists in plenty; almost every spot of our country, where are conditions congenial to their development, has become more or less familiar to those who take an interest in their study—in the damp shady glen, the wildest wood, banks of streams, mountain-sides, whether it be in their sheltered easy slopes, or in the almost inaccessible craggy steeps. Indeed, so diligently have the Fern explorings been carried on, that numerous curious and beautiful varieties, highly prized in our gardens, have been found. But, however laudable and agreeable Fern-growing may be, yet it is to be regretted that it leads to the extinction of some of our rarest native species. Even the more common are becoming scarce in localities within easy reach; great quantities being yearly

* Want of space prevents me entering more fully into the history of geographical distribution.

consigned to the London markets. Since Ferns have become so popular, those who patronize horticulture, from the members of the upper classes who construct large hothouses, to the humbler patron, or the hard-working mechanic who prides himself on his possession of a Wardian case, are desirous to obtain such information as will enable them to cultivate them successfully, either in the Conservatory, Wardian case, or Rockwork in the open air. To supply this, several books on the subject have been published. Indeed, if an example may be taken, besides the extraordinary degree of familiarity so rapidly attained in the knowledge of Ferns by all classes, the constant increase of collections gives sufficient evidence. Nevertheless, in publications which, through their cheapness, are within the reach of every Fern lover, frequent complaints are met with, tending to show that the result is not always satisfactory to the cultivator, especially among amateurs and persons of small means. These complaints are mainly attributable to the author not familiarizing his readers sufficiently with the habits of Ferns, the conditions under which they luxuriate in their native localities, and the necessity of imitating those conditions as far as practicable. It is, however, gratifying to observe that, in the progress of horticulture, the knowledge of the natural conditions of plants is more and more sought after, and appreciated with very good results, forming a lively contrast with the old times, when the *Trichomanes radicans* could only with difficulty be kept alive even by the most eminent horticulturists; a fact observed by the originator and promoter of the Wardian case system, under which this shade and moisture-loving Fern stands pre-eminent, even surpassing in size and luxuriance those in their native retreats. A familiar example of opposite nature to the preceding is our native species *Asplenium septentrionale,* confined to northern localities, generally growing on rocks and insinuating its delicate roots in the deep crevices and among the *débris* of irregular shelvings and prominences, not receiving any apparent injury from the rigour of winter or the scorching heat of summer. Although such is its hardy nature, it nevertheless does not flourish when taken from its native rocks and brought under artificial cultivation. What has been stated regarding the shade and moisture-loving *Trichomanes,* and the rock-loving *Asplenia,* may be viewed as the extremes in the nature of those Ferns that do not readily conform to ordinary cultivation; but, on the other hand, the greater number are not particular in their choice of place of growth, either in a wild state or under cultivation. Examples may be cited of species conforming to the most untoward and varying influences; for instance, the common hart's-tongue Fern (*Scolopendrium vulgare*) maintains itself in situations of the most opposite kind, having great predilection for the works of man, whether elevated in the air or sunk below the surface of the earth; such as stone, brick, or turf walls, embankments, hedge-banks and road-sides, pits, quarries, or deep open wells. In either place it multiplies freely, conforming itself to the various atmospheric changes to which those situations are liable. Such being the case, it is not surprising to find it assume different sizes and forms; in dry places it is only a few inches in height, while in open wells, such as may be seen in some of the nurseries and market-gardens near London, it produces fronds 2-3 feet in length. The *Asplenium Trichomanes, A. Ruta-muraria, Ceterach officinarum,* and *Polypodium vulgare,* are also wall and tombstone-lovers, and may be called our domestic Ferns.

Another remarkable example of a Fern making itself at home under extremes of temperature and moisture, is *Pteris longifolia,* a species having a wide range throughout the tropical and sub-tropical regions of both hemispheres. On the island of Ischia (Bay of Naples) it is found luxuriating within the influence of the hot vapours rising out of the cavities left by extinct volcanoes, growing in soft muddy soil at a temperature ranging from 140° to 160°. In our hothouses its spores vegetate abundantly upon all moist surfaces, and in the crevices of brick walls. Plants of it are nearly always to be found over the openings of hot-water tanks, and it has been seen in crevices of the walls outside hothouses, or even under iron gratings, where it could receive but little light, and where the temperature was often near the freezing-point. In the dry air of the Cactus-house plants of it have produced fronds from 2-3 feet in length.

In general the fronds of Ferns remain long in a perfect state; the exceptions to this rule are comparatively few, and these are chiefly supplied by the natives of climates alternating with seasons of heat and cold. But as many species are wanderers and conform to the effects of various climates, it is no wonder to see some of our native Ferns assuming the evergreen habit of their foreign allies, when grown in a temperate house; of which *Polystichum aculeatum*, *Lastrea dilatata*, and *L. Filix-mas* may be cited as examples, as also *Asplenium marinum*. In 1820 I found plants of it, having fronds from 2–4 inches in length, growing in a cave facing the German Ocean, on the east coast of Scotland; of these, two plants have been grown at Kew from that time; of late years, one in the Temperate and the other in the Tropical House. These became fine cæspitose plants, with fronds varying from 1–1½ foot in length; the greatest length being attained by the plant in the Tropical House, even assuming the character of a species native of the West Indies and Tropical America, and quite unlike the original plant; thus showing that although at home in the cold, sunless cave, it can well appreciate a better fed and warmer abode. Exceptions to this rule are some alpine species, Ferns in that respect being analogous to our ill success in growing many alpine flowering plants. This is no doubt partly owing to the difference of atmospheric density, and the varying influences of temperature and moisture common to the sea-level of this climate.

The species of cold climates truly deciduous, produce their fronds from an underground creeping sarmentum, of which *Pteris aquilina*, *Sitolobium punctilobium*, *Onoclea sensibilis*, *Anchistea virginica*, *Lorinseria areolata*, *Leucostegia immersa*, and *Phegopteris aurita* are the principal examples. On the other hand, instances of tropical species periodically losing their fronds without any apparent cause, such as by undue excess of heat and moisture, are *Phymatodes oxyloba*, *Pleuridium palmatum*, *P. venustum*, *Drynaria propinqua*, several species of *Davallia*, and a few others. Their rhizomes remain perfectly naked during the winter, and seem to be constituted for a season of dry rest.

These obversations are sufficient to show how certain species will grow under the most opposite conditions, and that a knowledge of them materially assists our efforts of cultivation, which I shall proceed to treat of under two heads,—special and natural.

II. Special or Pot Cultivation.

WITH the exception of those species sufficiently hardy to bear the cold of this climate, the whole family of Ferns can be artificially and most luxuriantly grown, under but two different scales of temperature, and with as little difference in other respects. There can be no better instance than that afforded by the immense collection now at Kew,* where, in the Tropical Fern-house, a great many species from various climates are placed under an average temperature of 60° to 70°. In this house† are species from nearly all the tropical and sub-tropical countries of the world, and, being in one compartment, they are all alike subject to the same amount of atmospheric moisture as of heat. However, in respect to moisture supplied to their roots, the amount is varied according to the nature and requirements of the plant. The same remarks apply to the Temperate House, of course with a reduction

* May 1864. † Length 130 feet, width 34 feet.

of moisture as well as temperature, the latter ranging from 40° to 50° in winter, that being the only period of the year when artificial heat is required. In this house* are growing, in the greatest luxuriance, species from Australia, New Zealand, China, Japan, North India, elevated portions of Ceylon, South Africa, North and South America, and other elevated regions within the tropics. The whole collection is grown in common deep or shallow pots, pans, and tubs; the latter, however, are only used for the large species of Tree Ferns and *Angiopteris*; and though such a system is the least natural in appearance, yet it is most convenient and found generally consistent with the prevailing fashion for in-door horticulture; this mode of cultivation, being mostly adopted, merits our first consideration. In the tropical and sub-tropical localities, where heat and moisture are abundant, by the manner in which they grow and the various positions they occupy, soil is of but little importance, except for affording their roots the means of obtaining permanency of position. This is evident by the remains of native soil adhering to the roots of imported plants, showing that Tree and other large Ferns in some places grow in stiff adhesive, red clay. To imitate this soil in pot culture is attended with no success, and it is remarkable to see how quickly the roots of newly-imported plants take to the fine loam and peat in which they are potted. In cultivation, however, the great beneficial influence of the natural atmosphere is not obtained, and the soil is therefore of great importance, necessitating caution in its

* Length 82 feet, width 13 feet.

selection. Taking a view of the large number of Ferns now in cultivation, their requirements as regards soil, and mode of growth, we easily recognize two classes—*terrestrial and epiphytal.* This division is important. For the first class, by far the most extensive, it is necessary to use the soil of a finer or coarser consistence, according to the delicate or stronger character of the plant. In it there is a more varied character of vernation, by which their terrestrial character is indicated and very easily recognized. Natural affinity of course is not taken into consideration. However, with tolerable accuracy in this respect, the genera *Phegopteris, Lastrea, Nephrodium, Adiantum, Asplenium,* and *Sitolobium* give examples of the chief variety of vernation indicative of the terrestrial nature; viz., erect, cæspitose, decumbent or sub-hypogeous, or, as in *Nephrodium unitum,* and *N. pteroides,* which, although with a true sub-scandent sarmentum, they, however, prefer the firm soil of the terrestrial group, a compost of two-thirds peat and one loam, with abundance of sand, according to the size of the plants for which it is to be used.

In potting Ferns, an over-depth of soil should be avoided, as well as a great depth of drainage. The one promotes stagnancy, the other gives an opportunity for the most vigorous roots to descend among its particles, and oftentimes they become too dry when the soil is apparently moist enough, and the tips of the foliage are in consequence injured, and the plant disfigured. This applies more strictly to species of small and delicate structure, such as *Asplenium firmum, A. dentatum, A. decussatum,* &c., and for such the shallow pot is best adapted, the width of which being considerably greater than the depth, gives a good extent of surface, and renders a deep drainage unnecessary. Such a kind of pot, of course, is most suitable for all species of decumbent or creeping habit, and those producing rapidly a great number of offsets.

The best time for repotting (or shifting, as it is termed) Ferns, more especially those of the terrestrial division, is at the end of February, or during March, as soon after that time they commence growth; but any time of the year, except winter, will suit them. The operation of potting is a matter of little difficulty, nor are the necessary utensils or materials very varied—a firm potting-bench, several different sizes broad-pointed firming-sticks, and a garden trowel. It is also necessary to have a few extra pots of different sizes, of both the deep upright and shallow flat kinds, also a quantity of material for drainage, which may consist of broken soft brick or hard knobs of old mortar; but for general purposes, potsherds, broken into different sizes, are most commonly used, and well known by the name of "crocks." The removal of the plant, with its ball of soil and roots, is readily effected by spreading the left hand over the surface of the soil, allowing the plant, when not too large, to come between the fingers; then, by inverting the plant and pot, and giving the rim of the pot a gentle but sharp tap with the right hand on the edge of the board; if the roots are in a healthy state, the ball will slip easily out of the pot. In all cases, a thorough drainage is indispensable, and to ensure this, the draining material should be of two sizes, the larger size at the bottom, with a concave large crock over the hole of the pot, the concave side downwards: there should be two, three, or four holes at the side, quite at the base, made in all pots larger than eight inches in diameter, the number of holes of course according to the size of the pot. This is the more essential when the pots are placed on smooth benches, as of stone or slate, the capillary attraction between the pot and bench retaining the superfluous water, and causing much stagnancy. After properly arranging the drainage, a small quantity of fibry material should be placed on it to keep the new soil from mingling with it. No plant should be repotted in a wet state, nor should it be allowed to become dry enough to cause it to flag, as this will not only give an immediate and severe check, but in giving water after it is potted, it will percolate the new soil (which should always be used in a moderately moist state) without entering the dry ball; if, from bad drainage or exhaustion of the soil, it has become impure, so much of it and of the old inactive roots should be removed. In such cases, as small a pot as possible should be used, which will admit the preserved roots freely, being carefully laid out by the fingers amid the new soil; for if in too large a pot, the soil becomes sour before any vigorous roots have entered and promoted drainage, as they are tardy to perform their functions vigorously after having been necessarily so much disturbed. However, when in a healthy and vigorous state, such space should be given it that will admit the fingers or both hands freely by its sides (according to the size of the plant) in inserting it in the new pot, where it should be so placed that its crown, if of the erect vernation, will be nearly level with the rim. The soil should be carefully introduced, and made firm round the sides by pressure of the firming-stick and a few thumps of the pot (if easily handled) on the potting-bench, if not, the stick must suffice; a space of about half an inch, more or less, according to the size of the pot, should be allowed at the surface to contain as much water as would moisten the whole ball.

The size of the plants will depend upon the amount of space and the number of species in the collection. For an amateur's collection in a small house, very fair specimens may be grown in pots from eight to twelve inches in diameter, after the shift into the largest-sized pot, and with good management the plant will not require any repotting for two or three years. By that time the running and cæspitose kinds, such as *Adiantum* and *Gleichenia*, will have become exhausted in the centre; the ball will, therefore, require division by passing a sharp knife through it, taking care not to injure the young growths next the sides of the pot; the most healthy portions to be selected for repotting, to become the new representative plant of the collection; and, if proper care is taken, the fronds will suffer but little injury. In operations of this kind, some gardeners entirely shake out the soil, cut away the whole of the fronds and roots, in order to make the plant, they say, come up strong. This may not do much injury to certain plants, such as bulbs, tubers, fuchsias, and such-like plants that rest in winter; but for Ferns it is a great mistake; it so weakens the plant that it takes a year or more to be worth looking at, and, indeed, some never recover. In *Adiantum* and other genera, the roots are naturally of a dark, or even quite black colour, and it is known to have been considered by some cultivators, not well versed in the nature of Ferns, as dead, and accordingly they are totally removed, even although with evidence of the plant being healthy and vigorous. The healthy state of the plant is readily explained, for on examining the numerous points of the black roots, each will be seen to be of a pale colour, which are the active feeders of the plant; and, in shifting, great care should be taken not to bruise or injure them.

The same directions apply to the gigantic Tree Ferns and species of *Angiopteris*. The soil for them, however, should be of a coarser nature when of a size to require the largest-sized pots or tubs; in no case is it desirable to give more than three inches extra space at one shift, even in the largest pot or tub. In all pots above one foot in diameter, the drainage should be elevated in the centre. When the shift is large, an inverted pot should be placed in the new pot, placing a layer of drainage round it, the height of the pot to be such that the crocks of the old ball (which must not be taken out) rest on the top of the new drainage or pot. By this the outer circle of bottom roots are not pressed together, or crushed by the weight of the ball, which is sure to be the case when the ball is set on a level surface, it also keeps the ball to its proper height. This mode of drainage has been the constant practice at Kew with all large shifts, not only for Ferns but *Proteaceæ* and all fine fibrous-rooted plants, the nature of which is to extend outwards and downwards through the new soil, ultimately forming a web of roots against the sides of the pot. In the ordinary practice of potting, it is a rule with many gardeners to remove all the old drainage crocks, the centre thus becomes a mass of soil, which in time becomes compact, inert, and useless, liable to become stagnant, and in time causing the plant to sicken. This is avoided by preparing and retaining the old drainage. This mode of drainage is, however, only necessary for plants required to be permanent in botanical collections. Its utility in prolonging life is verified by some yet existing plants of *Proteaceæ* at Kew, some of which are forty, and others above sixty years of age. What has here been stated may be considered as a general rule for the greater number of the Fern family, such as are usually found in ordinary collections of a hundred species, more or less. But in this extensive family there are many possessing some peculiar nature, and a few examples of these require to be specially noticed.

The group which has been termed *Epiphytal*, consists of the genera *Davallia*, *Goniophlebium*, *Drynaria*, *Phymatodes*, *Pleuridium*, *Platycerium*, *Lomariopsis*, *Polybotrya*, and others of like habit, characterized by a true and highly-developed rhizome or sarmentum, generally creeping, and adhering by their fine fibrous roots to the surface on which they grow. Their positions in their natural homes are generally on more or less perpendicular surfaces of moist and shaded places of rocks or ordinary soil, where decomposing vegetable matter abounds as a surfacing. They are often found on trees, and with many species of *Polybotrya* and *Lomariopsis* this seems to be their true position, and often by their vigorous growth, the trees are clothed with them in the manner of our common ivy. It will be understood by this that an open loose soil is essential for the whole group; good fibry peat is therefore all that can be desired, and on account of their creeping habit, as great an extent of its surface as possible is necessary to be gained in potting. To this end the shallow pan is indispensable; by an ordinary deep pot being inverted in the pan and covered with a layer of peat, of a thickness according to the size of the pan or pot used (at a proportion of two inches thickness for a pan of one foot diameter) carried up over the inverted pot in a conical form, the whole pressed firmly together. The rhizomes are fixed to its surface by pegs, and they will soon attach themselves by their own roots. The height of the cone will be according to the character of the plants; for *Pleopeltis, Anapeltis, Niphobolus,* &c., a cone of from 6–10 inches high, with a base from 1 to 1½ foot in width, will make handsome specimens. The genus *Oleandra* comes under *Epiphytal,* the rhizomes elongating rapidly, and in their natural places adhering to trees and moist rocks. *O. neriiformis* is, however, truly terrestrial, producing roots from its woody erect rhizome after ascending a few inches from the soil. The other species in cultivation, however, cannot be well suited by the conical mode, or being trained on straight sticks. An open netted wire cylinder, about three feet high, and about six inches diameter, answers the purpose exceedingly well. The cylinder is filled with peat, and by fixing the rhizomes round the bottom of the cylinder, the roots will soon adhere to the peat through the meshes, which being kept moist, a rapid growth is stimulated, and the cylinder soon becomes furnished with beautiful projecting fronds. If a greater height is desired, another cylinder may be added to the first. This mode is equally useful for *Stenochlæna, Polybotrya, Lomariopsis,* &c. In respect to species in which the rhizomes elongate slowly, the upper part of the cylinder is bare and unsightly for a time. This may be obviated by having a short cylinder and heighten it by adding short lengths when required. The genus *Elaphoglossum* is generally *epiphytal,* and as some of its larger growing species, such as *E. callæfolium* and *E. latifolium,* being decumbent cæspitose in vernation, it is not necessary to have the soil raised much above the rim of the pot, which should be of the flat kind, and three inches depth of soil is quite sufficient. Many smaller species of this genus are most lovely objects, and some of them are not yet in cultivation. Of the whole Fern Family the *Platycerium* may be considered the most grand, beautiful, and extraordinary; and it is thoroughly typical of this *epiphytal* group. Its natural position of growth is sometimes on moist rocks, but usually on the trunks and larger branches of trees. The spores becoming lodged there, germinate, and, sending out spongy fibrils, a little plant, like a circular disk, adhering to the tree, analogous to a foliaceous lichen, is formed, each succeeding disk (frond) becoming larger and overlapping the preceding one. In time the older ones loose their vitality, and by this mode of growth envelop, or nearly so, that portion of the tree whereon they grow in a dense, thick, spongy mass, among which the roots insinuate themselves and receive nourishment. As equivalent to this, in cultivation blocks of wood are mostly used; but they are objectionable from their constant liability to breed fungi and harbour insects. On that account pots are preferable, not of the ordinary shape, but with less difference in the diameter at the base than at the top, and with a wide opening at the side, extending about a third of its diameter from the rim down to within a sixth of its depth from the base (this will leave the pot in much the same form as a scoop). Thorough drainage should be given, and a material of very fibry peat will suit. The crown of the plant should be placed about the centre of the side opening from where the fronds will be produced. In time, the sterile fronds will spread in all directions, but mostly upwards, and quickly obscure the pot, and the fertile ones will hang loosely downwards. Their position in the house should be elevated, and, if possible, against a wall or partition. *Platycerium alcicorne* increasing rapidly by offsets, requires a considerable extent of surface; consequently rough sandy peat, arranged in a conical manner on a shallow pot, to which the plants will soon attach themselves, is most suitable, and which, if fancy leads, may be suspended from the roof of the house by a strong wire.

For the large-growing species of *Drynaria*, *Goniophlebium*, and *Phlebodium*, it is not necessary to raise the soil much above the level of the pot; their fleshy rhizomes soon reach the margin, to which they cling, and if standing on a moist surface, or near water, they creep down the sides of the pot.

In *Davallia pyxidata*, *D. ornata*, and others, the rhizomes are what may be termed aërial, rising considerably above the soil, often extending to a distance beyond the edge of the pot, and, as they in these cases do not produce roots, the rhizomes in time (three or four years) become weak, producing small fronds; it therefore becomes necessary to remove the older parts and re-pot the younger portions in fresh soil.

Very interesting and natural examples may also be had by placing common cylindrical red chimney-pots or drain-pipes in a pan wider by two inches than the base of the cylinder,—this space to be filled with soil and planted; the rhizomes will soon become attached to the surface of the cylinder and in time will cover the whole, forming a handsome pillar of fronds. In order to maintain a proper degree of moisture, a pan of water may be fixed inside the cylinder, and if a piece of woollen cloth is placed in the water, in contact with the cylinder, a constant moisture will be kept up congenial to the plants. The top of the cylinder is also useful for placing on it some of the pendulous species; for instance, *Phymatodes geminata, Goniophlebium dissimile, G. neriifolium, Campyloneuron angustifolium, Elaphoglossum Herminieri.*

In *Schellolepis subauriculata* and *S. verrucosa* the fronds are long and pendulous; this necessitates the plants being placed in an elevated position, such as suspended from the roof of the house in shallow wire baskets, the inside of which should have a lining of sphagnum moss, which assists to retain moisture, and also gives a clean and neat appearance. Plants of these two species thus treated have at Kew produced fronds 12 feet in length, hanging down in a very graceful manner. There are several other Epiphytes of special interest, such as the species of *Vittaria* and *Haplopteris,* which hang down from trees like bunches of grass, as also the remarkable *Ophioglossum pendulum*, which may be likened to ribands or bands hanging loose and waving with the wind, often many feet in length. These plants succeed in a small quantity of soil, firmly fixed in pots, with pieces of soft stone or potsherds, and the pot hung against a shady wall or pillar. For this purpose the pot should have a flat back, with the front rim lower than the back, so as to allow the fronds to hang quite free of the pot.

Neottopteris Australasica, and a few *Aspleniums* nearly allied, such as *A. sinuatum* and *A. crenulatum*, of precisely the same mode of growth, are of erect fasciculate vernation. Their roots being of peculiar mossy and delicate nature, they are not adapted for deep insinuation of stiff soil, but are rather what may be termed *aërial*. Two-thirds of their mass is produced above the surface of the soil. Substantial but open material is therefore required, of very rough, fibry peat, and porous, broken bricks, or soft sandstone, in equal parts; very little pot-room is necessary; a shallow pot of 18 inches diameter, with such material, will support a plant of two dozen fronds, and none less than 3 feet 6 inches long and 8 inches broad, with a stem a foot high, and as much through, principally composed of its mossy roots forming a spongy mass. As an instance of the long life under regular treatment may be cited the original plant of *Neottopteris Australasica*, which was imported in 1825, and is now (1864) a magnificent plant, in perfect health, having received but few shifts the whole of the forty years.

This is, however, far surpassed in size by the magnificent *Neottopteris musæfolia*, which is described by a Penang correspondent in the following words:—"I saw two fine specimens of the *Birds'-nest Fern;* each had between forty and fifty perfect green leaves; the average length of the leaves was six feet, and from one foot to fourteen inches across in the broadest part. They were growing on each side of a doorway; when I was walking up to them I thought they were *American Aloes*."

The remarks that have been previously made respecting the nature of the rock-loving *Aspleniæ* are equally applicable to others that inhabit rocky places, such as certain species of the genera *Notholæna, Myriopteris, Cheilanthes,* &c.; some are rocky-coast plants, others are mountain, being found in elevated situations within or near the tropics, where they are subject to the heaviest showers and most powerful sun, their surfaces being generally furnished with beautiful scales or woolly covering, which resist in a great measure the action of the sun. They all seem very impatient of moisture under cultivation, especially in winter. A material composed of finely-broken and mixed sandstone, bricks, old mortar, and a small quantity of sandy loam, suits them, placed in a position of the coolest shade, with abundance of moisture in summer, and in winter very little water, just sufficient to keep the soil slightly moist; a comparatively dry atmosphere and as much light as may be obtainable at that dull season, with a temperature not below 36°. As a general rule, the above may be considered applicable to all small-growing rock species of temperate regions, including the natives of this country. The species of the genus *Gymnogramma*, especially those with farinose fronds, commonly called Gold and Silver Ferns, are also very susceptible of moisture; they should never be syringed, or water allowed to fall on their fronds, as the farina, being loose, is disturbed by the water, and running down, gives the appearance as if the plants were smeared with dust. They, however, differ from the preceding, requiring more light, and the temperature of the Tropical House. The species of *Gymnogramma* vary very much in habit, as regards size and circumscription of the fronds, *G. trifoliata* having fronds from three to four feet high, while in *G. chærophylla* and *G. leptophylla* they are fragile, and average from two to six inches in length. These two species are peculiar in being, with the exception of *Ceratopteris thalictroides*, the only known truly annual Ferns. *G. chærophylla* grows freely; its spores vegetate abundantly throughout the house, often as a weed. *G. leptophylla* is, however, not so free in its growth. When its fronds decay, the pot should be covered with a piece of glass, and put in a dry place until the proper season arrives in spring, when the application of moisture will cause the latent spores to vegetate. *G. flexuosa* differs from the whole of the genus in having prelonging fronds, which climb in a rambling manner over bushes, like those of *Lygodium*, noticed in another page: it will probably hereafter form the type of a distinct genus. The beautiful genus *Lindsæa*, of which no less than sixty species are described in Hooker's "Species Filicum," are, with few exceptions, natives of the tropics of both hemispheres; in my Catalogue of 1857, only two species are recorded as being cultivated; but, within these few years, the number has increased to fourteen, the greater part having been introduced from Guiana by Mr. Backhouse, of York, and on account of their slow increase, plants of them are yet very rare: they are natives of generally the lower regions of tropical vegetation, growing in open places amongst herbage of small plants and grass, or on the skirts of woods, and sometimes under the shade of trees, the soil being very poor, sometimes almost nothing but sand or stony *débris*, in which their sarmentums are partially hypogeous, the soil and air never being much below the temperature of 80°, and although almost daily subject to the influence of tropical thunder-showers, yet, on account of the nature of the soil, the surface is never over saturated. Finding that they do not flourish under the medium temperature of a Tropical Fern House, it therefore is necessary that a special part of the house, on the principle of a Wardian Case, should be adapted, so that a moist air of 80° may be steadily maintained, and the plants occasionally sprinkled overhead, taking care that no superfluity remain in the soil, which should be no more than moist.

Like *Lindsœa*, the curious and interesting genus *Schizœa* does not readily conform to cultivation; plants of *Schizœa elegans* have often been freely imported from Trinidad, and although tried in various ways in high and moderate temperatures, it cannot be said they have yet become established. Under the Wardian Case, the native imported fronds remain for a considerable time fresh, and sometimes new fronds show themselves, but fail to come to maturity. In a letter lately received from Mr. Prestoe, in Trinidad, he informs me that the *Schizœa elegans* grows in solitary patches in loamy soil, covered with three or four inches of leaf soil, in company with *Adiantum* and other Ferns, as well as much *Cyperaceæ*, the whole forming a dense undergrowth in woods. From this it appears there is no great peculiarity from other Ferns as to its place of growth, and that our ill success arises in consequence of the nature of the plant, being difficult to re-establish after being taken from its native soil, of which there are many instances; the common *Pteris aquilina* being a familiar example.

The numerous and beautiful species of *Hymenophyllæ*, called Filmy Ferns, merit particular attention. They are natives of both tropical and temperate regions, and where they abound are generally found in shady moist woods, clothing the lower parts of the trunks of trees, especially Tree Ferns, or on dripping rocks, or surface-soil of the deepest ravines, rarely in exposed situations. In order to surround them with a moist atmosphere, adaptations must be resorted to, such as moveable cases upon the Wardian principle, hereafter to be described, and which are essentially necessary for the cultivation of this delicate tribe. Such, indeed, is their delicacy, that if once allowed to become dry on the surface of their fronds for any length of time, a rusty and shrivelled appearance will soon follow. Having, as already stated, succeeded in cultivating *Trichomanes radicans*, it becomes obvious that other species could also be made subservient to cultivation; of which there is ample evidence to be seen at Kew, where there are forty cases* of different sizes, arranged on a stone shelf on the north side of the large Fern House, in such a position that they

* May, 1864.

receive but little sun in winter, and are densely shaded in summer, all filled with patches of fine fronds, of a number of different species, varying from the delicate hair-like *Trichomanes tricoideum*, not more than two inches high, to the robust *T. anceps* and *T. radicans*.

They are grown in square shallow pans and boxes, well drained in the ordinary way, and having about two inches of peat soil mixed with nearly half its bulk of sand and small broken potsherds; but soft sandstone is best. For the creeping sorts the soil should be raised in the form of a mound, and for those that have long-extending sarmentums, if soft stone cannot be had, it is desirable to invert a pan or common deep pot, covering it with a layer of soil, as already explained, to which the plant will cling, and soon form a green hillock: junks of wood answer the purpose; but in a moist, close, and warm atmosphere, fungi and insects breed, and in a short time the wood decays, causing unnecessary disturbance of the whole mass of the plant.

The singular genus *Lygodium*, and its ally *Lygodictyon*, grow naturally in firm soils, generally amongst trees and bushes, their wiry, flexile, climbing fronds growing over and involving everything within their reach in the most intricate complexity. In most Ferns, the whole of the divisions of the fronds are formed in the nascent or bud state, and are unfolded as the fronds elongate; when the whole of the developed parts are unfolded, the frond ceases further extension. This is, however, not the case in *Lygodium*, *Salpichlœna*, *Pellea flexuosa*, *Gymnogramma flexuosa*, *Odontosoria aculeata*, and a few others, the fronds of which are of indefinite extension, their apices continuing to grow and produce lateral pinnæ, in every way analogous to the development of branches and leaves in woody plants. The climbing and rambling nature of these plants necessitates some kind of support when cultivated in pots, which may consist of open wire trellises of a cylindrical form, varying in height from three to six feet or more, which, if fixed to a pot ten to twelve inches in diameter, the plants will, with careful training, become handsome specimens; they are also well adapted for covering trellis against wall or pillars.

The fronds of the much admired genus *Gleichenia* are also indefinite in extension, and some of the smaller species, when growing amongst bushes, assume a climbing habit; but in *G. furcata*, *G. dichotoma*, *G. Cunninghami*, and *G. flabellata*, the fronds are rigid and erect, produced from a more or less stout or slender sarmentum, which is either superficial or creeps a little below the surface of the soil; therefore shallow pans or boxes are best suited for these plants; and with attention specimens of considerable size may be attained, as, for instance, at Kew a plant of *G. flabellata* measured twelve to thirteen feet in circumference and four and a half feet high, consisting of a thicket of fine fan-like fronds.

It may be expected that in such an extensive family some species would be found bearing the appellation of aquatics, but such is not the case; for although many species grow in wet places, such as *Osmunda regalis* and *Acrostichum aureum*, both of which love water, but also flourish even in dry places, the only Fern really entitled to be called a water Fern being *Ceratopteris thalictroides*, and which is also singular in another respect—in being one of the few Ferns that are only annual. It is widely dispersed throughout the tropics, growing in wet places, often flooded; its sterile, viviparous fronds floating on or below the surface of the water, as may be yearly seen in the Victoria Lily tanks at Kew. Being annual, care must be taken to preserve spores, which in the spring should be sown in a shallow pan of loamy soil made wet like mud, kept moist; and when the plants are of sufficient size, the pan may be either filled with water, or be placed about an inch deep in a tank. *Acrostichum aureum* is mentioned as growing in wet places; for instance, in Jamaica and other of the West-India Islands it is described as taking the place of the European *Typha latifolia*, attaining the height of seven or eight feet. By imitating its natural condition, placing it in a pan of water, or in a tank, specimens have been grown at Kew to the height of six feet; but although it enjoys and luxuriates in water, it also grows in dry places, its height then not exceeding one to two feet, and often with simple fronds.

The delicately beautiful *Selaginellas,* on account of the large share of notice they receive and meet from the admirers of the Cryptogamic family, deserve here an especial notice. Originally the species were included under *Lycopodium;* but have been separated on account of differing in the character of their spore-cases. The species are numerous, and their habit and free growth mark them as a very distinct group from true *Lycopodium.* With a knowledge of a few species, two very distinct modes of growth are easily understood; the upright or climbing ones, such as *S. Africana* and *S. filicina,* are examples, and the decumbent or creeping species, such as *S. serpens*, *S. uncinata*, *S. Galiottii*, &c., are examples. The latter delight in a light soil, composed of fibry peat and well-decomposed leaf-soil, free from pieces of wood (as it generates fungi), with a little sand intermixed, and require a depth of only about two inches in ordinary round pans, with ample drainage. If, however, good specimens are desirable, with an arrangement on a bench or low shelf, square pans about twice their height at the back as in front, are in every way preferable, and as at Kew, arranged alternately with the cases of *Hymenophyllæ*, with which they harmonize. The former-mentioned species, on account of their robust habit and strong rooting character, require the shallow pot, and a similar soil as for the last, but with a small proportion of light loam. The magnificent *Selaginella lævigata*, if supported by wires against a wall, becomes, in a remarkably short time, the loveliest object to be found in a Fernery. The species of the *Rosulate* section differ from the rest of the genus in having erect, fasciculate vernation, the frondules rising from a central developing axis, and, as in *S. lepidophylla*, spreading out nearly horizontal, and overlapping one another, forming a beautiful green rosette, about six inches in diameter; when dry, they turn upwards and inwards in an involute manner, the whole mass of the plant thus forming a firm ball, which, on becoming moist, again expands; and even perfectly dead plants, when expanded, seem as if alive. In that respect being analogous to that of the insignificant cruciferous plant called the *Rose of Jericho* (*Anastatica hierochuntica*): hence the balls of this *Selaginella* are frequently met with in curiosity-shops under the above name. This hygrometric property seems to indicate that this and its allied species are subjected to occasional drought in their native localities. But although they may frequently undergo this process in nature with impunity, it is, however, not desirous to be too often repeated on plants artificially cultivated. Shallow pots, four to five inches in diameter, best suit this section; and being firm holders, it is necessary to insert pieces of soft sandstone or potsherds amongst the shallow soil, so as to fix the plant firmly in the pot. The extremely delicate nature of the tissue of these little plants necessitates that a position the shadiest and moistest in a Fernery should be assigned to them. *S. Willdenovii*, *S. filicina*, and others are termed deciduous from their fronds decaying altogether in the early spring; new fronds, however, soon begin to grow.

The species of true *Lycopodium* are also numerous, some being terrestrial and others epiphytal, hanging from the trees like various-sized cord, one to four feet in length. Of the terrestrial, five are natives of this country; they, as well as several beautiful species that have from time to time been introduced from North America, refuse to become domesticated under ordinary treatment; to which must be added the wide-spread and beautiful tropical species *L. cernuum*. Equal ill success attends the epiphytal species; for although frequently imported, they are yet but poorly represented in our collections; they adhere to trees by producing many fibrous roots, which ultimately become a large spongy mass.

The extra height of roof required for the growing of good examples of Tree Ferns excludes them from many amateur collections. This has also been the case at Kew for the last twenty years, especially as regards Tropical Tree Ferns, many fine plants succumbing to the make-shifts that of necessity had to be resorted to after they had attained a certain height; but by beginning with young plants, they may be grown for a number of years in houses of the usual average height of ten to twelve feet, as also the large fronded tree-like *Lastrea villosa*, *Litobrochia podophylla*, *Asplenium striatum*, *Hemidictyon marginatum*, and many others of like habit. The latter, at Kew, in a 20-inch pot, produced beautiful fronds, seven feet in height, and which might, with encouragement, soon be made to produce them equal to those of native growth—fourteen feet. But in order to get rid of the inconvenient and unsightly look of large pots and tubs, it is best to adopt for these plants the system of natural cultivation explained further on.

In the "Species Filicum" about one hundred and twenty species of Tree Ferns are described; but, according to Mr. Moore's "Index Filicum," the number amounts to nearly two hundred. They are widely distributed, chiefly within the tropics. They love shade and solitude, and are generally found at elevations of from three thousand to five thousand feet in the humid regions. In the southern hemisphere they, however, extend much beyond the tropics, their southern limits being New Zealand, Norfolk Island, New South Wales, and Tasmania, where they grow at a lower elevation than within the tropics. On Mount Wellington, in the latter island, *Dicksonia Antarctica* is found in the greatest abundance, at an elevation of from one thousand five hundred to two thousand feet, attaining the height of from eighteen to twenty feet, growing only in damp places, generally gullies, where the sun rarely penetrates, and where they are sometimes covered with snow, and in summer the atmosphere loaded with vapour. This suggests that, with proper selection of situations, they might live in the open air in the south and west of England, as also the mild climate of Argyleshire, where shaded ravines and gullies may be found similar to those of Mount Wellington.

With few exceptions, Tree Ferns readily conform to cultivation. *Alsophila capensis, A. excelsa, A. Australis, Cyathea dealbata, C. medullaris, Dicksonia antarctica,* and *D. squarrosa,* grow freely in the Temperate House; the lofty *Alsophila glauca,* the beautiful tessellated stem of *Cyathea arborea,* with its crown of fine fronds, and the broad shining fronds of *Hemitelia horrida,* assume a grand appearance in the Tropical House. In general, the stems of Tree Ferns are of sufficient size to warrant the name of trees; but in many the thickness is more apparent than real, the diameter of the woody centre being often only a few inches, but in many cases covered with successive productions of out-growing aërial roots, which become hard and wiry, and by their interlacing, form a compact mass; the points of these roots are, however, the active feeders, and if a layer of fresh soil is occasionally placed round the base of the stem, their growth will be promoted, and vigour given to the plant. The most fastidious Tree Fern to cultivate is *Dicksonia arborescens,* a native of St. Helena. It was first introduced to this country in 1786, and many times since, but refuses to become established, either in a tropical or temperate house. This species, as well as the whole of the species of true *Dicksoniæ*, are characterized by the base of the stipes being clothed with a more or less coating of beautiful, articulated silky hairs; in some species of *Cibotium*, the quantity is so great, that in the Sandwich Islands it is collected, and ship-loads of it sent to California and Australia for stuffing cushions, beds, &c. Another remarkable species of this alliance is *Cibotium Barometz*, a native of China and other parts of Eastern Asia; its fronds attain the height of twelve to fourteen feet (even in this country); they rise from a thick decumbent caudex, which is densely covered with silky hairs, as above described; lying on the ground, it has the appearance of a woolly-clad animal. The stories told about it to early travellers led them to describe it as an animal with flesh and blood, but fixed to one position, from which it never moves; hence the story of the now fabulous *Barometz*, or *Vegetable Lamb*. This plant is of easy cultivation; if placed on soil slightly raised, a few years' growth will produce very good specimens of the "Lamb."

The rare *Schizocæna sinuata* is an exception to the general rule that characterizes Tree Ferns, the stem being slender, not exceeding an inch in thickness, attaining the height of three or four feet, and bearing a fascicle of simple fronds. This remarkable Fern is a native of Ceylon; it grows in shady places in woods where a degree of coolness prevails. A few years ago plants of it were received at Kew, and it was found necessary to place them in a large Wardian Case, in which they flourished.

Although the plants belonging to the order *Marattiaceæ* are ranked with Ferns, they nevertheless differ much in habit, chiefly as regards the nature of the roots, which are thickened and fleshy, and generally penetrating deeply into firm soil. In their general character they present much resemblance to the roots of *Cycadeæ*, *Cyclantheæ*, and other allied endogens; also in venation the remarkable genus *Stangeria* connects them through *Danæa* with *Cycadeæ*; and their compound fronds are represented in the same family through the new genus *Bowenia*. The species of *Marattia* and *Angiopteris* grow freely in a loamy soil, requiring plenty of water. Some species of *Angiopteris* require much space, for although the caudex does not rise much above the ground, yet a plant at Kew produced fronds that reached the height of twelve feet, spreading outwards, forming a diameter of thirty-four feet. The species of *Danæa* and *Kaulfussii* are, however, not such free growers; they require a moist and high temperature.

I have now gone through a few of the principal genera requiring special kinds of treatment; the next consideration is good management as regards temperature, watering, airing, and keeping free from insects. Much depends on the size and nature of the house; the larger it is, the less fluctuations of temperature take place, especially in a house of the ordinary construction of glass roof and sides, whether lean-to or span. The maintenance of a proper condition of the air in plant-houses depends much on the nature of the interior fittings. White or polished, reflecting and radiating surfaces, should be avoided as much as possible, such as iron pillars, rafters, spandrels, polished slate or smooth stone, iron or stone floors, &c. These kinds of material are often introduced in superabundance, for the purpose of giving consequence and dignity to the house, thus laying the foundation for conditions unfavourable to plants. The shelves or benches may consist of dark-coloured porous stone or rough slate, with an edging of smooth slate two inches deep, so as to form a shallow trough, which should be filled with dark-coloured sand or fine-sifted coal-ashes, the whole pressed down, so as to form a firm smooth surface, on which the pots are to be arranged. By this means a degree of moisture will be retained more congenial to the plants than when standing on stone or slate. An edging of about three inches of *Selaginella hortensis* or *S. denticulata* gives a neat appearance, and if allowed to run between the pots, the whole becomes more congenial than a naked surface. White sand or broken quartz is frequently used, but it soon becomes dirty, and has a harsh and dry appearance.

In the arrangement of the plants some degree of order must be observed, so that every specimen plant of a species should be seen to advantage, and if the collection is extensive, and the house has different aspects as regards light and shade, then it is desirable to arrange the plants according to their habits and requirements; thus, at Kew, the division *Eremobrya* occupy the whole of the south side of the house; this division being less sensitive to the effects of the sun or deficiency of water than the more extensive division *Desmobrya,* which, on account of their thin texture, very quickly suffer from dryness or too much exposure to the sun; therefore the north side of the house is best for them.

The natural direction of the fronds is the most pleasing, whether upright, pendulous, or spreading.

In the latter case, sometimes their own weight makes them fall lower than is convenient; it therefore becomes necessary to support them by inconspicuous stakes. Care must, however, be taken not to raise the fronds above their natural position, and to avoid making them have a stiff formal appearance.

In "Theories of Horticulture," it is said that abundance of light and a free circulation of air are indispensable for the cultivation of plants under glass. In my long experience, I have found, as a rule an abundance of light and a too free circulation of air prejudicial to the good cultivation of plants in general. With regard to Ferns, they love a quiet, dull atmosphere; the light afforded by a roof glazed with the ordinary transparent glass, admits fully double the amount of light necessary during eight months of the year. Sufficient means are, however, necessary for admitting air, in order to keep down the temperature in summer; but care must be observed in admitting it; if not, the increased ventilation in lowering the temperature will cause a rapid dispersion of the essential moist atmosphere at a time when most required. Shading must then be resorted to, this being the chief purpose which it serves, besides protecting the plant's foliage from the too powerful rays of the sun. It may consist of canvas blinds on rollers; two moderately thick ones are much better than one very thick, especially when *Trichomanes* and *Hymenophyllum* are grown; one should be permanent during the height of the summer, the other to roll over it on occasions of full sunshine. Although canvas is here spoken of as best for shading, and is extensively used, still it is troublesome, and in the end expensive, which makes it desirable to resort to some other method; such as by using several kinds of dulled or rough glass.

In former years, the fruit and plant-houses at Kew were glazed with a very dark-green glass called Stourbridge-green, and which was patronized by the late Mr. Aiton. Fine crops of fruit were produced under it, also the tropical plants in the Botanic Garden flourished without the aid of canvas or shade of any kind. Not many years ago, solitary squares of this glass might be seen in the roofs of the old hot-houses, which strongly contrasted with the modern clear glass. My experience with this glass led me to recommend green glass for the Palm House, which was adopted; but the modern-made green tint does not appear to be so fixed a colour as that of the old Stourbridge-green.

The Palm House in the nursery of the late Messrs. Loddiges, at Hackney (now things of the past), affords another example of successful plant-growing without the aid of canvas or other moveable shading material. On the late Mr. George Loddiges being consulted respecting the glazing of the Kew Palm House, he was asked if they shaded theirs; his reply was, "Oh, no; our thick rafters and sash-frames, with sooty glass, just afford the amount of light necessary for the plants." Under this roof, in an atmosphere of stillness and gloom, Palms, Ferns, Orchids, and numerous other tropical plants, grew in the greatest luxuriance, which, with the proverbial solitude of the place, and when viewed from the elevated platform, gave the idea of a ravine in a tropical forest.

To judge by the above examples of Kew and Hackney, it appears that fruits and plants were successfully grown under roofs very different in appearance from the elegant clear glass roofs of the present day, which, nevertheless, necessitates some kind of invention for temporary shading in summer; but it is to be hoped this will be superseded by further improvements. Double glazing is now spoken of as answering the twofold purpose of saving shading and fuel; but as there appear to be different opinions on its merits, further experience is therefore desirable before it can be safely recommended.

With regard to watering, it is difficult to explain in words, or lay down a rule, what constitutes the extremes of dry and wet soil in a flower-pot, feeling is perhaps the best guide; if, on taking a pinch of soil between the finger and thumb, the particles will not adhere, then it is too dry; if it adheres, and falls like a dry wafer, then the plant is not in immediate want of water; but, on the contrary, should the soil stick to the fingers like a wet wafer, then the whole is in a bad condition, and should the plant be weakly, it will get worse if allowed to remain in the sodden soil; for the correctness of this test care must be taken not to be deceived by the surface-soil, for it may be either dry or wet, and the contrary lower down. Another test of the state of soil is to give the side of the pot a smart tap with a hard piece of wood; should the sound be hollow, with a little tone, then the ball is dry; if, on the contrary, there is only the dull sound of the tap, without any tone, then the ball is moist and tight in the pot. To those in the habit of watering the same plants for a lengthened time, this test is of service, but it must not altogether be depended upon. The quality of the water has great effect on the health of the plants.

It should always be soft, or, if hard, should be aerated in open tanks; rain-water is best, but if collected from the roofs of hothouses, care must be taken that it does not become impure by the decomposition of the paint and putty, which is of frequent occurrence, and in time it so impregnates the water as to cover the foliage of the plants, when often syringed, with a thin film, that gives them a dull look; care should also be taken that the temperature of the water should not be less than 50°, especially in winter. Syringing is made an operation of great importance in guide-books on cultivation, and is readily accepted by the unskilled amateur as necessary, and being amusing, is often carried out in the extreme, to the injury of the plants. An amateur remarked not long ago, "Now we have got such a nice lady's syringe, it is quite a delight to use it." Another remarked, "I every day regularly water, syringe, and sponge my pet *Adiantums*, but with all my attention they are getting of a brown colour, surely your plants (at Kew) must have great attention to keep them so green." The lady syringer is told that many of the plants before her have been in the same pots for several years, and that they get water when dry, and are never syringed or sponged, unless for removing insects when they appear.

From what has now been stated, an amateur's successful cultivation of Ferns simply depends on proper soil and potting, careful watering, placing in a quiet, moist atmosphere, keeping down temperature in summer by shading, and to avoid currents of dry air, and in winter to maintain the proper temperature, according to the nature of the plants, by means of hot-water pipes. For the latter purpose, in small houses, temporary stoves of gas and charcoal are used, often to the utter destruction of the plants.

To assist in maintaining a moist atmosphere besides the usual moist surfaces of the pots, it is desirable to have water-troughs on the top of the hot-water pipes, as also to place pans of water on or near the pipes; also in hot weather, the paths and other evaporating surfaces should be wetted every morning and evening. An occasional syringing may be given, which should be done when the out-door air is charged with vapour; wetting plants when the air is dry causes a sudden evaporation, and a reduction of temperature on the surface of the frond is the consequence. This is especially the case with broad, smooth, glossy species, such as the genera *Phymatodes*, *Neottopteris*, *Asplenium*, *Diplazium*, &c.; occasionally the whole of some fronds or part of a frond becomes black. In order to raise vapour, a practice prevails with many cultivators to throw water over the hot-water pipes (or flues); there is no objection to this, providing the air out of doors is warm, but in the winter season it is often done in the evening, in order to counteract the dryness of the air, caused by the extra heat of the pipes required at that season; the house then becomes filled with hot vapour, which coming in contact with the glass of the roof condenses and falls in a shower of cold drops on the plant; at the same time the temperature of the house rapidly falls, thus causing black fronds, spots in orchids, and such-like complaints.

Ferns, like other plants, are liable to be more or less infested with insects; the soft membraneous kinds being the most subject, while, on the contrary, the smooth-fronded kinds, such as the whole of the division of *Eremobrya*, *Elaphoglossum*, &c., are, it may be said, quite exempt. The mealy bug, three kinds of scale, and thrips, are the chief pests, and should be carefully watched. The oblong brown scale is generally the most common, and very quickly overruns the under side of the frond, forming lines along the midribs. Many kinds of fluid mixtures, powders, and other nostrums, have from time to time been advertised for their destruction, but it too often happens what kills insects also injures the plant, and makes it look unsightly. Fumigation with tobacco is the grand preventive against the breeding of plant-insects; very slight fumigation destroys the winged male insect of all the Coccos family, and if frequently repeated, the whole in time will become extinct. When the mealy bug gets a head, it is readily got rid of by syringing; not so the scale, it requires to be loosened with a soft brush, or blunt-pointed peg, which, if the weather permit, should be done out of doors, and the plant afterwards syringed, so as to clear away all loose scale and eggs. Scale insects are adherers and suckers only; not so the less conspicuous insect the thrip, which is a small, slender, black, shining insect, with large prominent eyes, and quick in its movements, herding in groups, generally on the under side of the frond, voraciously feeding on the cuticle, which they soon destroy, permanent injury being often done without any indication of their presence till too late; they must, therefore, be carefully watched, and, when seen, at once destroyed by pinching with the fingers.

and syringing the plant. They are generally found on species of *Adiantum, Pellæa, Platyloma, Doryopteris, Hemionites cordifolia,* and other genera having fronds of a smooth, chestnut-brown colour. Within the last twenty years a small, white-winged insect, like a midge, has made its appearance (supposed to have been first introduced with imported plants to Kew). They congregate on the under side of the fronds, and, when the plant is moved, dart off like a flock of white pigeons. At first it was supposed to be harmless; but such is not the case, as it has been found to feed on the cuticle like thrips; but they are not so easily caught. Repeated tobacco fumigation destroys it.* Red spider seldom attack Ferns; when such is the case, it is a sure sign that the air of the house is too dry.

Few plant-houses are exempt from the well-known cockroach, which when once introduced is one of the greatest of pests. They increase and multiply most rapidly, and are most voracious feeders, eating and gnawing the fronds of Ferns, old and young, often completely in one night spoiling fine plants; therefore war in all its forms must be waged against them. They are night marauders, hiding and breeding in crevices of masonry in dark places, generally near the furnaces, or hottest parts of the heating-pipes, and are even found lurking amongst the drainage, in large pots, from which they issue at dusk to commence their ravages. Many expedients are resorted to for their destruction, such as attacking them in their day

* This insect has been described and figured in the *Gardeners' Chronicle,* of 1856, by Mr. Westwood, under the name of *Aleyrodes vaporariorum.*

abode, by the appliance of boiling-water, sulphur fumes, or exploding gunpowder: when such agents can be used, they deal with them quickly and wholesale. Various kinds of traps are used, which, with poison, will, if daily attended to, completely extirpate them. But it must be borne in mind that, although the whole, old and young, may be got rid of in the course of a fortnight by poison — the effect of which is greatly increased by the living eating the poisoned dead,—eggs are however left, which will soon produce a new generation that must not be allowed to arrive at maturity.

Under the ordinary varying atmosphere of hothouses, insects seem not to be affected, for if their extirpation is not attended to, they will be found in more or less abundance all the year. Not so the sooty mildew,* a fungus covering the upper surface of the leaves of plants with a black, sooty coat, and for their sudden appearance, like that of the grape mildew, the potato disease, and other sporadic plagues, no satisfactory causes have as yet been assigned. The pest now under consideration may be called one of these plagues; in some years it is not seen, while in another it soon overruns and quickly covers Ferns, and other plants, in hothouses. The broad-fronded species of *Aspidium*, *Meniscium*, *Goniopteris*, *Angiopteris*, &c., are very subject to its attacks. Books on mycology name and describe these pests, but not how to prevent them; and books on horticulture instruct how to get rid of them; the principle of which seems to be dusting with sulphur, washing,

* *Fumago foliorum*, Fries.

and syringing, the latter being the only remedy for the black mildew.

This concludes my observations on pot-cultivation, and in order to save repetition, it must be understood that the process as regards soil, watering, airing, &c., is equally applicable to natural cultivation, which I now proceed to describe.

III. Natural Cultivation.

THE natural cultivation of Ferns consists in growing them without the aid of garden-pots. This is accomplished by placing them, as far as artificial appliances will permit, under conditions and influences analogous to that of their native wilds; for that purpose uneven irregular surfaces are best adapted, whether natural or artificially formed; and as Ferns are generally called rock-plants, which many truly are, therefore, in order to imitate rocks, the surface on which they are grown is made irregular and covered with rude stones, on or between which the Ferns are planted; hence the term Rockery is applied to this kind of cultivation.

For all species of Ferns, either hardy or tender, this system of culture, which admits of their being planted out, is far preferable, to whatever extent it may be desirable to practise it, whether in the sheltered nook, in the open air, or in stoves or greenhouses, either partially or entirely devoted to it. For by it is the finest health imparted to the plants. The rockery bank which once occupied the Temperate Fern House at Kew, together with the fine fronds of tropical species grown on rockery in the Palm House, afforded excellent examples of the superiority of this mode of growing plants over that of pot culture; by proper arrangement from ordinary level positions, their delicate hue and elegant form of outline may be seen to much greater advantage, as the upper surfaces of their fronds are generally presented to the eye more fully. And where this system is largely carried out, if a raised platform or footway be erected considerably above them, the advantage is still greater. The great beneficial influence this system has on the plant is mainly consequent on the more uniformly moist atmosphere, so congenial to all Ferns, which can be successfully maintained from the great extent of the more natural evaporating surface of the soil, and material, with which the rockery is composed; and as they are allowed to establish themselves at freedom in it, there is a similar uniformity of temperature at the root, as well as of moisture. As is well known, in large masses of earthy and other solid material, its temperature does not fluctuate with that of the house, at least, as in the case with ordinary-sized pots of soil, where the absorption and radiation of the heat takes place so much more rapidly. It may further be mentioned, that, in a house entirely devoted to the cultivation of Ferns under this system, its evaporating surface does not wholly consist of the principal mass of material composing the rockwork; there is the large extent of wall-surface, which, if properly constructed with rough and absorbent material, is continually giving off its moisture, besides, from that reason, affording an excellent opportunity for Ferns of more or less scandent habit to assume their native character in luxuriant profusion, by growing on it as on natural rocks. As regards the design of any structure intended to be wholly or partially devoted to the cultivation of Ferns, it is of course a good deal a matter of taste and convenience, as the system is available in almost any kind of structure, large or small, but varying in the degree of economy in its construction.

To this end, the first consideration should be to construct in a manner that will insure the necessary amount of shade and moisture and (when required) heat, with as little auxiliary assistance as possible. Therefore if a lean-to house be adopted, the aspect should be west or north-west, with the back consisting entirely of wall. Should a span-roofed structure, however, be preferred, which indeed for all purposes is much the best (and which will be here treated of as a general example), its aspect should be north and south, by which a shadier and consequently a moister side of the house will be obtained, suitable for Ferns of more or less delicate structure, such as the genera *Trichomanes* and *Hymenophyllum*. A sunk house, in the manner of a ditch or railway-cutting, is most suitable, and will give ample height for the growth of Tree Ferns, without the necessity of having high side walls, and thus exposing a lessened portion of the house's surface to the action of climatic vicissitude; consequently a warm moist atmosphere may be maintained with but little assistance from artificial heat. As a source of heat the hot-water apparatus is to be preferred, on account of its economic utility, and where practicable the pipes should be hidden as much as possible with the rockery material. Perkins's system of heating is perhaps the best for a house of this kind; it consists of coils of small pipe, which can be arranged in any part of the house, and are readily detached and again fixed at another point when found necessary. The first deposit for the rockery should be rough and somewhat absorbent, such as irregular brickbats, masses of stone, loosely disposed to admit of a ready dissemination of the heat amongst its parts, and then follow with the material proper. Apertures for the escape of the heat should be constructed at frequent intervals, drain-pipes being very suitable for this purpose, which, by the mass being kept properly moist, will allow the heated air to circulate in a congenial steamy condition throughout the house.

In the design and arrangement of the principal mass of material composing the rockwork, it is, as in the design of the house, a good deal a matter of taste and convenience as regards size; whether in representing in miniature a rugged mountain side, or deep glen, with its clear pool, reflecting the elegant forms of the Ferns growing near it; or may be a miniature valley, with undulating surfaces gradually rising on either side, with projecting rocks. There are numerous kinds of material more or less suitable for constructing a rockery either indoors or out, and but a few localities where some of these may not be easily procured. The chief desideratum is to have such that is of porous consistence, and generally with rough and ragged surfaces. It is useless to be particular in the selection of various kinds of stone on account of their fine quality, either in texture or colour; for where Ferns properly thrive all such will be speedily obscured by their luxuriant growth; consequently any brick-field affords very good material that would contribute largely in the general structure, in the way of conglomerated masses of brick, or bricks artificially amassed with Portland cement to form miniature rocks, rugged cones, rough pillars, or perpendicular surfaces. However, when procurable, the absorbent and roughly quarried soft stone should enter largely into the composition of the rockery, as it is pre-eminently useful, not only in contributing so largely to the natural appearance of the whole, but, when the more delicate members of the Fern family of creeping habit are grown, it is invaluable in affording a moist surface for their tender rhizomes to cling to. It is seldom that wood of any kind is found of much use, as it is so soon reduced by decomposition, and fungi produced in contaminating profusion. The soil to be used in the rockery-house should be selected and applied according to the principle already given in pot-culture. The style of the Fern-rockery having been decided on and executed accordingly, the next step will be to arrange the plants about it; and to carry this out satisfactorily is not the least difficult part of the whole undertaking, as it is not only essential to provide positions for each individual plant, but such that will prevent any injurious interference of its neighbour, either by foliage or root.

Should the roof require support (and if not, should it be consistent with elegance and regularity), pillars composed of rough porous stone or brickwork may be constructed, and, if admissible, united above as archings, which, when clothed with the luxuriant foliage of scandent and pendulous Ferns, intermixed with the broad foliage of creeping *Aroideæ*, growing from their crevices and recesses, would greatly enhance the beauty of the whole, and give a most accurate and durable example of the manner in which many species of Ferns and *Aroideæ* clothe the trunks of trees in tropical forests; also several species of epiphytal woody plants may be introduced with good effect, such as *Tanæcium parasiticum, Marcgravia umbellata, Norantea coccinea, Hoyas,* and other creepers not subject to be infested with insects. The species of *Phymatodes, Pleopeltis, Davallia,* and several allied genera, as also *Polybotrya* and *Stenochlæna,* will luxuriate over the moist surfaces and form a compact mass; and, if planted near the wall, direct their rhizomes upwards, and the plants will quickly assume a most elegant appearance. The grand and beautiful *Schellolepis verrucosa,* and its allies, should be placed in elevated positions, as their fronds are long and pendulous. Elevated positions will also be found for the species of *Platycerium,* the grand *Drynaria Heraclea,* and *D. coronans:* the latter is remarkable in its thick rhizome, progressing in a circular direction round trees, which, with the erect, broad, rigid fronds, gives the idea of a coronet or crown; this circular direction appears to be normal. A fine plant at Kew, in the course of eight years, grown in a shallow pot, formed a circle more than a foot in diameter. For that portion of the terrestrial group composed of stronger growing species, such as *Phegopterideæ, Pterideæ, Asplenieæ,* &c., for which firmer soil is necessary, the ordinary surfaces of the rockery offer most suitable positions, arranged there in a manner as their sizes indicate, so as to be in harmony with those of smaller size around them. It is perhaps essential to note that the *Gymnogramma, Cheilanthes,* and their allies, require a place in the Fernery, the lightest and best drained; therefore the higher part of the rockery is best suited for them. In such a Fernery as this now before the mind's eye, convenient positions, in the form of dripping crypts and recesses, may be provided for the delicately beautiful and easily grown *Trichomanes* and *Hymenophyllum*, without the aid of glass cases that are necessarily used, as already explained, under pot-culture.

Where Tree Ferns would be grown, the proper situation for them is the lowest part of the house, in the valley, where the greatest head-room would be obtained, and most constant shade and moisture about their stems, which will encourage the essential growth of the aerial out-grown roots. The path in the valley may be straight or winding between the Tree Ferns, the intervening spaces to consist of raised mounds or hillocks, to be planted with the smaller kinds of Tree Ferns and other large-fronded Ferns. The different species of *Selaginella* should occupy the intervening spaces between the plants, which will add greatly to the natural appearance of the whole and assist in promoting a genial atmosphere.

To whatever extent a house of this nature may be carried for tropical Ferns, the same must be allowed for the extra-tropical kinds, and being sunk, the necessary amount of heating power required will be just sufficient to keep out frost. Keeping the house shut during the winter season, so as to maintain a quiet, still atmosphere, greatly assists in repelling frost, indeed for such houses ventilation is only necessary to keep down high temperature during summer; free admission is, however, beneficial when the external air is still and moist, when even the tropical division may be freely left open during the nights of summer, but taking care to prevent currents of dry air. I have now stated the chief points to be observed in forming and arranging an indoor natural Fernery, the principles of which are capable of application to houses of the ordinary construction and usual average size, or to any extent that means will allow; even to realize the grand idea of the celebrated Loudon, who, more than thirty years ago, speaking in favour of span-roofs, said, "There would be no difficulty in covering ten acres of Kew Gardens with glass by a series of span-roofs." Although since then much has been done at Kew, by the erection of lofty plant-houses, still the area covered falls far short of ten acres; but as natural cultivation is now patronized, and with the improving age and desire for novelty, let us hope that the time is not far distant when Kew may have at least one acre converted into a straight or winding Fern-valley covered with glass, the highest part not to exceed thirty or thirty-five feet above the centre of the valley, that being sufficient height for such Tree Ferns as may perchance withstand the vicissitudes of thirty years careful treatment.

In a house of this kind, furnished with means of maintaining a temperature of 60 in the coldest weather, situations in it would be found suitable to the good growth of not only *Ferns* and *Aroideæ*, but also *Bromeliaceæ*, *Orchids*, and other kinds of plants of an epiphytal nature.

For an open-air Fernery a sheltered situation should be chosen, and if possible within the influence of shelter and shade, but not under trees; a sloping bank, a natural or artificial hollow, such as an old gravel-pit or sunk fence, may be turned to good account. The number of species of exotic Ferns sufficiently hardy to bear the cold of our winters is very limited, much less than is generally believed, the chief of them being natives of North America, which, with the British species, amount to about 80 in all; therefore the extent of space required for a simple collection of hardy Ferns is small; an area of 60 feet by 10 will be ample, and which may be either on the side of a slope, or between two slopes with a path in the centre and open at one end only; or it may be a raised oblong mound with a path all around it. If not confined to a strictly scientific collection, and space to be had, an interesting natural glen may be formed by introducing other plants conspicuous for their large foliage.

It has been already stated, that although our small-growing native and several exotic species, such as the rock *Asplenicæ*, *Woodsia*, &c., endure the extremes of heat and cold in their native localities, they nevertheless suffer by full exposure under artificial cultivation; it is therefore necessary to plant them on the rockery, so that they can readily be protected in winter; an inverted flower-pot, placed over each plant and covered with leaves, or some loose material analogous to the covering of snow of their native country, is a good protection against being injured by excess of moisture or severe cold. Where the rockery is to be carried out on a large scale, then caves and chasms may readily be constructed, and with a natural or artificial head of water they can be kept moist by the constant dropping and spreading spray; thus making appropriate positions for the natural growth of *Adiantum Cappilus*, *Hymenophylleæ*, &c. But when a full collection of the many varieties of *Scolopendrium*, *Asplenium*, *Lomaria*,

&c., are grown, then a pit, or frame with glazed sashes, is the most convenient mode of keeping them. The width of the pit should not exceed 5 feet, 3 feet or more high at the back, to slope to about a foot at front, its length regulated according to the number of plants. They may be either grown in pots, arranged to slope to the front, or a bank may be formed with small rockery stones, neatly arranged with the plants between them, and if the mason work for the frame or pit is contrived to be hidden, the whole will then be in character with the general rockery. By this an interesting little bank may be formed, and by proper arrangement of the plants the variety lover can at a glance see the difference that marks one favourite from another. The glass protection will stimulate the plants into early growth, and finer fronds will be made, but care must be taken not to expose the young fronds to the harsh drying winds of spring, or late frosts, which often do great injury to out of door Ferns; by midsummer, the glass sashes may be entirely removed, due attention being paid to watering and shading.

If the soil and subsoil of the situation chosen for a hardy Fernery is stiff and retentive of moisture, and the neighbouring trees, walls, &c., are clothed with *Musci* and *Lichens*, it is favourable for Ferns without the aid of much rockery; but if the soil is light and sandy, then rockery is most essential for retaining moisture, not only for Ferns, but also all kinds of small herbaceous plants. Formerly there were great masses of rockery in the Botanic Garden at Kew, but modern taste has swept the whole away, which, with the removal of trees, walls, and old shrubberies, has caused the garden to become drier, as is manifested by the disappearance of above nearly forty species of *Mosses, Hepaticæ,* and *Lichens;* the effects of this, with other causes, have been hurtful to the hardy Fern collections. In forming a new rockery, after the plants have become once established, as few changes should take place as possible, either with the plants or stones on which *Musci* should be encouraged to grow; and in order to give a decorative effect, *Sedums, Sempervivums, Saxifrages,* &c., may be placed on the projections and shelvings of the stones, the whole bordered by a line of spring-flowering plants, such as *Iberis sempervirens, Alyssum saxatile, Arabis alpina,* and other plants of that nature.

It may be also mentioned that the vicinity of water is a favourable position for a Fernery; where such does not already exist, an artificial piece of water should be made in conjunction with the rockery, varying in size from a small basin to any extent: it may be circular, oblong, straight, or winding, with irregular projections. In ordinary cases its width need not exceed ten to twelve feet; its margin should consist of a bog or border, not less than three feet wide, having a slope to the edge of the water. This border will afford the means of growing a great many curious sub-aquatic and bog plants, and the projections will be excellent positions for the different species of *Osmunda;* water-lilies occupying the centre.

The want of such a rockery and aquarium has been much felt at Kew; for with all the great cost for lakes, ponds, and water-works, yet no arrangement has been made for growing a scientific collection of bog and aquatic plants, either hardy or tropical.

IV. Cultivation of Ferns in Ward's Cases.

IT is now thirty years since I was invited by Dr. Ward to visit him at his house in Wellclose Square, for the purpose of seeing plants growing in cases and glass jars, so closed as to be considered air-tight. Knowing, as I did, the common practice of growing plants under hand and bell-glasses, I therefore could not appreciate what I had gone to see until I was made aware that the plant-loving residents of such smoky and soot-falling districts of the metropolis, as that of Wellclose Square, could grow rare and delicate plants equal to those at Kew. An account of this method of growing plants appeared in the *Companion to the Botanical Magazine* for 1836, and in April, 1838, the celebrated philosopher Mr. Faraday delivered a lecture at the Royal Institution on the subject, which may be considered as the advent and introduction of Wardian cases, under which a large portion, and decidedly the most beautiful of the Fern family, are now successfully cultivated in the sitting-rooms of the town-confined lovers of natural objects. In 1842 Dr. Ward published a small work on the subject, giving a history and details of management, which renders it unnecessary for me to say more regarding the early history of Ward's cases. The principle on which the system is founded, consists simply in shutting up air in glass cases, in such a manner that it is not readily influenced by changes of the external atmosphere. The case also contains several inches depth of moist earth, that gives off moisture to the absorbent, undisturbed air above it, which becomes more or less saturated, as dependent upon changes of temperature, and thus becomes proper for the growth of plants.

The case may be constructed of any shape or size, according to taste or means; it can be square or round, an octagon or hexagon; the roof may be a dome, span, or sloping, but by all means avoid a flat one; and be the shape whatever it may, the design should in every point be neat, and not of such an ornamental description as to be more attractive than the plants. A very good, interesting collection may be grown in one 3 ft. long by 1½ ft. in width, and 2½ feet high; it will have the best appearance if the sides are constructed with single squares of glass; but if divided, it must not be into more than three, as many divisions spoil the effect; the ends should be made to open, to enable any dressing or removing of old fronds to be done conveniently, and a small piece of perforated zinc should be inserted in the apex of the roof at each end, which will assist in preventing the almost universal complaint, that the plants cannot be seen for condensed water on the inside of the glass; this is caused by the variation of temperature. If the case stands in the sun or becomes warm inside during the day, and retains the heat, whilst the temperature of the room may fall considerably during the night, the cold air, acting upon the glass, condenses the warmer vapour inside and obscures the plants. By the introduction of the above-named remedy the temperature is more equally balanced, and the plants are always to be clearly seen. Should the air outside become very hot and dry, it will be advisable to close the ventilators for a short time during the day. The glass case should be entirely independent of the soil-box, but to fit the inside, resting in a groove or rabbet. The box should be about 6 in. deep, and may be made of zinc, brass, or wood lined with gutta-percha or zinc; but metal of any kind in contact with the earth or air in which plants are grown is not genial to either their roots or foliage. I have always found the plants succeed best in a neatly-made wooden box lined with pitch, having a small tap or cock in one corner of the bottom, for letting away any excess of water; but this will not be necessary if proper attention is observed in supplying the plants with a sufficient amount of water at one time, which, in consequence of little or no evaporation taking place, will be seldom required. This knowledge can only be gained by practice; many amateurs' failures with Ward's cases being caused mostly by giving too copious waterings at certain stated intervals.

The height of the stand must be regulated according to whether the plants are to be viewed in a sitting or standing position; for the former the ordinary height of a table is a sufficient guide, and for the latter a few inches higher, so that the plants can be seen through the side glass rather than from the top. In preparing the box for the plants, about one inch of its depth should be filled with sand or other drainage material, such as is already explained in pot-culture—but in Ward's cases this is only necessary as a precaution against an over-supply of water,—the rest filled with soil, which should be good fibry peat and silver sand, intermixed with pieces of sandstone broken small, or (if peat cannot be procured) good light loam. Having proceeded so far, two systems of planting the case present themselves. The first plan is to plant in the soil, which should be raised in the centre; the number of plants will depend on the size of the box, and care must be taken not to plant them in too crowded a manner, the distance apart depending on the size and nature of the plants, but in no case should they be closer than six inches. The second method is to have the plants established in 4 to 6-inch pots, plunging them in the soil sufficiently deep to hide the rims of the pots. The last system has one decided advantage, namely, should a plant die or does not succeed, it can be removed and replaced without disturbing its neighbour. It must be understood that the above mode of planting relates to plants with erect vernation only, but those with creeping rhizomes, that form cæspitose tufts, require to be planted each on a separately raised hillock, of which, if the case is large, there may be a series, and for the creeping *Hymenophylla* lumps of porous stone are very suitable; fine patches may be obtained in this way, not only of *Hymenophylla*, but also of the smaller species of *Niphobolus*, *Anapeltis*, the beautiful *Humata pinnatifida*, *Davallia pentaphylla*, and others of like habit. After the Ferns are planted, some small-growing *Selaginella*, such as *S. apus*, may be pricked in over the surface of the soil, in pieces a few inches apart, which will soon cover it, and give to the whole a neat and finished appearance, and also materially assist in maintaining the proper state of moisture in the air of the case and about the plants. The planting being now completed, the soil must be brought to a uniform state of moisture; this is best accomplished by two or three moderate waterings at intervals of half an hour, and when thoroughly drained, the case may be closed and placed in position. To make cases, ornamental pieces of marble, shells, &c., are often introduced; but this should not be tolerated, as they do not harmonize with the occupants of such a structure.

For *Hymenophylla* it is necessary they should be sprinkled overhead occasionally; to enable this to be done, the tops of all small cases should be moveable, and in large ones a hinged pane is required. A sponge, or small thumb-pots filled with water and placed out of sight, will greatly assist in keeping a moist atmosphere, which is so essential for the health of these filmy-leaved plants. If the cases stand in a room where a fire is regularly kept in the winter, a great many tropical species may be grown, and in situations where they stand exposed, such as balcony windows, halls, &c., care must be taken, as winter approaches, that the soil does not become frozen, such being very detrimental to even the hardiest Ferns. Hot bricks and bottles filled with hot water have been resorted to as a preventive against frost, as also for maintaining a proper temperature for tropical species; but unless the bottom of the case is constructed for that purpose, and arrangements made for this mode of heating, to be strictly and regularly attended to, it had better be dispensed with. Should the case be exposed to the direct rays of the sun during the summer, it must be shaded, and care taken that the temperature inside does not become too high; it should not exceed 70°. This will be much assisted by placing the shading material at some distance from the case. Subjoined is a list of species suitable for cases of ordinary dimensions; it must, however, be understood that a Ward's case may consist of a bell-glass, containing a single plant, up to a shut-in window, or area covered with glass, where larger-fronded species may be grown.

Humata heterophylla.
Davallia bullata.
—— pentaphylla.
—— Canariensis.
Leucostegia hirsuta.
—— chærophylla.
—— pulchella.
Polypodium pectinatum.
—— Schkuhrii.
Lepicystis sepulta.
—— squamata.
—— rhagadiolepis.
Goniophlebium appendiculatum.
Lopholepis piloselloides.
Anapeltis stigmatica.
Pleopeltis nuda.
Niphopsis angustatus.
Drymoglossum piloselloides.
Hymenolepis spicata.
Phymatodes longipes.
—— glauca.
Pleuridium juglandifolium.
—— venustum.
Selliguea caudiformis.
Niphobolus Lingua.
Drynaria propinqua.
Elaphoglossum piloselloides.
—— vestitum.
Hymenodium crinitum.
Rhipidopteris peltata.
Microstaphyla bifurcata.
Psomiocarpa apiifolia.
Stenosemia aurita.
Gymnopteris quercifolia.
Platycerium alcicorne.
Xiphopteris serrulata.
Gymnogramma tomentosa.
—— Calomelanos.
—— Martensii.
—— chrysophylla.
—— Peruviana, *var.* argyrophylla.
—— pulchella.
Llavea cordifolia.
Hemionitis palmata.
—— cordifolia.
Dictyoxiphium Panamense.
Ceratopteris thalictroides.
Meniscium simplex.
Goniopteris asplenioides.
Nephrodium molle.
Cyrtomium falcatum.
Fadyenia prolifera.
Aspidium Pica.
Hypoderris Brownii.
Trichiocarpa Moorii.
Polystichum mucronatum.
—— Lonchitis.
—— triangulum.
—— anomalum.
—— denticulatum.

Lastrea elegans.
—— concinna.
—— strigosa.
—— vestita.
—— podophylla.
—— erythrosorum.
—— Mexicana.
—— æmula.
—— glabella.
—— Shepherdi.
—— hirta.
—— sancta.
—— hispida.
—— deparioides.
Woodsia hyperborea.
—— polystichoides.
—— mollis.
Arthropteris albo-punctata.
Nephrolepis pectinata.
Nothalæna brachypus.
—— trichomanoides.
—— Marantæ.
—— sinuata.
—— sulphurea.
Myriopteris lendigera.
—— myriophylla.
—— elegans.
—— tomentosa.
—— vestita.
—— hirta.
Cheilanthes viscosa.
—— fragrans.
—— tenuifolia.
—— microphylla.
—— Alabamensis.
—— multifida.
—— argentea.
—— farinosa.
—— pulveracea.
—— capensis.
—— radiata.
—— pedata.
Cincinalis nivea.
—— flavens.
—— pulchella.
Pellæa geraniifolia.
—— intramarginalis.
—— hastata.
—— atropurpurea.
—— Calomelanos.
—— ternifolia.
—— cordata.
Platyloma Brownii.
Adiantum reniforme.
—— caudatum.
—— lucidum.
—— macrophyllum.
—— villosum.
—— pulverulentum, *var.* rigidum.
—— fovearum.
—— curvatum.
—— affine.
—— formosum.
—— hispidulum.
—— setulosum.
—— fulvum.
—— tenerum.
—— sulphureum.
—— Chilense.
—— Capillus.
—— Æthiopicum.
Onychium Japonicum.
—— auratum.
Pteris Cretica, *var.* albo-lineata.
—— longifolia.
—— crenata.
—— heterophylla.
—— semipinnata.
—— argyrea.
—— tricolor.
—— scaberula.
Litobrochia denticulata.

Litobrochia leptophylla.
Doryopteris sagittifolia.
—— pedata.
—— collina.
Blechnum Lanceola.
—— cognatum.
—— occidentale.
—— longifolium.
—— campylotis.
Doodia blechnoides.
—— caudata.
Lomaria Patersoni.
—— lanceolata.
—— L'Herminieri.
Asplenium Hemionitis.
—— alternans.
—— formosum.
—— Brasiliense.
—— tenellum.
—— erectum.
—— marinum.
—— firmum.
—— obtusatum.
—— brachypteron.
—— Belangeri.
Asplenium viviparium.
—— viride.
—— fontanum.
—— flabellifolium.
—— pinnatifidum.
—— macilentum.
—— fragrans.
Diplazium Zeylanicum.
Antigramma rhizophylla.
Odontosoria tenuifolia.
Microlepia cristata.
Gleichenia dicarpa.
—— rupestris.
—— alpina.
—— speluncæ.
Hymenophyllum species.
Trichomanes species.
Féea spicata.
Lygodium palmatum.
Anemia tomentosa.
—— fulva.
—— adiantifolia.
Anemidictyon Phyllitidis.
Mohria thurifraga.
Todea hymenophylloides.

V. Propagation of Ferns.

THERE is considerable variety in the way in which Ferns reproduce themselves, and we only aid nature when we attempt to increase any particular species; such is usually effected, and certainly most numerously, by sowing their *spores* or *seeds*, as commonly called, which are contained in cases, as already explained at page 51; they consist of atomic particles, which, under favourable conditions as regards light, can be seen discharging from the spore-cases like puffs of smoke; therefore Fern seed is said to be invisible, and to have the power of rendering persons invisible. "I have the receipt of Fern seed and walk invisible."* Be that as it may, each particle is by the aid of the microscope seen to have definite forms varying in different genera, being globose, oval, or angular, smooth, plain, striated, or echinate, presenting very beautiful microscopic objects. Although these spores (see plate I.) are so small, they nevertheless are endowed with an extraordinary power of retaining their vitality; and being easily wafted by currents of air, readily account for the wide geographical range of many species of Ferns. When naturally or artificially placed under favourable conditions, the spore vegetates by expanding in the form of a simple oblong cell, from which other cells are successively produced, ultimately forming a thin green membrane, called the *Prothallium*, which lies nearly flat, and by the aid of fine spongioles attaches itself to the surface, and when arrived at full size it is of a reniform cordate shape, or sometimes bilobed, or obcordate, varying from about ¼ to ½ inch in diameter, having much the appearance of a small foliaceous lichen. But before proceeding to describe the method of sowing and rearing, it is important that the seed sower should be made acquainted with the remarkable discovery made, in 1848, by Count Leszczyc Suminski, that Ferns possessed organs analogous to stamens and pistils of flowering plants, and that these were produced on the *Prothallium*. In the progress of growth peculiar cells

* Shakespeare.

are formed on its under surface, of two kinds, one called *Antheridia*, and the other *Archegonia;* the first (of which there are generally between 30 and 40) containing round cells, called *sperm cells*, which contain vermicular spermatozoids that issue from the cells when arrived at maturity. The cells of the *Archegonia* differ considerably in their structure from the *Antheridia;* they contain an embryonal sac, which at a certain period protrudes from the cell, and with which the moving spermatozoids are said to come in contact, and thus produce fertilization; soon after this the embryo sac is changed, a bud is formed, and ultimately a young Fern Plant is gradually developed, and in time the *Prothallium* decays. This singular phenomenon excited considerable interest, and has been carefully investigated by the late Professor Henfrey and other eminent microscopists.*

Although the above is only an outline of the principal points of this curious subject, yet it will be sufficient to draw the attention of Fern seed sowers to the subject, and materially assist them in understanding, and arriving at the cause of the many anomalies in the good and ill success attending the raising of Ferns from seed. Much has, however, yet to be investigated before any rule can be arrived at; such as why the *Desmobrya* division does, as a general rule, produce plants from seed in the greatest abundance, even to some species becoming the weeds of the hothouse; while, on the other hand, those of the division *Eremobrya* are comparatively few, and may be considered the exception and barrenness of this rule. The differ-

* Henfrey, in *Linnæan Transactions*, vol. xxi. 1853.

ence in the fertility of the spores of these two divisions is very remarkable; this is, it must be remembered, as found with the plants under cultivation, and as with many true Ferns we find, from experience, the spores of the latter seldom germinate in less than eighteen months after having been sown. On the contrary, in the *Desmobrya* group, many require but two or three weeks, and occasionally certain species of *Gymnogramma* and *Cheilanthes* will germinate in as many days. From the irregularity, however, in this respect it is in a great measure governed by the state of, and conditions which surround, the seed; from repeated trials with spores of *Brainea insignis*, *Prothallia* were produced in forty-eight hours. Consequent on this irregularity, it is impossible to state what time of the year is most suitable for sowing to ensure success, especially for imported spores, which should be sown as soon as received, provided a proper condition can be given them. Therefore extra attention must be paid when it unfortunately occurs that the young Ferns are in their *Prothallium* stage during the winter season, to keep them from damping and the growth of conferva and musci. When it is desirable to increase any particular species by spores, some precaution is necessary to secure them in a proper ripe condition, and without spores of other species growing near being mixed with them. The fructified frond should be taken as soon as the spore-cases on the plant begin to open, and placed between sheets of paper in a moderately dry place for a few days, at the end of which time there will be escaped spores in abundance. These should be immediately sown in pots prepared in the following manner;—four to six-inch pots are generally the most convenient size; they should be half filled with good drainage material, with the smallest particles at top, when another fourth of depth should be occupied by fine soil, half sand and peat, one quarter loam, with a sprinkling of finely-broken sandstone or soft brick slightly pressed down on the top; it should then be watered and time allowed for the whole to become uniformly moist; then the spores to be very thinly distributed over it, the whole covered with a bell-glass or a piece of glass same size as pot, to be placed on its rim, allowing a space of about one inch between it and the surface of the soil. In order to keep the whole moderately and constantly moist, the pots should be placed in pans of water of half an inch depth, care being taken not to allow the soil to become over saturated; and whenever any copious condensation takes place on the glass, it should be carefully sponged off. As the spores germinate, and the *Prothallia* become crowded, so as to touch each other, they should be immediately thinned, and if it is desirable to save the thinnings they can be removed in little clumps on the particles of brick or sandstone to other pots prepared as for spores. It is not, however, with all our care in sowing different species in separate pots, that the species sown come up in the pot in which it was sown: plants of it may be found in other pots, or in different parts of the house on moist surfaces. This is easily accounted for, as the least motion of the air carries away the spores while in the action of sowing, and indeed all superfluity of spores may be with profit distributed over the whole house, the moist walls often affording abundance of young plants. It also often happens that a good crop of *Pteris aquilina* is the result, its spores being no doubt latent in the soil used for filling the pots. To avoid this it is advisable to bake the soil before using, and to bring it again to its proper moist state by the use of water that has been boiled.

It is said that spores retain their vitality for a number of years; in my experience I have no direct proof of this; but several remarkable instances of plants making their appearance without the spores of the species having been sown, or even an Herbarium specimen having been seen in this country. In the instance of *Lomaria Patersoni,* a species originally discovered in Tasmania, which spontaneously made its appearance at Kew in 1830, only one specimen was at that time said to be in the possession of Mr. Brown, at the British Museum, which I never saw; and Allan Cunningham informed me that he never found the plant, and was very much surprised when he saw it growing at Kew. This in time gave specimens to many Herbaria, and living plants to botanic gardens. A similar instance was that of *Doodia blechnoides,* which made its appearance at Kew in 1835. Other instances might be quoted, such as the appearance of *Asplenium stipitatum,* of which two plants spontaneously made their appearance about twenty years ago, and I at first supposed they had originated from the spores from a specimen in my Herbarium of a Luzon plant named by me *Neottopteris stipitata;* but in time it became evident that the two plants were quite distinct from it, and, like the *Lomaria* and *Doodia,* I had never seen native specimens. By what means the spores that produced these plants came to Kew it is impossible to say. In 1829 I found a plant of *Ceterach officinarum* growing in a crevice of masonry on one of the towers of the New Palace at Kew (since taken down). As this Fern is not found wild near London, it would be useless to speculate where this solitary spore came from; it seemed however to have found a proper nidus in the crevice, enabling it to germinate and resist all untoward influences, to pass through the *Prothallium* state and become a plant.

Polypodium vulgare and *Asplenium Ruta-muraria* may be considered our domestic Ferns; for many years a plant of *Polypodium vulgare* grew on the brick wall separating Hyde Park from Kensington Gardens, and there it remained till the wall was taken down. These few instances of isolated appearances of Ferns readily explain the wide geographical distribution of some species over the surface of the earth.

It has been shown that the spores of many species germinate quickly and abundantly, and become fully developed *Prothallia*, yet it often happens that no plant bud is formed, and in time the *Prothallia* decay; the cause of this has always been supposed to be undue moisture or some atmospheric action not sensible to us, as this has always occurred in certain species of special interest, such as *Brainea insignis*, the spores of which, as already stated, germinate readily; yet we have not succeeded in obtaining young plants, not even one *Prothallium* being seen to make a plant bud. Without special microscopical examination of the *Prothallia* we can only speculate on the probability, that, as in flowering plants, the whole of the spores of some Ferns (such as *Brainea*) are unisexual or may even be entirely destitute of both *Antheridia* and *Archegonia*, in either case deficient of the elements necessary for the production of a plant bud.

I have at page 65 mentioned that a few intermediate forms of the genus *Gymnogramma* had been raised from spores, which are considered *sports* by some and by others *hybrids;* the latter can only be admitted on the supposition of two *Prothallia* of two different species growing so contiguous to each other that the spermatozoids of one *Prothallium* have the power of passing and fertilizing the *Archegonia* of the other, and thus produce a hybrid, as in flowering plants. Another point of some practical importance is, that in general only a single plant bud is formed on each *Prothallium.* This may be supposed to be owing to the vital function of the *Prothallium* not being able to support more, in that respect analogous to only one ovulum being fertilized in ovaries of many flowering plants. Admitting that, then, how are we to explain that on removing the plant bud a new bud is formed; and even as many as eight to ten have been obtained from *Prothallia* of *Hymenodium crinitum,* each of which, by proper care, becomes a plant. Then again experiments have shown, that by dividing the *Prothallium* from the base upwards with a sharp instrument into two or even four parts, each part produces a plant bud. Seeing this, it is reasonable to infer that *Prothallia* have the power of producing plant buds, analogous to leaves of *Begonias* and other plants; but whether such is the case, or each bud is the result of the action of spermatozoids on latent *Archegonia,* is not known. Then again we have the remarkable instance of the great profusion of plants produced by the spores of all farinose Ferns, such as species of *Gymnogramma, Cheilanthes, Notholæna, Cincinalis, &c.,* and also of the smooth ebenous genera *Pellæa, Platyloma, Doryo-*

pteris, and *Adiantum*, while comparatively only a few of the smooth-fronded species of the division *Eremebrya* produce plants from spores. This subject yet requires much experimental investigation before satisfactory reasons can be assigned for what is here stated.

The majority of Ferns that do not increase by spores, often, however, readily do so by other means, such as by offsets, and viviparous buds, or bulbils produced on the upper surface, on the apex of the fronds, or in the axils of the segments, which, when placed under favourable circumstances, become plants. Ferns of cæspitose vernation will occasionally produce buds or crowns laterally on the old caudex, which may be readily separated for propagation with a sharp knife; when the vernation consists of a creeping rhizome, such may be cut in pieces of whatever length desirable, with a bud or growing point in each piece, and, as with the separated lateral crowns, should be placed in as small a pot as convenient, with soil suitable to their kind (selected according to the rule already given for establishing plants), and the whole subjected to an extra close atmosphere till thoroughly established. Up to the present time attempts to propagate Ferns by separated portions devoid of any previously joined bud have proved fruitless; although by some a solitary instance in *Scolopendrium vulgare* is considered sufficient evidence to the contrary, as portions of the base of its fronds, if separated, inserted in soil, and kept close and moist by the aid of a bell-glass, will readily strike; so also with some of the abnormal forms of the same genus, if portions of the margins of their fronds are treated in the same manner. Upon close examination, however, previously formed embryo buds are observable and considering, too, its close affinity with the *Aspleniums*, the most proliferous (in the formation of bulbils) of all Ferns, this cannot be wondered at. To propagate Ferns by the buds produced on their foliage is most easy. As soon as the bulbil plants have attained a size to be handled conveniently, they should be carefully taken off and pricked out in pots filled with moderately fine soil, and kept covered with a bell-glass till thoroughly rooted, when they may be potted off in single pots, as required.

Species with long, slender, hard sarmentum, such as *Gleichenia*, do not root readily when separated; indeed, large plants have been entirely destroyed by too free division of their sarmentum; to prevent this, layers are resorted to, which is accomplished by fixing prolonging sarmentums over small pots filled with soil, which, when well rooted, can be separated with safety, and without injuring the specimen plant. Again, in regard to the division *Eremobrya*, they are not only remarkable in the sterility of their spores, but also in not producing viviparous buds; however, the readiness with which small portions of their rhizomes form plants, and the already described tenacity of life, seem to make them independent or to render less need of perfect spores or bulbils.

In concluding this treatise, I deem it necessary to explain, that, in consequence of the woodcuts occupying more space than was calculated for, and in order to keep the book within a limited size, it has become necessary to considerably reduce the original manuscript on Cultivation. It is, however, hoped that what is now given in the preceding pages will be sufficient to show the nature of Ferns, and the methods adopted for propagating and preserving them in the collections of this country.

NOTE.

In the preceding article on Cultivation it has been shown that all Ferns are capable of being cultivated in this country under one of three conditions as regards temperature, determined by the nature of the native climate of the different species, which, as regards the species enumerated in the preceding catalogue, I classify as follows :—

1st. Hardy, in the open air.

North and Central Europe, including Great Britain and Ireland, North America, North Asia.

2nd. Temperate House.

South Europe, Madeira, South Africa, North India, China, Japan, Australia, South of the Tropics, Tasmania, Norfolk Island, New Zealand, Chili, Mexico.

A few species of those countries are hardy,—such are marked H. after the name of the country; some others improve by a higher temperature,—such are marked Tr. after the name of the country.

3rd. Tropical House.

West Tropical Africa, including St. Helena and Ascension, Mauritius, Ceylon, India, Malacca, the

Malayan Archipelago, including Penang, Singapore, Java, and the Philippines, Borneo, the Polynesian and other Islands of the Pacific within or near the Tropics, Sandwich Islands, Tropical America, including Venezuela, New Granada, Panama, Peru, Guiana, Brazil, West Indies.

Some species from high altitudes within these countries will thrive in the Temperate House,—such are marked T., after the name of the country.

A LIST OF AUTHORS AND BOOKS

QUOTED IN THIS WORK.

N.B.—When no special book is quoted after an author's name, his writings on ferns are to be found in botanical and other scientific journals of their time, and which are too numerous to notice in this work.—Those with an asterisk are living authors.

**Agardh.* *J. C. Agardh*, Professor of Botany at Stockholm; author of a "Monograph on the genus *Pteris*."

Ait. *W. & W. T. Aiton*, father and son, Directors of the Royal Botanic Gardens, Kew, from 1760 to 1841; authors of 1st and 2nd edition of "Hortus Kewensis," 1793 and 1813.

A. Rich. *A. Richard*, a French botanist; "Voyage de l'Astrolabe (Botanique)."

**Arn.* *Walker Arnott*, Professor of Botany at the University, Glasgow (see *Hook. et Arn.*).

**Bab.* *C. C. Babington*, Professor of Botany at the University, Cambridge; an eminent British botanist.

**Backhouse.* *James Backhouse & Son*, nurserymen, York; importers and cultivators of Ferns.

Bauer. *Francis Bauer*, a celebrated botanical painter (see *Hook. et Bauer*).

Beauv. *Palisot de Beauvois*, a French botanist; author of "Flore d'Oware;" figures and descriptions of plants in the Bight of Benin, 1810.

Bernh. *J. J. Bernhardi*, Professor of Botany at Erfurt.

Blume. *C. L. Blume*, Director of the Botanic Garden, Batavia; "Enum. Plant. Jav.," 1830; "Fl. Jav.," figures and descriptions of the plants of Java.

Bolt. *J. Bolton*, an English botanist; "Fil. Brit.," figures of British Ferns, 1790.

Bory, J. B. G. *Bory de St. Vincent*, a French traveller and botanist.

**Brack.* *W. D. Brackenridge*, Botanist to the United States' exploring expedition; Descriptions and figures of the Ferns of the expedition.

Braun. *A. Braun*, Professor of Botany, Berlin; "Monograph on Selaginella."

Br. *R. Brown*, the most celebrated of botanists; "Prodromus Floræ Novæ Hollandiæ," 1810; "Observations on Ferns in Wallich's Plantæ Asiaticæ Rariores;" Horsfield's "Plantæ Javæ."

Br. *P. Browne*, author of a "History of Jamaica," 1756.

Brongn. *A. Brongniart*, a French botanist.

Burm. *J. Burmann*, a Dutch botanist, and writer on plants of India, Ceylon, &c.

Carm. *Captain D. Carmichael*, a Scotch cryptogamic botanist.

Cav. *A. J. Cavanilles*, Professor of Botany, Madrid.

**Colenso.* *Rev. W. Colenso*, a New Zealand botanist.

Col. *A. Colla*, a collector and namer of Chilian ferns.

Cunn. *A. Cunningham*, a celebrated botanist and traveller in Brazil and New South Wales from 1815 to 1830.

Dec. *Aug. Decandolle*, a celebrated French systematic botanist.

Desf. *M. Desfontaines*, a French botanist and traveller in Barbary; "Flora Atlantica."

Desv. *N. A. Desvaux*, an eminent French botanist; author of several papers on Ferns, from 1808 to 1814.

De Vriese. *G. H. De Vriese*, a Professor of Botany, Leyden; "Monograph on the genus *Angiopteris*."

Dick. *J. Dickson*, an English cryptogamic botanist.

Don. *D. Don*, Professor of Botany, King's College, London; "Prodromus Floræ Nepalensis."

Dry. *Jonas Dryander*, librarian to Sir Joseph Banks; a writer on Ferns in the "Linnæan Transactions."

**Eat.* *D. C. Eaton*, an American botanist.

Ehrhart. *F. Ehrhart*, a German botanist.

Endl. *Endlicher*, Professor of Botany, Vienna; "Prodromus Floræ Norfolkicæ."

Eng. Bot. *English Botany*, "Figures and descriptions of British plants," by Sir J. E. Smith and Jas. Sowerby.

Eschw. *F. L. Eschweiler*, a German botanist.

**Fée.* *A. F. A. Fée*, Professor of Botany, Strasburg, an eminent pteridologist; "Genera Filicum," descriptions and figures of the genera of Ferns, 1850-1852; "Memoirs on *Acrostichum* and other Genera."

Fisch. *Dr. Fischer*, a Russian botanist, and Director of the Imperial Botanic Garden, St. Petersburg (see *Lang. et Fisch.*).

Fl. d'Oware (see *Beauv.*).

Forsk. *Peter Forskahl*, a Danish naturalist and traveller in Arabia; "Flora Ægyptiaca."

Forst. *John Reinhold Forster*, botanist to Captain Cook's second voyage; "Figures of Ferns in Schkuhr's 'Cryptogamia.'"

Gal. *H. Galeotti*, a German botanical collector in Mexico.

Gard. Chron. *Gardeners' Chronicle*, a weekly journal. New garden ferns described by T. Moore.

Gardn. *Dr. G. Gardner*, a botanical traveller in Brazil, and Director of the Botanic Garden, Ceylon.

Gaud. *M. C. Gaudichaud*, a French botanist; "Plants of Freycinet, Voyage de l'Uranie," 1817-1820.

Gill. *Dr. Gillies*, a Scotch botanist and collector of plants in Chili.

Gmel. *J. G. Gmelin*, a Russian botanist and traveller in Siberia; author of a "System of Plants" and "Flora Sibirica."

**Gray.* *Dr. Asa Gray*, Professor of Botany, Havard University, United States.

**Grev.* *Dr. R. K. Greville*, an eminent botanical artist (see *Hook. et Grev.*).

**Griseb.* *A. H. R. Grisebach*, a German botanist; "Flora of the West Indies," 1864.

Haenk. (see *Presl*).

Ham. *Dr. Francis Hamilton*, an Indian botanist.

**Hance.* *Dr. H. F. Hance*, an English botanist, and writer on Chinese Ferns.

Hedw. *J. Hedwig*, a German cryptogamic botanist.

H. et B. & H. B. K. *Humboldt, Bonpland, and Kunth.* The two first famous travellers and botanists in South America. *Kunth*, a German botanist.

**Hew.* *Robert Heward*, a zealous botanist, and writer on Ferns of Jamaica, in the "Magazine of Natural History."

Hitch. — *Hitchcock*, a North American writer, "Silliman's Journal."

Hoff. *G. F. Hoffmann*, a German botanist, and writer on cryptogamic plants, 1784.

Homb. et Jacq. *Hombron et Jacquemont*, French voyagers to the South Pole; "Voyage au Pol Sud, &c.," History of the Voyage and Plants.

Hook. *Sir W. J. Hooker*, Director of the Royal Botanic Gardens, Kew, from 1841 to 1865; an admirable descriptive botanist and eminent Pteridologist; "Exotic Flora," figures of rare plants, 1823; "Icones Plantarum," figures of 1,000 rare plants—many Ferns; "Second Century of Ferns," 100 figures of rare Ferns; "Filices Exoticæ," figures of 100 Ferns; "Garden Ferns," 64 figures; "Journal of Botany," 4 vols.; "London Journal of Botany," 7 vols.; "Journal of Botany and Kew Miscellany," 9 vols.; "Species Filicum," 5 vols., descriptions of all known Ferns with upwards of 300 figures, 1844-64.

Hook. et Arn. *Hooker and Arnott*, "Botany of Capt. Beechey's Voyage."

Hook. et Bauer. *Hooker and Bauer*, the "Genera Filicum," figures of the genera of Ferns, illustrated by Bauer, 1838.

Hook. et Grev. *Hooker and Greville*; "Icones Filicum," figures of rare Ferns, 230 plates, 2 folio vols., 1831.

**Hook. fil.* *Dr. Joseph Hooker*, Director of the Royal Botanic Gardens, Kew; appointed 1865; "Flora Antarctica," "Flora of New Zealand," "Flora of Tasmania."

Hoppe. *J. C. Hoppe*, a German botanist, and collector of plants.

Hort. Gardens; *Hort. Ang.* English gardens; *Hort. Berol.* Berlin garden; *Hort. Linden.* Horticultural Garden, Brussels.

**Houlst.* (see *Moore et Houlst.*).

Huds. *W. Hudson*, an English writer on British plants.

Humb. *A. v. Humboldt*, a celebrated traveller and philosopher (see H. B. K.).

Jacq. *Nicolas Joseph & François Jacquin*, father and son, eminent Austrian botanists; "Icones rariorum," figures of rare plants.

**Johns.* *J. Y. Johnson*, a writer on Madeira ferns.

**J. Sm.* *John Smith*, Curator Royal Botanic Gardens, Kew (retired 1864); "Gen. Fil.," an arrangement of the genera of Ferns in Hooker's "Journal of Botany," 1841; "Enum. Fil. Philipp.," an enumeration of the Ferns collected by H. Cuming in the Philippine Islands; "Seem. Bot. Voy. Herald," an enumeration of the Ferns in Seemann's "Botany of the Voyage of the 'Herald';" "Cat." Ferns cultivated at Kew, in appendix to *Botanical Magazine*, 1846; "Cat. Cult. Ferns," catalogue of Ferns cultivated in British gardens in 1857.

**Karst.* *Dr. Karsten*, a German botanist and traveller in Tropical America; "Flora Columbia."

Kaulf. *G. F. Kaulfuss*, Professor of Botany, Leipsic; "Enum. Fil.," Enumeration of the Ferns collected in Chamisso's Voyage.

Klot. *Dr. Klotzsch*, a German botanist.

Kunze. *G. Kunze*, Professor of Botany, Leipsic; "Analecta Pteridographia," figures and descriptions of Ferns, 1834; 'Schkuhr's Cryptogamia," continued, 1841–51; numerous other papers on Ferns.

Labill. *J. J. Labillardière*, a French navigator sent in search of

M. La Perouse; "Nov. Holl. Plant. Spec.," figures and descriptions of New Holland plants, 1804-1806; "Sertum Austro-Caledon.," figures and descriptions of plants in New Caledonia.

Lag. *M. Lagasca*, Professor of Botany, Madrid.

Lam. *Jo. Bapt. Monet de Lamarck*, a celebrated French naturalist and compiler.

Lang. et Fisch. *G. Langsdorf*, a Russian botanist. "Icon. Fil." figures of Ferns, chiefly Brazilian, 1810.

L'Hérit. *C. L. L'Héritier*, a French botanist.

Lieb. *Liebmann*, a German botanist.

Lieb. *Liebold*, a German traveller and botanist, and collector of plants in Mexico.

**Linden.* *J. Linden*, a nurseryman at Brussels, and traveller in Mexico and Peru; a collector and importer of rare plants.

Lindl. *Dr. John Lindley*, an eminent systematic botanist, Professor of Botany, University College, London; "Lindl. & Moore: Nature-printed British Ferns."

Link. *H. F. Link*, Professor of Botany, Berlin; "Enumeration of the Ferns of the Berlin Garden."

Linn. *Car. Linnæus*, the celebrated Swedish botanist, and founder of modern botany.

Lodd. *Conrad Loddiges & Son*, nurserymen and great cultivators of Ferns at Hackney.

Lour. *Loureiro*, a Portuguese botanist and traveller in Cochin China; "Flora Cochinchinensis."

**Lowe.* *E. J. Lowe;* "Lowe's Ferns," figures and descriptions of exotic Ferns, in 9 vols.

Mart. et Gal. "Figures and descriptions of Mexican Ferns," by H. Galeotti and M. Martens.

Mart. *C. F. P. Martius*, Professor of Botany in Munich; "Icon. Crypt.," figures and descriptions of Brazilian Ferns, 1820.

**Metten.* *Dr. G. Mettenius*, Professor of Botany at Leipsic, and eminent writer on Ferns; "Figures and descriptions of the Ferns in the Leipsic Garden;" "Monograph on the genus *Asplenium*," &c.

Mey. *C. Meyer*, a German botanist.

Michx. *A. Michaux*, a French botanist and traveller in North America; "Flora Boreali-Americana."

**Miq.* *F. A. G. Miquel*, a Dutch botanist.

**Moore.* *T. Moore*, Curator of the Apothecaries' Garden, Chelsea; "Index Filicum," an alphabetical list of all names of Ferns; various books on British Ferns (see *Lindl.*).

**Moore et Houlst.* *T. Moore and W. Houlston;* "Descriptions and Wood-cuts of Cultivated Ferns," published in Ayer's "Magazine of Botany."

Muhl. *Muhlenberg*, a North American botanist.

**Newm.* *Edwd. Newman*, author of works on British Ferns.
Nutt. *D. Nuttall*, a North American botanist.

Plum. *Car. Plumier*, a French botanist and traveller in the West Indies. "Plum. Fil.," figures and descriptions of Ferns, chiefly of the French West India Islands, 1666.
Poir. *M. Poiret*, a French botanical compiler.
Presl. *C. B. Presl*, Professor of Botany, Prague; a famous Pteridologist; "Reliqua Haenkæana," figures and descriptions of Ferns collected by the traveller Haenke; "Tent. Pterid.," a new arrangement of the genera of Ferns, 1836; "Epimeliæ Botanicæ," figures and descriptions of Ferns, being an addenda to the preceding.
Pursh. *Frederick Pursh*, author of a "Flora of North America."

Radd. *J. Raddi*, a German botanist, who travelled in Brazil; "Fil. Bras.," figures and descriptions of Brazilian Ferns, 1825.
Raoul. *M. M. E. Raoul*, a French botanist and writer on New Zealand Ferns.
**Regel.* — *Regel*, Director of the Imperial Garden, St. Petersburg, 1866.
Retz. *A. J. Retzius*, a German botanist and writer on plants of Ceylon.
Rheede. *H. van Rheede*, a Dutch botanist; "Hortus Indicus Malabaricus," a large work on the plants of Malabar, 1703.
Reichenb. — *Reichenbach*, a German botanist; "Flora Germanica."
Reinw. — *Reinwardt*, Professor of Botany, Leyden, and traveller and collector of plants in Java and other Malayan islands.
Remy in Gay. *Remy*, an authority in Cl. Gay's "Flora of Chili."
Rich. *L. C. Richard*, a writer on the plants of Guiana.
Roth. *A. G. Roth*, a celebrated German botanist and writer on Ferns.
Roxb. *Dr. Roxburgh*, an English botanist; "Observations on St. Helena Ferns."
**R. T. Lowe.* *R. T. Lowe*, a writer on Madeira Ferns.
Rudge. *E. Rudge*, author of a work on the plants of Guiana.

Schk. *C. Schkuhr*, a Dutch botanist; "Crypt. Schk.," cryptogamia; figures of Ferns; "Crypt. Supp.," supplement of the above, continued by Kunze.
Schlecht. *D. F. L. Schlechtendahl*, Professor of Botany in Berlin; *Schlecht. Adumb.* "Adumbratio Plantarum," figures and descriptions of South African Ferns.
Schott. *Heinrich Schott*, Director Royal Gardens, Vienna; "Schott Gen. Fil.," the genera, illustrated by figures.
Schreb. *J. C. Schreber*, a German botanist.
Schum. — *Schumacher*, a writer on plants of Guinea.

**Seemann.* *Dr. Berthold Seemann*, an eminent botanist and voyager; "Botany of the expedition of the surveying ship *Herald*."

Sibth. *Dr. Sibthorp*, an English botanist and traveller in Greece; author of "Flora Græca."

Sieber. — *Sieber*, a celebrated German botanical traveller and collector in various parts of the world.

**Sim.* *R. Sim*, a nurseryman and celebrated grower of Ferns at Foot's Cray, Kent.

Sloan. *Sir Hans Sloane*, a traveller in the West Indies; author of "History of Jamaica," and founder of the British Museum.

**Sm.* *Sir James Edward Smith*, a celebrated British botanist and writer on botany in various journals; first President of the Linnæan Society, and purchaser of the "Linnæan Herbarium."

**Sowerby.* *E. Sowerby*, a British botanist; author of a work on British Ferns.

Spenn. — *Spenner*, a German botanist.

Split. *F. L. Splitgerber*, a Dutch botanist; "An enumeration of the Ferns of Surinam."

Spreng. *C. Sprengel*, Professor of Botany, Halle; "Syst.," a systematic enumeration of plants according to the Linnæan arrangement.

Spring. — *Spring*, a German botanist and writer on *Lycopodiaceæ*.

**Stansfield.* — *Stansfield*, a nurseryman at Todmorden, Yorkshire, celebrated as a discoverer and cultivator of numerous varieties of British species of Ferns.

Sw. *Olaf Swartz*, a Swedish botanist and traveller in Jamaica; "Synopsis Filicum." an enumeration and description of all known Ferns since 1806.

Thunb. *P. Thunberg*, a Dutch traveller and botanist in Japan and Cape of Good Hope; "Fl. Jap.," the Flora of Japan, 1784.

**Thwaites.* *G. J. Thwaites*, Director Botanic Garden, Ceylon; author of "Flora of Ceylon."

Tuckerman. *E. Tuckerman*, a United States botanist.

Vahl. *Mart. Vahl*, a German botanist.

**Van Houte.* *L. Van Houte*, a celebrated Belgian nurseryman; author of periodical botanical works.

**Veitch.* *James Veitch & Son*, nurserymen, London, celebrated importers and cultivators of rare Ferns.

Vent. *E. P. Ventenat*, a French botanist.

Vogler. — *Vogler*, a German botanist.

Wall. *Dr. Nathaniel Wallich*, Director Botanic Garden, Calcutta. MSS. catalogue of the plants contained in the "Wallichian Herbarium" at the Linnæan Society.

**Watson.* *Hewet Cottrell Watson*, an eminent British botanist.
Webb. *P. B. Webb*, an English botanist; "Flore des Canaries."
Weis. *F. W. Weis*, a German cryptogamic botanist.
Willd. *C. L. Willdenow*, Professor of Botany, Berlin; "Species Plantarum;" Enumeration of plants in the Berlin Garden, 1809.
**Wollast.* *T. Wollaston*, an English botanist and writer on Ferns.
Wulf. *F. H. Wulfen*, a German botanist.

Zenk. —*Zenker*, a German botanist and writer on plants of the Nilgheery.

INDEX

OF

GENERA, SPECIES, AND SYNONYMS.

N.B.—The names in italics are synonyms. Where two pages are given, the second refers to Cultivation.

Number of adopted species in first edition (1865)......	1,084
Ditto in Appendix (1877)	226
Total	1,310

INDEX OF SPECIAL TERMS

DESCRIBED IN ORGANOGRAPHY.

WYMAN AND SONS, PRINTERS, GREAT QUEEN STREET, LONDON.

www.ingramcontent.com/pod-product-compliance
Lightning Source LLC
Chambersburg PA
CBHW020904210326
41598CB00018B/1770

* 9 7 8 3 3 3 7 0 7 0 0 4 5 *